P9-CKT-924

ETERNAL EPHEMERA

NILES ELDREDGE

ETERNAL EPHEMERA

ADAPTATION and the ORIGIN of SPECIES from the NINETEENTH CENTURY through PUNCTUATED EQUILIBRIA and BEYOND

COLUMBIA UNIVERSITY PRESS NEW YORK

Columbia University Press
Publishers Since 1893
New York Chichester, West Sussex
cup.columbia.edu

Library of Congress Cataloging-in-Publication Data

Eldredge, Niles.
Eternal ephemera : adaptation and the origin of species, from the nineteenth century, through punctuated equilibria and beyond / Niles Eldredge.
pages cm
Includes bibliographical references and index.
ISBN 978-0-231-15316-4 (cloth : alk. paper)
ISBN 978-0-231-52675-3 (e-book)
1. Darwin, Charles, 1809–1882. On the origin of species.
2. Punctuated equilibrium (Evolution) 3. Evolution (Biology)—Philosophy.
4. Emergence (Philosophy) I. Title.

QH398.E43 2015
576.8'2—dc23
2014022519

Columbia University Press books are printed on permanent and durable acid-free paper.
This book is printed on paper with recycled content.
Printed in the United States of America
c 10 9 8 7 6 5 4 3 2 1

FRONTISPIECE: Southern South America, showing locales important to Darwin on the *Beagle* voyage, 1832–1835. (Artwork by Network Graphics)

COVER IMAGE
Plate 3 from Joachim Barrande, *Système silurien du centre de la Bohême (1852)*.
COVER DESIGN
Lisa Hamm

To Michelle, who has given me everything

"Like the Shadows of the Clouds in a Summer's Day"

In this vast host of living beings, which all start into existence, vanish, and are renewed, in swift succession, like the shadows of the clouds in a summer's day, each species has its peculiar form, structure, properties, and habits, adapted to its situation, which serve to distinguish it from every other species; and each individual has its destined purpose in the economy of nature. Individuals appear and disappear in rapid succession upon the earth, and entire species of animals have their limited duration, which is but a moment, compared with the antiquity of the globe. Numberless species, and even entire *genera* and tribes of animals, the links which once connected the existing races, have long since begun and finished their career.

—ROBERT E. GRANT, INAUGURAL ADDRESS, UNIVERSITY OF LONDON, 1828
MENTOR TO CHARLES DARWIN, EDINBURGH, 1827

CONTENTS

PREFACE

Robert Grant left little iron-clad evidence that he was the radical, pro-Lamarckian thinker that in fact he was. Were it not for Charles Darwin's brief passage describing Grant's outburst as they were walking together one day, few would have placed Grant among the small but influential group known as the "Edinburgh Lamarckians." Gentlemen did not often publicly express their pro-evolutionary views in Great Britain in the 1820s.

The excerpt from Grant's inaugural address to the public on the occasion of his being appointed professor at the University of London, used as the epigraph to this book, is the closest thing to pure poetry in the annals of evolutionary biology. Thanks to Darwin, we know of Grant's evolutionary convictions, so when we read his words, there can be no doubt of their true meaning.

Individual organisms, and indeed even species, though they tend to persist little changed over millions of years, are in the end evanescent "like the shadows of the clouds in a summer's day." And yet they are all linked together in ancient skeins of life—individuals, species, and larger groups, many of which are lost in the deep annals of time. Together they are the "eternal ephemera," whose origins and extinctions, and their adaptively forged roles in the economy of nature, are the original, and still continuing, subject matter of evolutionary biology.

This is the story of the search for rational, non-miraculous understanding of these eternal ephemera. It is, at the same time, an exploration of my own struggles with these issues. I have learned only late in my career, decades after "punctuated equilibria" and related themes, of the long-lost connections to my own intellectual forbears of the early nineteenth century. The evolution of evolution, as it were.

ACKNOWLEDGMENTS

In the acknowledgments to his magnum opus, *The Rise of Anthropological Theory*, Marvin Harris said that "when an author has completed his book, it would seem a simple task to thank those who helped him in its preparation. In actuality the task is not an easy one at all."

Very true. Acknowledging everyone who played a role in showing me the way all along is nearly impossible. But I'll give it a shot.

I was an undergraduate at Columbia College from 1961 to 1965. I decided as a freshman to remain in academic life. Initially, I thought that Greek and Latin would be an appropriate major for me. But I also met Michelle Wycoff as a freshman; we were married in 1964, and she has been the major influence in my life all along the way. She has been indispensable, in particular with her help and encouragement on this book, which she has copyedited and made readable. I can also say that she is very glad that I am finally finished with it! Thanks, babe—I love you so much!

Michelle knew several of the Anthropology Department faculty members—among them Marvin Harris. She often baby-sat for the Harris kids, and I got to know the family right away. In those days I thought anthropology was more a matter of Louis Leakey–style hominid fossil hunting than Margaret Mead–style cultural anthropology. (She was also at Columbia and the American Museum and always very nice to me. Marvin, of course, was a cultural anthropologist—though hardly in Margaret's style!) I started taking anthropology courses and, under Marvin's aegis, joined a small group of greenhorn undergraduates in the summer of 1963. We lived in and around Arembepe, Brazil, in the days before paved streets, running water, or electricity, save on two special feast days each year.

In many ways that Brazilian experience in 1963 was one of the most important formative experiences of my life—and certainly of my subsequent career in evolutionary biology. Among many other things, I decided while there that my future path would lie more along the fossils that studded the sandstone "reef" at Arembepe than in pestering people about their private lives in my baby-talk Portuguese.

Returning to New York that fall, I enrolled in a paleontology class taught by John Imbrie. Imbrie's courses, plus the summer (1964) research opportunity under paleontologist Roger L. Batten at the American Museum of Natural History (AMNH), sealed the deal: paleontology was it! By the end of my college sojourn, I eventually amassed more credits in geology than in anthropology.

I stayed for graduate school in the Department of Geology from 1965 to 1969. My main mentors/advisors were Roger Batten and the great Norman D. Newell, whom I had met only in passing in the corridors of the AMNH. In those days, American Museum curators often had dual appointments at Columbia, and Roger and Norman were no exception. In the end, Norman became my official advisor.

I also studied with Bobb Schaeffer, Edwin (Ned) H. Colbert, and Malcolm McKenna. They, too, had joint appointments at the AMNH and Columbia—Ned and Malcolm in geology, and Bobb in biology. Columbia faculty who were especially important to my care and feeding as a nascent professional paleontologist with a special passion for evolution were Marshall Kay in geology, and Walter Bock and Frederick Warburton in biology.

In many ways, though, it is who you go to school *with* that matters at least as much as who your teachers were. Kids on a mission tend to teach one another, banding together in important ways as they individually and collectively grope their way to completing their doctorates and entering the daunting world of professional academe.

And in this critical arena of fellow students, I was especially lucky. Many of us have worked together over the years as colleagues and will be encountered again in these pages. They included Harold B. (Bud) Rollins, Stephen Jay Gould, Joel Cracraft, Eugene S. Gaffney, and John C. Boylan.

In due course, I got my degree and started work "at the top" in 1969: I was lucky enough to be appointed as assistant curator in what was then called the Department of Fossil Invertebrates at the American Museum of Natural History. I held that position (though I did get promoted!) for forty-one years, retiring from what is currently the Division of Paleontology in 2010. While working at the AMNH, I of course had many friends and helpful colleagues, including in those early days Bob Adlington, Frank Lombardi, and Bea Brewster; and later Iris Calderon, Bushra Hussaini, and Steve Thurston.

A special note of profound thanks goes to Sidney Horenstein. Sid was there to greet me the first day I showed up at the door in 1964. We have worked together on two exhibitions: the Hall of Biodiversity in the 1990s, and *Darwin* in the first decade of this present century. Most recently, we have written *Concrete Jungle*, an exploration of the role of cities both in environmental destruction and biodiversity loss, and the promise of cities as the last best hope for life on this planet. Perhaps unsurprisingly, being life-long New Yorkers, we chose our city as a microcosmic exemplar (as Steve Gould might have called it) of the world's great cities. Thank you Sid!

Among the many stimulating and sometimes provocative colleagues I have worked with at the American Museum, in addition to Margaret Mead in the earliest days, I think Gareth Nelson was among the most stimulating (and sometimes irritating) intellects I have ever run across. I thank Gary for his continual prodding and challenging.

I am glad to say that two of my fellow students have also been valued colleagues on the American Museum staff down the years: Gene Gaffney and Joel Cracraft. And I am very glad to acknowledge the presence of Neil Landman, ammonite specialist extraordinaire, for keeping the flame of invertebrate paleontology very much alive at the AMNH.

Most recently, I have become close friends with termite specialist Kumar Krishna, a very wise man. Hashing over issues both scientific and personal in both New York and Kumar's native haunts in India has helped me immeasurably as I have been working on this manuscript. Thank you so much Kumar!

And finally, among my American Museum colleagues, I come to my dear friend and trusted and valued colleague, Ian Tattersall. We have known each other since the early 1970s, and on an almost daily basis (save when Ian was on one of his global peregrinations!) we have been in constant touch, going all the way back to the days of the Blarney Castle. Ian is also a wise man—though he often denies it. We have written together. And we have hashed over the internal politics of the museum as well as the issues confronting the world as it has spun on its axis over the past forty-plus years. Thank you Ian, for being yourself!

Among all my friends and colleagues, none have been more important to me than Steve Gould and Elisabeth S. Vrba. Once again, a more complete account of our work, separately and together, comes toward the end of this narrative. But they deserve special attention here, as working with them separately and as the "Three Musketeers" for a time in the 1980s has been so critical to the development of my own perspectives in evolutionary biology.

My career would simply not have been the same had one of the entering graduate students at Columbia in 1963 not included Stephen Jay Gould. Steve was a sort of older-brother role model—with his penchant for theory, and his incredibly workaholic-like approach to his studies and his career. He set the very best example I could have had—and we had a long and productive working relationship that is in many ways the highlight of my career.

Likewise, Elisabeth Vrba, whom I first met in 1980, has always had the very highest intellectual standards. Hers was the most creative mind, hands down, in the realm of species ("taxic") evolutionary biology in the 1980s—as I endeavor to make clear in my narrative. More than once she kept me on track as my focus on issues would occasionally flicker. And several times she taught me what to think about certain issues—even if she did not particularly agree what I ended up thinking (see the section on "species selection").

To Steve and Elisabeth, my profound respect and deepest admiration. Many thanks for all that you both have accomplished over the years.

I also want to acknowledge the profound influence that biologist Stan Salthe and philosopher Marjorie Grene had on me—especially in the 1980s when we were developing hierarchy theory.

And so I come to the production of this, my personal equivalent to Marvin Harris's *Rise of Anthropological Theory*. My narrative links the early days of truly scientific examination of evolution with events that occurred starting in 1935. The beginnings revolved around the simple, yet profound, question: What is the natural (that is, non-miraculous) explanation for the origins of species in the modern world? It is thus a book about species: what they are, how they originate, live out their ultimately ephemeral lives (often over millions of years!), and eventually die; and how all this relates to the general process of adaptation.

My narrative by no means is a general history of evolutionary theory. But it does nonetheless cover a lot of territory. And for this I have needed a lot of help.

I would never have thought (or dared) turn my paleontological evolutionary gaze on the deep recesses of the history of evolutionary biology had I not been invited by the administration of the American Museum to lead the development of an exhibition on Charles Robert Darwin. Though I was initially daunted, it turned out to open so many conceptual evolutionary doors to me that I felt compelled to continue reading Darwin and his predecessors, going all the way back to Lamarck in 1801. I am profoundly grateful to have had this opportunity, and I thank the museum's administration for entrusting me with this task.

It turns out that our show *Darwin*, still on display in various venues around the world, has been a tremendous success. I know nothing of how to actually make a three-dimensional display, and the success of our *Darwin* show derived in largest measure from the museum's superb Exhibition Department, headed by David Harvey.

But it was historian David Kohn, who has devoted his professional life to the intricate subject of Charles Darwin, who made the project come to life. David, the talented writers of the Exhibition department, and I developed the narrative text. And David was the key figure, along with Darwin's great-great-grandson Randal Keynes, to specify the list of "must-have" objects we needed to borrow for the exhibition. More on all this anon, but I want to acknowledge particularly David Kohn's continued coaching on how to think about all things Darwinian even long after the exhibition was up and running. David gave me the confidence to invade the turf of professional historians!

I am also deeply grateful to the staff of the AMNH Library, especially Director Tom Baione, Barbara Mathe, and Mai Reitmeyer. Profound thanks for all your help over the years, including the acquisition and online posting of key documents leading up to the publication of "punctuated equilibria."

I am also deeply grateful to colleagues Rolando González-José, Teresa Manera, Rodrigo Tomassini, Rodrigo Medel, Michel Sallaberry, and Roberto Nespolo for organizing and hosting a 2008 excursion that gave me a first-hand appreciation of the critical localities and environments of southern South America visited by the young Charles Darwin from 1832 to 1835. I am further very grateful to Dr. Michael Rosenbluth for his generous support for this critically important travel experience.

For specific advice on this present volume, or issues surrounding it, I thank historians David Kohn, David Sepkoski, Jon Hodge, and Martin Rudwick. Fellow paleontologists Bud Rollins, Gil Klapper, Ian Tattersall, Bruce Lieberman, Stefano Dominici, and William Miller III have been helpful for their corrections and advice and encouragement—and insistence from many, if not all, of them that I should try harder to improve the writing!

I also thank philosopher Telmo Pievani and biologists John N. Thompson, Dan Brooks, Ilya Tëmkin (also a paleontologist!), and T. Ryan Gregory for all their helpful comments. I also acknowledge the helpful comments from two anonymous reviewers for Columbia University Press. And I thank my editors there, Patrick Fitzgerald and Bridget Flannery-McCoy, for all their help and encouragement.

My agent, John Thornton, and I have been working together since the early 1980s when John signed me up, when he was editor at Washington Square Press, to write one of my earliest books, *The Monkey Business* (1982), on creationism. John is unflappable, wise, and most capable—and it is indeed a pleasure to thank him here publicly.

And finally: this narrative is told for the most part in the words of its cast of characters—the protagonists of this prolonged drama. I am but one of them. My thanks to them all for having created and developed the themes of adaptation and the "origin of species" recounted in these pages.

ETERNAL EPHEMERA

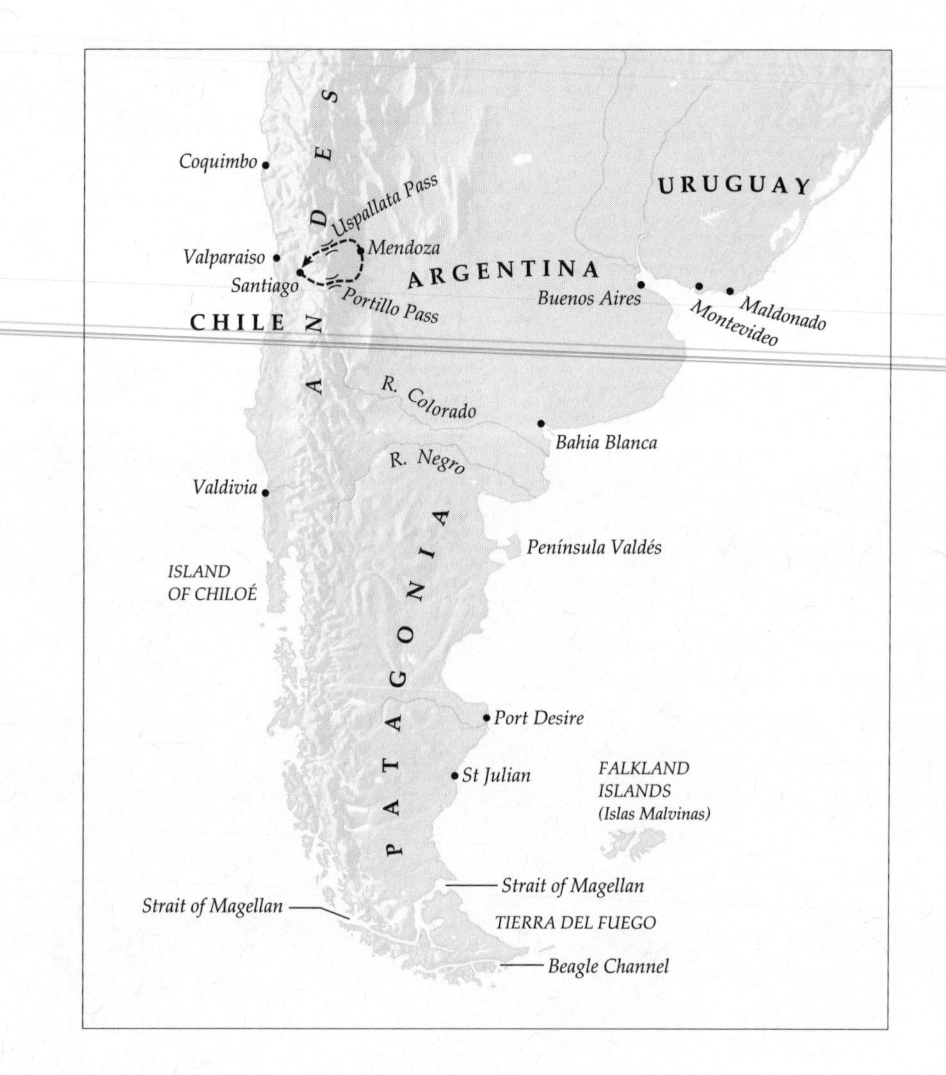

Southern South America, showing locales important to Darwin on the *Beagle* voyage, 1832–1835. (Artwork by Network Graphics)

INTRODUCTION

Approaching Adaptation and the Origin of Species

The last time I saw the great biologist Ernst Mayr, he was in his nineties. He was clutching a martini at the end of a long meeting at which we both had spoken. He seemed happy, and I told him he clearly was still having fun. Ernst put on a mock scowl, and growled, "Come on, Eldredge, you know as well as I that evolutionary biology is hard work!"

And so it is. I have been an intensely engaged evolutionary biologist since the mid-1960s. I still love it—especially the ideas themselves, if not always the infighting that all such pursuits inevitably entail. But there is also the element of a game to it, something like "monkey-in-the-middle," in which people jockey for position and ideas are contested as if they are a ball to be tossed, grasped—or dropped.

More substantively, some thinkers have seen evolution itself as a sort of a game in which the "goal" is to maximize adaptive, hence reproductive, success. They in fact equate evolution purely with the process of adaptation through competition and some form of selection. This approach is perhaps especially effective (or at least most commonly encountered) in theories of material cultural evolution. But it has been a dominant theme throughout much of the history of evolutionary biology as well.

But alongside adaptation, there has been this persistent, almost pesky, theme of the "origin of species." Starting with Jean-Baptiste Lamarck's work of 1801 on marine fossil shells and their relation to modern species, questions on the nature and modes of origin and extinction of species have been out there on the evolutionary table. Indeed, the origin of the species of the modern fauna and flora was the original subject matter of evolutionary biology. Adaptation as a topic requiring analysis in and of itself did not come along until several decades later.

Once adaptation came to the fore, starting with Darwin's Transmutation Notebooks of the late 1830s, the question then became: What is the relation between the origin of species and selection-driven adaptive change? Are new species strictly a byproduct of adaptive evolutionary change? Or do species exist and have natural causal origins, histories, and deaths through processes involving, not simply a fallout of, selection-driven adaptive change? Much of the history of evolutionary biology has involved wrangling over the nature of species and the relationship between speciation and adaptation.

In the waning days of the 1970s, I was invited to give a lecture at the University of Rochester in upstate New York. I already had a decade of employment as an evolutionary paleontologist at the American Museum of Natural History under my belt. I had developed the early versions of punctuated equilibria and published the final version with my co-author Stephen Jay Gould in 1972. By the mid-1970s, the paper had gained a lot of notoriety, and I was receiving many invitations to speak at universities in the United States and abroad.

Punctuated equilibria was based on two empirical patterns: geographic species replacement in the living biota, which had prompted a renaissance of geographic speciation theory by people like the geneticist Theodosius Dobzhansky and Ernst Mayr; and stasis, the very common observation that, once they first appear, species seem to be relatively stable, often for millions of years. There was little hint of the slow, steady, gradual change that Darwin, by the time he went public with his ideas, had insisted must be the general rule.

The paleobiology program at Rochester was very strong, staffed as it had been by the likes of David M. Raup, Jack Sepkoski, and Dan Fisher—the latter two former students of my friend from graduate student days a decade earlier, Stephen Jay Gould. But my invitation actually had come from the evolutionary biologists at Rochester, who were immersed in the study of genetics, systematics, and ecology of living species—rather than fossils.

I had been in the business long enough to know that, on most campuses, most of the time, the paleobiologists residing in the geology department (by then beginning to call themselves more fashionable names like "Earth and Planetary Sciences," reflecting the revolution in plate tectonics) had at best only a nodding acquaintance with the systematists, ecologists, and evolutionary biologists of the biology department (which, by then, and prompted by the molecular revolution, had begun to call themselves names like "Ecology and Evolution," as they began to split off from the newly minted molecular biologists). And vice versa. I had long since discovered that evolution interests people from many different walks of academic life: biology and paleontology, to be sure, but also anthropology, psychology, history, and philosophy of science—even comparative literature. For the most part, people from different departments didn't seem to know—let alone work with—each other all that well. If anything, Rochester was more open and communicative across disciplinary lines than most universities were at that time.

I had given my lecture title in advance: "Alternative Approaches to Evolutionary Theory" (1979), the same title as a paper I had recently completed. The campus newspaper announced the event, but the title had somehow become garbled: the ad for my talk now read "Alternatives to Evolutionary Theory," explaining why the rather modest-size classroom in which I spoke was crowded to overflowing, as twenty or so rather clean-cut kids (who no one from either biology or geology had ever seen before) showed up just as the lecture was about to begin. They were, of course, creationists; they kept politely silent, and softly shuffled out the door when my disappointing talk was (finally!) over with.

My actual talk dealt with what I had come to see as something of a rhetorical schism in ways of thinking about evolution. I was trying to turn the schism into a dialectic—a dialogue that might shed light on the entire subject if only evolutionary biologists (in the broadest sense of the term) would acknowledge that different disciplines made certain different assumptions about what evolution is; about what exists in the biological world that can reasonably be said to "evolve"; and regarding the fundamental nature of the processes that underlie evolution.

Although I was yet to learn Ernst Mach's admirably concise definition of science (something like "the description of the entities of the material universe and the interactions between them"), and was still decades from learning the great John Herschel's much earlier characterization to the exact same effect, it did always seem pretty obvious that science dealt with what philosopher Mario Bunge (1977) had already been calling the "furniture" of the universe, and the things that happen to that furniture.

And what are those entities that lie at the core of modern evolutionary discourse? Genes and species, each with their subsets like "alleles" (and more in these post–molecular revolution days) and their frequencies within "populations" (subsets themselves of species). Looking upward, some evolutionary biologists (paleontologists, but also "neontologists") would also consider "higher taxa" (such as genera, families, and the like) for the most part to be groups of genealogically interconnected species. Higher taxa are the large-scale results of the evolutionary process.

Missing from the gene/species duality, of course, are organisms, and more particularly their anatomical, physiological, and behavioral (phenotypic) characteristics. These are generally held to be the expression of the underlying genetic information, almost invariably assumed to be adaptations forged by natural selection working on genetically based variations within populations within species. But organisms themselves have ontogenies, births, and deaths. Organisms change throughout their lifetimes, but do not themselves evolve. Their genes and their phenotypic expressions evolve. And species evolve.

And it made sense, in the division of labor that inevitably arises in all human endeavor (certainly in all branches of the academic world) that some people, hence some research traditions, and perhaps even entire departments, would focus on one subset of phenomena, in the interests of efficiency, and following the funding of emerging new technologies. So some people were geneticists, until subsets of genetics began to develop. And some geneticists of course could consider the nature and role of entire species (Theodosius Dobzhansky, as we shall see, played a huge role in that respect). But it was quite possible to study the changing distributions of alleles in populations, whether on a sheet of paper (using the mathematical analytical techniques developed primarily by Ronald Fisher [1930], J. B. S. Haldane [1932], and Sewall Wright [1931, 1932] in the 1920s and 1930s); in experimental conditions using fruit flies, mice, and, nowadays, *E. coli*; and also in the field (as Dobzhansky himself did—and as still goes on in myriad studies such as John N. Thompson's (1994, 2005, 2013) co-evolutionary work, and Peter and Rosemary Grant's (2011) work on the Daphne Major Island population of the Galápagos finch species *Geospiza fortis*). Though some of these studies contemplate species as a whole (including their origins), most do not, focused as they are on the dynamics of gene frequency change mediated by natural selection, sexual selection, genetic drift, and a host of molecular causes.

And that, of course, is as it should be. But what was bothering me back in the late 1970s was that something deeper seemed to be going on: that many evolutionary biologists seemed to be acting as if species, in a sense, did not really matter—or, in a sense, even truly exist. By that time, Richard Dawkins's *The Selfish Gene* (1976), one of the most influential books on evolution in the latter half of the twentieth century, had appeared, and one could easily read that book and come away with the idea that genetic change through natural selection is all there really is to the process of evolution. Indeed, Dawkins took Darwin's original concept of natural selection, in which the distribution of heritable features of organisms within species could change owing to the differential success in the "struggle for existence" (as Darwin had put it) of individuals with various different traits leading to their differential reproductive success, down a notch. Dawkins seductively argued that it is the genes themselves (or, really, the information that they contain) that are competitively vying for representation in the next generation.

Thus in the 1970s evolution was seen by many biologists almost purely as a process of adaptive change through natural selection. Nor was this an entirely novel gambit: evolutionary biologists could rightly cite Charles Darwin himself, whose overwhelming message left to his intellectual descendants in all editions of *On the Origin of Species* was one of inexorable evolutionary change through natural selection through long periods of geological time. Darwin had essentially included species in the pile that accumulated on his cutting room floor. That Darwin's thinking as a younger man had been in reality far more complex than this cartoon caricature of his ideas was not really known to biologists, as modern Darwinian scholarship really did not come to the fore until the 1980s.

But there was another way of envisioning the organization and evolution of biological entities: specifically, species seen as discrete entities with births, histories, and eventually deaths. Adaptation generated through natural selection was very much a part of this theoretical perspective. But the origin of discrete species is a phenomenon that itself needed causal explanation. It also needed to be folded into general evolutionary theory. This is the tradition of evolutionary biology that I absorbed in graduate school—even though my friends and I, as budding paleontologists, were firmly ensconced in the geology department.

At Rochester, then, I was talking about the dichotomy between two lines of evolutionary thinking: one that denied the significance of the existence of discrete species and the possible connections between their origins and processes of adaptive change; the other that held that the origin of discrete species is intimately connected with adaptive change in evolutionary history.

And I was beginning to see that the implications of stasis, which, along with geographic speciation was the other component of punctuated equilibria, were potentially profound. Stasis is the empirical refutation that most adaptive change accumulates slowly within species as time goes by. The two views were in fact antithetical, to the point of mutual exclusion. What I didn't know then was that Darwin had reached that very conclusion in 1839 when he was thirty years old.

When the gist of these ruminations was published in 1979 as "Alternative Approaches to Evolutionary Theory," I called the two views the "transformationist approach" and the "taxic approach." It was the only paper of mine that Ernst Mayr wrote to me for a reprint. What matters is the circumstances under which adaptation through natural selection occurs. And I readily admit that it is not intuitively obvious that the two possibilities need be mutually exclusive. Darwin didn't think so when he first saw both possibilities. Nor did Dobzhansky and many others who reinvented the concept of geographic speciation beginning in the 1930s.

But Darwin had come to think so, and, based on punctuated equilibria, so had I by the late 1970s. Darwin chose what we called "phyletic gradualism" as the predominant way natural selection forges and modifies adaptations. In contrast, we chose geographic speciation as that arena in which selection-generated adaptive change mostly occurs.

And thus my entire narrative is about the development of a natural causal explanation for the origin of the species of the modern fauna and the addition of the phenomenon of adaptation generated by natural selection—thus setting up a conflict that flared up again in the twentieth century after Darwin had seemingly laid matters to rest.

The first half of this narrative begins with the story of the development of evolutionary theory in France, then Italy, and finally, by the 1820s and early 1830s, in Great Britain. It is a story inspired, at base, by Isaac Newton's insistence that there be natural causes (sometimes called "secondary" causes in deference to the prime causal actions of the Creator) for natural phenomena. Newton, by all accounts, was a deeply religious person, yet he was someone who saw no inherent conflict between his religious beliefs and his search for natural causal explanations for the phenomena he studied.

I tell the essentially adaptation-free story of the earliest decades of the scientific study of "transmutation," long since called "evolution," in chapter 1. The initial focal point for all early evolutionists was the search for a natural causal explanation for the origin of species alive today.

It is remarkable that the two basically polarized positions that have permeated evolutionary thought throughout the last two hundred years were established by the first two naturalists to seriously study the subject. Both men based their ideas on empirical data drawn from a comparison of fossil mollusks with the species still alive along the coasts of their respective countries. First, there was the Frenchman Jean-Baptiste Lamarck, who, in 1801, painted a picture of slow, steady, inexorable change of species through time. In Lamarck's "system," species do not so much become extinct as much as they transform slowly into their descendants, up to and including the species of the modern biota.

Little more than a decade later, in 1814, the Italian geologist Giambattista Brocchi published a rather different set of ideas on the origin of modern species (that's right . . . the first two scholars to lay the groundwork for modern evolutionary biology were both named John the Baptist). Brocchi said that species are like individuals: both individuals and species have natural causal explanations for their births and deaths. He denied Lamarck's vision of slow, steady, gradual change, seeing species appearances and disappearances as reflections of discrete events, births and deaths, by analogy with what happens to individual organisms such as us. Brocchi saw descendant species replacing one another, becoming progressively more and more like modern species in progressively younger rocks.

For the remainder of chapter 1, I examine the paper trail, primarily in Edinburgh in the 1820s, where the ideas of both Lamarck and Brocchi were considered in some detail. Most of the discussion was published anonymously in the *Edinburgh New Philosophical Journal*, founded in 1826 by geologist and medical school faculty member Robert Jameson. Patterns of geological replacement of species progressively more like, and eventually identical with, modern species became a commonplace catchphrase for British scientists not afraid to engage, however discreetly, with the concept of transmutation. The influential philosopher John Herschel also talked, in several places, of this pattern of progressive replacement of older species by their presumed descendants.

Both Jameson and the invertebrate biologist (and physician) Robert Grant were teachers and mentors to students, including Charles Darwin. Darwin read Lamarck's treatise of 1801 on invertebrates while a student at the medical school in Edinburgh, and on other evidence, including Darwin's own somewhat grudging words, there is little doubt that he was familiar not only with Lamarck's picture of evolutionary change, but with Brocchi's as well.

Darwin's exposure to scientific analysis, natural history, and transmutational thinking continued as he completed his formal education at Cambridge in the last few years of the 1820s. It was at Cambridge that Darwin read John Herschel's *Preliminary Discourse on Natural Philosophy* (1830). Herschel's book is a clarion call for the Newton-inspired search for natural causal explanations for natural phenomena, among which was the progressive, successive replacement of older, extinct species by more modern ones. Darwin once said that Herschel's *Preliminary Discourse* and Alexander von Humboldt's *Personal Narrative of Travels to the Equinoctial Regions of the New Continent* (1819–1829), were the two books that persuaded him to pursue a life in science.

In chapter 2, I move on to consider Darwin's experiences, primarily in South America, during the five-year journey of HMS *Beagle*. Through Geological Notes, Zoological Notes, Diary, and letters sent home to his Cambridge botany teacher, John Stevens Henslow, and family members, a picture of Darwin wrestling with the possibility of transmutation emerges from his earliest formative experiences (in 1832) with the fossils and living species he encountered.

First, Darwin thinks he has found evidence of the replacement of an extinct species by a descendant still alive in southern South America. He then, I think quite originally, begins to see geographic replacement of "closely allied" species of the modern biota (predominantly birds). And, as the icing on the cake, beginning in 1834 in the Falkland Islands (Malvinas), and especially, of course, in the Galápagos in 1835, he sees patterns of replacement of mainland species by similar ones found on islands, a pattern further broken down by replacement of species on different islands in the chain. Years later, Darwin opens the *Origin of Species* (1859) alluding to these patterns.

There was, however, an intriguing case of adaptation and transformation of characters in Darwin's notes, Diary, and some of his letters, involving a poisonous snake he collected in 1832. Beyond this, nothing much of a Lamarckian-style tracing of morphological transformation between closely related species appears in Darwin's *Beagle* writings.

When did Darwin become an evolutionist? Part of the solution to this perennial question is recognition that adaptation need not be the clinching evidence for the existence of a genuine evolutionary formulation, at least in these early days. Darwin was clearly toying with transmutational notions as soon as he started his field work on the *Beagle*. But when did he become convinced? I think the case is pretty clear that Darwin was a firm transmutationist in his first essay on the subject, written in Valdivia, Chile, and entitled, simply, "February 1835." This was written some six months before he reached the Galápagos. Before "February 1835," it is impossible to say.

Historians seem to agree that Darwin was a full-blown transmutationist by the time he wrote (in early 1837) the passages in the latter half of the Red Notebook after he was safely home. These passages are simply a rewrite of the gist of what he had to say in "February 1835."

Thus Darwin's earliest flirtation with transmutation was, on the whole, much more Brocchian than Lamarckian in tone and content. I close chapter 2 with Darwin's admission of precisely this in a letter written to Leonard Jenyns in 1844.

The rest of the first half of my narrative (chapter 3) looks at the late arrival of adaptation on the evolutionary conceptual scene. Darwin opens Notebook B in 1837 invoking the spirit of his grandfather Erasmus's book *Zoonomia* (1794, 1796) and, convinced that a natural law of adaptation is near at hand, Darwin seizes the moment and links adaptation directly with two scenarios: speciation in isolation, on the one hand, and gradual progressive change on the other. At first, Darwin sees no conflict between the two very different models. But after he formulates natural selection in Notebooks D and E, and after acknowledging the uncomfortable fact that paleontologists have failed to find any convincing evidence of gradual progressive change, by the end of Notebook E (1838) he acknowledges a conflict, and adaptation temporarily becomes a problem. At that point, Darwin feels compelled to choose between adaptive change in conjunction with the origin of species in isolation, or, instead, a picture of gradual wholesale transformation of species.

Darwin resolves the conflict by insisting that, lack of evidence to the contrary, gradual evolutionary change must be the general rule. I conclude this first half of chapter 3 with an exploration of why Darwin saw the two pictures as antithetical—a subject that recurs in the second half of the book when I look at the events in evolutionary biology beginning in 1935.

Basically, Darwin had what we later called "punctuated equilibria" firmly in his grasp, but dropped the ball, as he downplayed the importance of speciation-through-isolation in the generation and accumulation of adaptive change in evolutionary history. The remainder of chapter 3 examines how the importance of isolation in evolution, especially in conjunction with the origin and modification of adaptations, becomes more and more dim in Darwin's essays of 1842, 1844, mid-1850s, and finally *On the Origin of Species by Means of Natural Selection. Or, the Preservation of Favoured Races in the Struggle for Life* (1859, and the sixth edition of 1872, the one that still dominates the bookshelves).

Darwin left a lot on the cutting room floor, and the second half of my narrative can be seen as much as a restoration and elaboration of insights that were grasped even before Darwin as it can be seen as anything starkly new under the sun.

As a paleontology student in the 1960s, I of course knew next to nothing of all this early work that laid the foundation of evolutionary biology. I was hardly alone, as none of the earlier leading lights in evolutionary biology and paleontology of my era knew anything in detail about pre-Darwinian evolutionary thinking, nor of Darwin's early 1830s thoughts, never published in his lifetime, about evolution. I was only to learn of all this in the last decade of my career, after I had spent something like forty years learning about modern evolutionary concepts, working to improve evolutionary theory by bringing it more into synch with patterns of biological history that the fossil record seemed to be resonating so loudly and clearly, all the while trying to make a living and help support a family in suburban New Jersey.

I had, of course, been delving into Darwin's *On the Origin of Species* off and on throughout my career, starting as a college senior in 1965, when I took a copy to the lecture hall where Louis Leakey was scheduled to speak on the latest hominid fossil finds from Olduvai Gorge. I went early to grab a good seat. Leakey was very, very late, giving me the chance to delve into the *Origin*. I was overawed to begin with, afraid that I would not be able to wrap my mind around the great man's words. The Victorian prose was additionally unnerving, though I managed to shoulder on. But by page 15 of the first chapter, Darwin had already immersed himself and his unwitting readers in a discussion of the intricacies of domestic pigeon breeding. Ennui quickly replaced awe. I put the book down, and at long last Leakey walked into the room.

I finally read the *Origin* in graduate school, switching to the original edition, in a facsimile version with an introduction by Ernst Mayr, a book I still have and consult regularly. Especially when Gould and I were being accused of anti-Darwinian tendencies by the mid-1970s, it was important to be accurate in our characterizations of what Darwin in fact had said. But that was about the extent of my knowledge of pre-twentieth-century evolutionary biology.

Part II of my story deals with the sequence of events that led to the resurrection of that second line of thinking that Darwin had all but tossed aside.

Though the importance of geographic isolation to the evolution of new species had never been wholly lost sight of after 1859, it was not until Theodosius Dobzhansky resurrected this whole line of thinking in a short paper published in 1935, followed two years later by *Genetics and the Origin of Species* (1937, 1941, 1951), that speciation theory assumed an important position in modern, post-genetics evolutionary thinking. It was thus a geneticist (that is, not a paleontologist) who had, in effect, brought back notions of the reality of species and the importance of their origins, largely (if not exclusively) through geographic isolation.

This reinvention, revival, call it what you will, of the theory of the origin of species in isolation is the subject of chapter 4—the first chapter of the second half of my narrative.

Dobzhansky took the existence of species literally. He said that mutations at the level of individual organisms produce discontinuities: alternative discrete forms of genes, or "alleles." But at the level of populations, continuity is the rule, as allelic frequencies within populations could be modified gradually and intergradationally among generations, through the agencies of natural selection (a statistically deterministic process) and "genetic drift," the sampling error "random" process accounting for changes especially among populations within the same species originally proposed by Sewall Wright.

But Dobzhansky also pointed to a third, higher level: that of species themselves. Darwin had left the message that discontinuities among closely related species are largely a matter of extinction of intermediate forms. In sharp contrast, Dobzhansky argued that discontinuities between species are there for a reason: to focus species on "adaptive peaks." Dobzhansky thought that, while some variation is necessary within a species to maintain flexibility, to avoid extinction should environmental conditions change, nonetheless organisms are adapted to a set of environmental conditions that is typical for any given species. Were there no genetic gaps between species, such an adaptive focus would be difficult, if not impossible, for a species to achieve.

Five years later, in *Systematics and the Origin of Species* (1942), systematist Ernst Mayr followed up Dobzhansky's line of thinking, and the notion of geographic ("allopatric") speciation, an idea developed at least as far back as Darwin in the 1830s, was finally back on the table in evolutionary biology. Yet again, by the 1970s, the newer game in town ("selfish genes") seemed to imply that the existence and origins of species made little or no substantive difference to understanding evolution. In some quarters of evolutionary biology, by the 1960s and 1970s at least, species and speciation once again seemed to dim in importance.

Which brings us to chapter 5, devoted to the development of the theory of punctuated equilibria, fondly known around our house, and elsewhere, as "punk eek."

Schermerhorn Hall on the Columbia University campus in the 1960s housed five departments: Biology (top floors); Geology (appropriately on the ground floors down to the basement); Fine Arts and Archaeology (wedged incongruously in the intervening floors); and, in the building's "extension," Anthropology and Psychology. As an undergraduate in the 1960s, I mostly hung out in the Anthropology Department, toying with the idea of taking it up seriously under the cultural materialist Marvin Harris, who in fact became my academic role model. But by the time I graduated, I had already also started hanging around the Geology Department, having preferred collecting Pleistocene fossils to badgering people about the details of their personal lives in my rudimentary Portuguese while on a three-month trip to Arembepe, Brazil, in 1963.

I'll never fully understand why I thought that paleontology was the proper way to study evolution. It just seemed to make sense. After all, fossils tell us what was living before the species of the modern world evolved. They usually belong to species now extinct, the forerunners of what is living today. I was blissfully unaware that the study of evolution had long been firmly in the hands of biologists immersed in the living world and, for decades, especially in the intricate details of genetic systems. I had no clue that it had all begun with fossils and that our nineteenth-century predecessors began with the thought that the fossil record holds the key to understanding the origin of modern species. That was another thought we had to reinvent for ourselves.

But in the 1960s we paleontology students were keenly aware that genetics, ecology, and systematics (the core of evolutionary biology) focused on the nature and modes of origins of discrete species to a high degree. I used to commute between the Geology and Biology Departments on the Schermerhorn elevator, taking courses in both geology and biology, reading in both libraries, trying to understand how the disparate dialogues of these two rather separate worlds actually fit together.

One day a fellow graduate student in paleontology asked me what book I was carrying as we ascended to the lofty aerie of Schemerhorn's ninth floor (former home to Thomas Hunt Morgan's "fly room," where the elements of genetics were hammered out in the first two decades of the twentieth century). That was the very place that attracted the attention, and eventual presence, of that young Russian ladybird beetle systematist Theodosius Dobzhansky. The book I was holding that day on the elevator was Mayr's *Systematics and the Origin of Species* (1942), the opus that picked up where Dobzhansky had left off five years earlier, developing further the notion of geographic speciation and establishing its importance as a vital component of the evolutionary process.

Punctuated equilibria (my preferred term—as originally published; Steve Gould inexplicably changed it at some point to "punctuated equilibrium" as it perhaps is most commonly known these days) combines the near-universal pattern of stasis with the idea of the origin of species through geographic isolation. Ask virtually any paleontologist what happens to a species in the interval of time between their first and last recorded appearances in the fossil record. You'll be told that most of the known species appear not to change very much as they live out their "lives," often over millions of years. That's stasis.

And new species overwhelmingly appear rather abruptly in the fossil record. It is vanishingly rare to be able to trace the slow, steady, gradual modification of an ancestral species into an undoubted descendant, including tracing very young fossils slowly evolving into the appearance of still-living species. Thus the imagery caught in the name "punctuated equilibria": relatively quick spurts of evolution (the "punctuation") interrupting vastly longer periods of business-as-usual non-change of species as one chases them up cliff faces and pieces together their histories from disparate far-flung outcrops of rocks that, together, tell their evolutionary history.

Ernst Mayr once famously remarked (in the very book I was carrying on that elevator ride) that *of course* species are "real" entities: Why else have a theory of their origins? Certainly Gould and I thought (and I still do) that species are "real." Punctuated equilibria is a general characterization of the births, histories, and deaths of species. Species, to people like us, at any rate, are discrete in both time and in space. They are, as well, component parts of evolving lineages. Every species, as the overwhelming rule, is descended from an ancestral, progenitor species.

That Darwin grasped at least the gist of punctuated equilibria, before dropping it in favor of gradual progressive adaptive change, gives our work the slight aura of "reinventing the wheel." But I take the more optimistic view that speciation in isolation, the reality of species, and other topics were anticipated, if only to be promptly discarded, by Charles Darwin, thus leaving a vacuum. The phenomena remained essentially unexplained until the twentieth century, beginning with Dobzhansky's work in the 1930s. This suggests that these concepts centering on the origin of discrete species *had* to be reinvented. It further implies that, at a very fundamental level, they must be essentially right.

Prompted in part by the attention garnered by punctuated equilibria, theoretical analyses of the nature of species and their putative roles in an expanded evolutionary theory exploded in the 1970s and 1980s. These novel theoretical gambits, loosely gathered under the term "macroevolution," are the subject matter of chapter 6.

Independent of our work arose the "radical solution to the species problem" posed by biologist Michael Ghiselin: Ghiselin said that people tended to think of species as classes—that is, as collections of similar individuals bound together by traits held in common. Instead, they are entities with proper names—a position that appealed to me especially, as a paleontologist who had no trouble seeing species as real entities.

And of course, here again there is more than a hint of reinventing the wheel: Brocchi had proposed a formal analogy between individuals and species, at least in terms of their births and deaths. Indeed, seeing species as analogous to individuals was the initial insight that began the serious discourse in evolutionary biology that has led us down to the dialogues of recent times. Darwin used it as his entrée into the scientific contemplation of the history of life on earth. But when Ghiselin (1974) first proposed it (and as it was extensively elaborated by philosopher David Hull [1976]) it seemed to us all very new, bold, plausible—and very welcome.

If species are "real," discrete (and, I would add, for the most part evolutionarily stable) entities, species themselves can be seen as playing their own roles in evolutionary history. Once genetic isolation is established within sexually reproducing lineages, what happens later within a newly fledged lineage will have no effect on the collateral kin that also survive. New species embark on their own, separate evolutionary odysseys, variably spreading over the landscape and diversifying; or remaining localized and virtually invariant, or any combination thereof; and they may, in the course of their own "lifetimes" give rise to their own descendant species. And they and their descendant species may survive differentially more than their collateral kin; or they may die out sooner. Concepts such as "species selection" and "species sorting" (a term introduced by paleontologist Elisabeth S. Vrba) have been debated actively in recent decades.

Then, too, there is the delicious possibility that speciation may actually be a factor inducing adaptive change. There's a switch: instead of assuming, as is almost universally and traditionally the case, that it is adaptive change that "causes" speciation, it seems to some biologists these days that the reverse is more nearly the case: that it is the process of speciation that triggers adaptive change. Indeed, and once again, Darwin spelled out how that could be the case in Notebook B (1837), before essentially dropping the possibility in Notebook E (1839).

Stasis alone suggests that speciation may trigger adaptive change. At the very least we know empirically that morphological adaptive change occurs for the most part in and around speciation events, rather than in the much longer periods of species persistence with little or no lasting accumulation of change. Whatever variation and change through time an ancestral species may have developed prior to giving rise to its descendant, that change is seldom, if ever, going in the morphological direction of any eventual new daughter species that might later arise. It is exceedingly rare to find convincing examples in which geographic and temporal patterns of variation within a species turn out to be good indicators of the evolutionary future.

And now, as I write, some molecular biologists are claiming to see similar patterns in their data. Evolutionary genetic change is now emerging as closely correlated with branching points in evolutionary history, suggesting to some molecular geneticists that evolutionary change is actually contingent on episodes of evolutionary branching: speciation.

As Dobzhansky suggested, there is indeed a hierarchy of entities and phenomena at disparate spatiotemporal scales. Organisms, with their component genes, are parts of populations, with their own, continuous processes of selection and drift ("population genetics"); and populations are parts of species, which are genetically and reproductively discontinuous with closely related other species. This is the "genealogical hierarchy."

Ecological systems are also hierarchically arranged. And the interactions between components in both systems are, in the most general sense, what drives evolution. "Hierarchy theory," developed from 1980s to the present, recognizes the hierarchical structure of biological systems of varying spatiotemporal scales and dimensions, and examines how they interact to form a more complete understanding of the evolutionary process: the "how and why" the history of life on earth looks the way it does. The "sloshing bucket" theory I proposed a decade or so ago links scales of evolution of entities from local populations, through species and on up through higher taxa, with scales of environmental disruption, from local to global. The relation between extinction and evolution is profound.

All of this—geographic speciation, punctuated equilibria, species as individuals, species selection, species sorting, hierarchy theory, the sloshing bucket—struck especially those of us, primarily paleobiologists, as new and excitingly original. Here at last, we thought in our headier moments, paleontology, arguably for the first time, had real contributions to make to evolutionary theory alongside those who study the biology of living organisms. For my own part, it seemed that it had not been altogether a stupid move after all to try to study the evolutionary process by contemplating the long-dead fossilized remains of extinct species. And even some of those not especially enamored with our work did in fact welcome paleontologists to the "High Table" of evolutionary discourse.

The "data" of my story are the actual words of the players in the game, from Lamarck on up to the present day.

My narrative in the ensuing chapters, logically enough, begins with my take on the man who began it all: Jean-Baptiste Lamarck. It is of course open-ended, as life, certainly including the study of evolution, goes on. Thus my narrative does not coincide with the order in which I led the life, had the experiences, and learned my own professional lessons. I started in the middle and lived it up to the present time. Only later in life did I learn the details of the first half of the story.

I still think Ernst Mayr was having fun as he clutched his martini after giving a lecture at ninety-plus years. So, for that matter, am I.

I

BIRTH OF MODERN EVOLUTIONARY THEORY

1

THE ADVENT OF THE MODERN FAUNA

On the Births and Deaths of Species, 1801–1831

If evolution is the non-miraculous, scientific explanation for the origin of species, it will come as no surprise that the original problem that led to the development of evolutionary theory was the search for natural (secondary) causes to explain the origin of the modern biota: the living species of animals, plants, and fungi. It was a simple, hard-to-ignore fact that, as you climb vertical stacks of fossiliferous rocks, the fossils you find become progressively more modern in aspect, until, near the top of the sequence, in the youngest sediments, species still alive in the modern fauna begin to make their appearance.

Years ago, as a fledgling paleontologist, I used to wonder how it could possibly have been that, according to the historians of geology and biology I had read, early paleontologists and naturalists had evidently failed, if not to see, then at least to think seriously about, this pattern and what it might mean for an understanding of causal pathways leading to the advent of the modern fauna.

As it turns out, this general pattern of the progressive modernization of the fauna was indeed recognized by a diverse array of early natural philosophers, including some who remained opposed to the general idea of evolution (or at least Jean-Baptiste Lamarck's version of transmutation: Georges Cuvier, for example); others among the earliest to embrace transmutation (for instance, Robert Jameson); and the famous transmutational waffler (but skilled geologist and ecologist) Charles Lyell. As already noted, historians have tended to gloss over these early discussions of progressivism (or successionalism), trapped as most have been into thinking that evolution is fundamentally, perhaps even solely, a process of adaptive modification of organismal phenotypic characteristics.

THE OLD SHELL GAME

The fossil record of vertebrates has always held the most fascination for naturalist and layman alike, with the by-now dense fossil record of human evolution the secular equivalent of the Holy Grail. But it was the immensely denser, richer, and easier to find and to collect fossil remains of marine invertebrates in the younger sediments, typically distributed near the margins of continents, that led to the first empirical and analytical explorations of what might be the patterns and natural, non-miraculous causes underlying the appearance of new species, up to and including the advent of the modern fauna.

Earlier savants had advocated natural causal explanations for the history and diversity of life. Most famous, perhaps, was Charles Darwin's own grandfather Erasmus Darwin, who deserves recognition in his own right, and not just for the more successful accomplishments of his grandson. But as grandson Charles was later to complain in *Autobiography*, Erasmus's work was rather high on speculation and rather low on empirical evidence.

Systematics, the recognition and classification of natural groups of what came to be called "allied forms" by naturalists long before most of them subscribed to any notion of transmutation, was also a necessary, if not a sufficient, precursor to the emergence of what we would recognize today as modern evolutionary theory.

The Frenchman Jean-Baptiste Lamarck and the Italian Giambattista Brocchi shared not only the same first name, but far more importantly, the distinction of being the first to develop natural causal explanations of the origins of modern species that were both empirically and phylogenetically based.

Both Lamarck and Brocchi attempted to trace lineages among closely similar species considered to be members of the same natural group, especially among species within a genus. They were applying a phylogenetic perspective to the search for causal explanations of the origin of modern species. Nor was this insistence on focusing the discourse on closely similar, apparently allied species necessarily a no-brainer. For example, Charles Lyell, in the second volume of *Principles of Geology* (1832), estimated that species originate, and become extinct, roughly at the rate of one per year, measured in all the ecosystems worldwide. But his gaze was thoroughly and steadfastly non-phylogenetic. Believing that, for example, a new species of carnivore could not appear until the proper prey species had appeared, Lyell's notions of species origination and extinction were distinctly ecological, rather than phylogenetic, in character. Lyell's book was largely an extended refutation of Lamarck's ideas, and against the very idea of transmutation, though he did pirouette on the fence in one long paragraph, in which he acknowledged the pattern of the progressive "younging" of species as one approaches modern times in the fossil record.

Both Lamarck and Brocchi developed their earliest ideas through studies on Cenozoic (roughly 65 million years ago down to the youngest sediments) fossil mollusks. The advent of the modern fauna, as seen especially as the replacement of species within genera in marine mollusks, was the crucible in which the rudiments of evolutionary theory were born. And from its very inception, evolutionary process theory was divided into two camps, based on strikingly different claims on the very nature of patterns of stability and change of species in the fossil record: Lamarck and Brocchi made radically different claims about empirical patterns they saw revealed by their fossils.

Lamarck saw utter continuity and intergradation in time and space: all species are destined to change slowly into descendants through time, and, when the data were complete, would also be seen to intergrade geographically into other species of the same genus.

In sharp contrast, Brocchi regarded species as discrete and stable entities. Species have births, histories with little or no change, and eventually, inevitably, deaths programmed into them like the deaths of individual organisms. Old species do not change into descendants. Rather, they give birth to new, descendant species, just as organisms give birth to offspring.

It is one of the greatest ironies in the history of biology that Darwin, so often seen as the polar opposite of Lamarck in terms of the mechanisms they put forward to explain transmutation (natural selection versus the "inheritance of acquired characters"), came to insist that Lamarck was right when it comes to the patterns of evolution: species are bound to change through time, intergrading imperceptibly into descendants. While the fossil record seems to say the opposite, Darwin decided that the fossil record itself was at fault. That species do not seem to intergrade laterally, he decided, was that intermediate populations, subspecies, and closely related species had succumbed to extinction. The pattern of evolution Darwin left us with was Lamarckian—though his initial scientific thinking in evolution was pure Brocchian transmutation.

In the terminology I used in the late 1970s, then, Lamarck was thinking "transformationally," while Brocchi was thinking "taxically." The dichotomy was there from the inception. What follows is a bit more detail on Lamarck's and Brocchi's seminal views, and the critical demonstration that young Darwin was thoroughly familiar with both sets of ideas through his training in the 1820s—especially as a medical student in Edinburgh from 1825 to 1827, but also through his experiences at Cambridge in the years leading up to the *Beagle* voyage.

LAMARCKIAN TRANSMUTATION

Containing the earliest significant arguments proposing transmutation based on perceived temporal sequences of closely related species, Jean-Baptiste Lamarck's writings in the early nineteenth century stand firm as the earliest empirical, transmutational scientific work critical to the emergence of evolutionary biology as we know it today (figure 1.1). And though Lamarck is perhaps best remembered for the evolutionary ideas expressed in *Philosophie zoologique* (1809), his first important statements on the subject appear in the short introductory section on fossil mollusks in *Système des animaux sans vertèbres* (1801:403–411).

As becomes clear as we trace the influence of both Jean-Baptiste Lamarck and Giambattista Brocchi, primarily if not solely among the medically trained savants in Edinburgh in the 1810s and 1820s, both

FIGURE 1.1 James Hopwood Sr., *Jean-Baptiste Pierre Antoine de Monet, Chevalier de Lamarck.* (© National Portrait Gallery, London)

Lamarck's admirers and his detractors often mocked him for his exaggerated claims—not over process, but instead over his actual, putatively empirical, claims about the patterns of biological history he claimed to be generally true.

Consider, for example, the words of Charles Lyell in the opening page of volume 2 of *Principles of Geology*. Lyell equated transmutation strictly with Lamarck. (Later in the book, however, Lyell also discusses Brocchi by name. Lyell was virtually the only person to do so, save his own father-in-law.) Framing the problem of transmutation in explicitly empirical terms, Lyell ([1832] 1997) challenged his reader to "inquire, first, whether species have a real and permanent existence in nature; or whether they are capable, as some naturalists pretend, of being indefinitely modified in the course of a long series of generations" (183).

Lyell is sneering at what he goes on to characterize as Lamarck's blatantly absurd claim about the pattern of transmutation. I cite this passage here, anachronistically, simply because it so well typifies the way many natural philosophers felt about species. Whether creationists (like philosopher William Whewell) or those otherwise disposed toward the search for natural causal explanations of the origins of modern species (such as geologist Robert Jameson), pretty much everybody acknowledged the reality and apparent stability of species. What did Lamarck really say?

Lamarck was hired in 1793 to be in charge of invertebrates (animaux sans vertèbres) at the Natural History Museum (Jardin des Plantes) in Paris. It was not uncommon in these early days of natural history for appointments to be made to positions for which the nominee had no particular training or expertise: Charles Darwin's mentor at Cambridge, John Stevens Henslow, was already Professor of Mineralogy when he was named Professor of Botany in 1825, reflecting his growing interest in the subject. Henslow duly turned himself into a thorough-going botanist. Adam Sedgwick, who was similarly appointed Professor of Geology at Cambridge without any real track record in the subject, quickly became a leading geologist of his time. And so it was, two decades earlier, that Lamarck, taking his appointment seriously, set out to study invertebrate organisms, publishing *Système des animaux sans vertèbres* in Paris in 1801, or in "the eighth year of the Republic." Historian Janet Browne (1995) reports that Darwin read Lamarck's work on invertebrates while in medical school in the mid-1820s, and would have read the "text of his lecture" (83) on animal change through time.

It is in the introduction to the section "On Fossils" that Lamarck launches into a discussion of what fossils are and why they are interesting. On their own, he says, fossils, typically having lost their original color, have next to no intrinsic interest. But when fossils are seen as "extremely precious monuments" for the study of "revolutions" to which different places on the earth's surface have been subjected—and the study as well of the "changes that living beings have successively experienced there"—they become of the "highest interest" to naturalists (Lamarck 1801:406).

Lamarck (1801:407) goes on to say that several naturalists (evidently thinking especially of his colleague Georges Cuvier) have claimed that all fossils belong to the remains of animals or vegetables for which there are no living analogues. Cuvier was even then promoting his ideas of "revolutions" on the surface of the globe, the forerunner to modern discussions of mass extinction events. Cuvier was specifically concerned with proving that many species of large fossil mammals, such as the South American *Megatherium*, were not only extinct, but had no particularly close living relatives.

Not so, says Lamarck, for such naturalists "want to explain everything," and don't take the trouble to study the course that nature takes. And here Lamarck (1801:408) begins to walk an interesting tightrope— saying that in fact a small number of fossil species *do* have living analogues. And besides, among the fossil species without apparent living analogues, many belong to the same genera as are found in the modern oceans, differing more or less from their fossil relatives to the point at which they cannot be considered the same species.

The explanation of why only a few fossil species seem to be the same as living species, Lamarck tells us, is that most fossil species have changed in the course of time. He goes on to say that nothing is constant on the face of the earth, and "diverse mutations" occur, prompted by the "nature of objects and circumstances." Environmental change prompts changes in "situation, form, nature and appearance," and a diversity of habitats, a "different way of existing," followed by "modifications or developments in their organs and in the form of their parts," such that "every living being must vary insensibly in its organisation and form." The changes are propagated "through generation [reproduction], and after a long chain of centuries, not only will new species, new genera and even new orders appear, but every species will necessarily be varied in its organization and forms."

So Lamarck concludes that what is astonishing is not that so few fossil species have living analogues, but that there are any living species at all that are also known from the fossil record. And most of the relatively few fossil species with living analogues still extant must be among the youngest fossils known, for they simply have not had the time to change.

So one cannot really conclude, Lamarck says, that fossil species with no exact living analogues are in fact extinct. Humans may have driven some of the large fossil mammals to extinction, but this is not a matter that can be decided from the fossil record alone, as there are many regions on earth yet to be fully explored.

Hence Lamarck's tightrope: he wants to deny extinction, but has to admit that relatively few fossil species seem to be identical to still-living counterparts. He argues against external causation for extinction, arguing instead for a form of extinction-through-transmutation: the inevitable and insensible changes in form of species as the ages roll virtually ensures that no living species will be found to be exactly the same as its fossil counterparts. Though he and Cuvier actually agree that many fossil species have no evident exact living counterparts, Lamarck disagrees strongly with Cuvier on why this is so.

In 1809, Lamarck published *Philosophie zoologique*, easily the better known exposition of his transmutational views. Lamarck's words leave no doubt that Lyell, Darwin, and others who read him were correct to say that Lamarck saw the organic world in a constant state of flux within phylogenetic lineages, with constant, gradual change from one species to another through time, as well as in space. For example, Lamarck ([1809] 1984) wrote: "Let me repeat that the richer our collections grow, the more proofs do we find that everything is more or less merged into everything else, that noticeable differences disappear, and that nature usually leaves us nothing but minute, nay puerile, details on which to found our distinctions" (37).

Lamarck's notion entailed the destruction of the creationist view that species are inherently stable, discrete, forever unchanging, and are unconnected to any other species. The alternative view developed by Brocchi and adopted by a number of naturalists was that species are indeed real, and discrete, but are connected in phylogenetic series through a process of birth of descendant species from older ones. And unlike Lamarck, Brocchi and like-minded naturalists were perfectly willing to concede that species die—become extinct—through natural causes, though they disagreed on what those causes may be. And even a few (like Jameson) entertained a sort of mixture of Lamarckian and Brocchian views.

One final note on Jean-Baptiste Lamarck before moving on to Giambattista Brocchi. Lamarck died in 1829. He was eulogized by his long-time adversarial colleague Georges Cuvier (1836). The eulogy ("elegy") appeared in English translation in Jameson's *Edinburgh New Philosophical Journal*, most likely translated by Jameson himself. The standard interpretation is that Cuvier's eulogy is one of the unkindest on record. And it is indeed true that Cuvier does not shy away from accusing Lamarck of a hyperactive imagination and the development of ideas in the absence of hard evidence. For example, Cuvier says that "too great indulgence of a lively imagination has led to results of a more questionable kind."

Cuvier structures the eulogy around the twin themes of Lamarck's lasting empirical work and his great diligence in mastering new fields, on the one hand, and his penchant (according to Cuvier) of theorizing beyond the known facts, on the other. And we have already seen enough of Lamarck's writing on transmutation to agree with Cuvier (1836) that "M. de Lamarck could not fail to come to the conclusion that species do not exist in nature; and he likewise affirms, that if mankind thinks otherwise, they have been led to do so only from the length of time which has been necessary to bring about those innumerable varieties of form in which living nature now appears" (15). Lamarck indeed saw species in flux and not really existing in nature.

But (as pointed out by my colleague Stefano Dominici) this is simply not the whole story of Cuvier's eulogy to Lamarck. When Cuvier (1836) turns to Lamarck's work on fossil invertebrates, his tone changes as he writes:

> There is one branch of knowledge in particular to which he has given a remarkable impulse, the history, namely, of shells found in the bowels of the earth. These had attracted the attention of geologists from the time that the chimerical notion was exploded, which attributed their origin to the plastic force of a mineral nature. It was perceived that a comparison of such as belong to the different beds, and their approximation to those now living in different seas, could alone throw light on this anomalous phenomenon,—the deepest, perhaps, of all the mysteries which inanimate nature presents to our view. (20)

Cuvier, insofar as I am aware, was never a transmutationist in the sense that he ever admitted a need to understand the births of species in terms of natural causes. He was an early champion of the notion that species are real, and die through natural causes. But the births of species? According to historian Martin Rudwick, Cuvier was against Lamarck's ideas of transmutation, though not necessarily against natural causal explanations of the progressive modernization of fossil faunas. And here, in the elegy to Lamarck, Cuvier is acknowledging that Lamarck turned attention on the comparison of fossil with recent mollusks, thereby starkly posing the question of the origin of species of the modern fauna.

As they "approximate" to those now living, as one collects in progressively younger rocks up the stratigraphic column, light could be shed on this "anomalous phenomenon" that Cuvier pronounced "the deepest, perhaps, of all the mysteries which inanimate nature presents to our view."

A month or so after Jameson published the English translation of Cuvier's eulogy to Lamarck, John Herschel wrote to Charles Lyell on February 20, 1836, from Cape Town, commenting on volume 2 of *Principles of Geology* (1832). In a famous passage, Herschel wrote: "Of course I allude to that mystery of mysteries, the replacement of extinct species by others" (quoted in Babbage 1838:225–227; see also Kohn 1987:413n.59-2). Darwin (1839) saw a published version of that letter some two years later and remarked in Transmutation Notebook E that "Herschel calls the appearance of species, the mystery of mysteries. & has grand passage upon problem! Hurrah.—'intermediate causes'" (59; Eldredge 2005:8–9; Kohn 1987:413).

Twenty years later, Darwin (1859c) wrote in the second sentence of *On the Origin of Species* that "these facts [his observations on the living and extinct fauna of South America] seemed to me to throw some light on the origin of species—that mystery of mysteries, as it has been called by one of our greatest philosophers" (1).

Herschel had already played a big role in fomenting Darwin's initial enthusiasm for transmutation. Had Herschel's "mystery of mysteries" been triggered by Cuvier's eulogy for Lamarck? Whatever the case may be, Cuvier put his finger right on it when he underscored the importance of Lamarck's comparison between fossil and living species. He could not have liked Lamarck's conclusions. But in saying that Lamarck was the first to discuss the "mystery" by comparing progressively younger fossils with species of the modern marine fauna, he was really affirming that Lamarck was indeed the one who got evolutionary theory up and running in the new natural philosophy.

BROCCHIAN TRANSMUTATION

Who is this Giambattista Brocchi (figure 1.2)? And how can he possibly be a co-founder of modern evolutionary biology, and *the* founder of the line of thought that sees species as spatiotemporally real, discrete, stable entities, with births and deaths analogous to those of individuals? If he is that important, why has no one heard of him?

Well, not quite "no one." The historians Giuliano Pancaldi (1991) and, most recently, Martin Rudwick (1995, 2005, 2008) have certainly heard of Brocchi. Pancaldi coined the expression "Brocchi's analogy" in

FIGURE 1.2 Giambattista Brocchi.
(By permission, Museo Biblioteca Archivio, Bassano del Grappa)

reference to Brocchi's major conclusion that species are as real as individuals, and like individuals have births and deaths explicable through natural causes. And Rudwick (2005) has recently written that Brocchi's work "also suggested . . . that the *origin* of species, might have an equally natural, yet episodic, mechanism, analogous to the birth of individuals" (527). Most recently, my colleague Stefano Dominici (2010; see also Dominici and Eldredge, 2010), a geologist/paleontologist at the Museo di Storia Naturale, University of Florence, recognized the importance of Brocchi's work and has published extensive excerpts, translated into English, from Brocchi's monograph, concentrating on the "beautiful speculations" of Brocchi's theoretical discussions. What did Brocchi say?

Giambattista Brocchi's monograph *Conchiologia fossile subapennina* (1814) was the second important treatment of Tertiary molluscan fossils. Only Jean-Baptiste Lamarck's work on the fossils of the Paris Basin was earlier. Both men had compared their fossils with the known extant species of marine mollusks of their respective countries. They came up with rather different conclusions.

On Georges Cuvier's claim that no species known as fossils are still extant, Brocchi in part agreed with Lamarck (though not mentioning Lamarck explicitly by name), that species known as fossils are indeed still alive in the modern fauna. Brocchi claimed that Cuvier himself more or less knew that, or at least that the fossil record showed progressive approximation to the living fauna through time. Brocchi, addressing Cuvier's ideas on catastrophic extinction, wrote that "it does not account for the loss of fresh water shells and, what is more important, it cannot be applied to the loss of terrestrial quadrupeds, a matter on which he [Cuvier] has himself observed that all unknown species belong to the rocks older than those others that bear remains of known species or more similar to living ones" (Dominici 2010:593).

To Brocchi, the whole purpose of molluscan systematics is to shed light on the origins of modern species, and to explain their origins in natural (secondary) causal terms: "Since indigenous species are mixed with exotic ones, and those which we deem lost are together with others that nevertheless exist, we want to produce a system to reconcile facts of such variety, and that by satisfying all concomitant circumstances, tries to explain them without outraging reason, and in consonance with physics" (Dominici 2010:589).

"In consonance with physics" indeed. And to make his quest as plain as possible, Brocchi talks of the futility of analyzing lineages of extinct species that have no possible bearing on understanding the origin of the modern fauna, writing disdainfully of "busying ourselves to plot a distinct genealogy of some obscure descent since long gone" (Dominici 2010:588)—an evident reference to the study of genealogies of long-since extinct groups, such as the Mesozoic ammonoids.

But note the word "genealogy" here. Both Lamarck, and then Brocchi, were dealing with skeins of species that were so closely similar that the main debate was whether the species were the same, or whether they were closely related, but slightly different, species within the same genera. Naturalists since at least Carolus Linnaeus (the tenth edition of *Systema Naturae*, widely regarded as the starting point of modern systematic biology, was published in 1858) had long since become accustomed to talking of "natural groups" of "allied forms," but few if any of them (at least publicly) harbored transmutational views until Lamarck and Brocchi pointed to lineages within progressively younger fossil faunas, up to and including species still alive in the modern fauna, and sought an explanation of this "anomalous" "mystery" (Cuvier's words) in natural (secondary) causal terms. There can be no doubt that Linnaean-style systematics was the necessary forerunner to a full-blown search for an explanation of genealogies generally, and specifically of the advent of the modern fauna. That was the task Lamarck and Brocchi set for themselves.

But if Brocchi saw no immediate payoff for studying the phylogenies of wholly extinct fossil taxa, he also, to be fair, admitted that there is little use to simply cataloguing the shells of the modern Italian molluscan fauna: "I agree that wanting to describe all the shells of the sea, to sort them by order, genera and species does not lead to great consequences," except if they serve as a measuring stake to chart the progress of the advent of the modern fauna; Brocchi continues: "But if no one dared to treat in an academic way marine conchology, how could we usefully study fossil conchology which gives units of measure in geology and paves the way to so many beautiful speculations?" (Dominici 2010:589).

Beautiful speculations indeed—referring to Brocchi's own theory, so different from Lamarck's, of the dynamics underlying the patterns of stability and change, or the progressive appearance and disappearance of species. Before we look more closely at Brocchi's "beautiful speculations," though, we must look more closely at his thoughts on the nature of species, their appearances and disappearances, in the fossil record.

To begin with, Brocchi was dealing with younger rocks (hence fossils) than Lamarck. Lamarck's Paris Basin fossil faunas were Eocene in age (somewhere between 56 and 34 million years ago). In contrast, Brocchi's Upper Tertiary Miocene-Pliocene-Pleistocene strata were much younger, for the most part at somewhere during the past 5 million years. Thus it comes as no surprise that Lamarck could point to relatively few (though there were some!—having his cake and eating it too) species in the fossil collections that could be said to be still alive in the modern fauna. But Brocchi estimated that some 50 percent of the fossil species could be found still living in the offshore Italian waters.

Charles Lyell later became famous for proposing (in volume 3 of *Principles of Geology*) that Cenozoic rocks around the globe could be classified and correlated based on percentages of the species of the living fauna that could be identified in the fossil record. In the first proposal for the chronological subdivision of European Tertiary rocks, from the oldest to the youngest subdivisions, Lyell ([1833] 1997:394–398) estimated that only some 3.5 percent of known Eocene species survive into the modern fauna (in close agreement with Lamarck's minimalist claims), and roughly 18 percent for the succeeding Miocene. For the still younger "Older Pliocene," Lyell saw a range of from 33 percent to slightly over 50 percent survival of the fossil species into the modern biota; the rocks of Lyell's "Older Pliocene" include those studied by Brocchi, who estimated approximately a 50 percent species survivorship. And finally, 90 percent of species known as fossils in the "Newer Pliocene" (later known as the Pleistocene) are still present in the modern fauna.

Clearly Lyell, reputedly not a transmutationist until (finally!) Darwin published *Origin of Species* in 1859, had nonetheless been inspired by (or simply cribbed from) his predecessors—Lamarck, to some degree, and almost certainly Brocchi—to propose the famous "percentage of the Recent" relative dating of marine Cenozoic fossils, and thus the rocks in which they are found. And the anti-transmutationist Lyell was doing this based on the work of two people, Lamarck and Brocchi, whose "beautiful speculations" were distinctly transmutational in character.

Brocchi, as quite, almost explicitly, distinct from Lamarck, saw species as discrete, real, and pretty much stable entities. In what Stefano Dominici and I think must be a side-swipe at Lamarck's vision of species constantly changing, Brocchi wrote that "the alterations that take place in the animal machine and that are the symptoms of decline of the species, do not produce a large change in structure, what would be a true metamorphosis" (Dominici 2010:593).

In other words, species do not change all that much in time, though they may show some signs of aging. Brocchi's interpretation of fossil species is that they are discrete and stable. They do *not* display the sort of large-scale change that Lamarck claimed pertained to the fossil mollusks of the Paris Basin.

Which brings us directly to the core of Brocchi's "beautiful speculations." Biologists and historians of evolutionary biology tend to focus first on mechanisms: Darwin is said to have bested Lamarck because he had a more plausible mechanism of evolutionary change: "natural selection" as opposed to "inheritance of acquired characters." But this is not the whole story by far, as what really separates players in the game of evolutionary biology (now as much as in the past) are the often conflicting but all-important *empirical* claims of what evolutionary patterns should look like in any particular data set, be the data molecular, systematics of living biota, fossils, whatever.

We have already seen that Lamarck and Brocchi differed drastically on the claims of some of the most fundamental empirical patterns of biology: the constant and complete spatiotemporal flux of species claimed by Lamarck, contrasted with Brocchi's vision of species as discrete, more or less stable entities, with natural births and deaths creating lineages of descent leading right up to the appearance of modern species. And we have seen enough of Brocchi's words to know that, though the pattern of stability of species empirically matched the standard creationist claims of his day, he was looking for an explanation of the progressive "younging" of fossil faunas up to the advent of modern species in natural (secondary) causal terms (that is, "in consonance with physics"). Brocchi, in other words, was no creationist.

One tends to search in vain for causal mechanisms among early writers of a transmutational bent. Lamarck was an exception, granted; he was criticized not so much for his putative mechanism, but for his claims of constant flux between species through time and in space. Yet it is not accurate to say that early transmutationists (and here I am thinking especially of Brocchi) were more intent on establishing the reality of the progressive origin of one species from another through time than they were for causal mechanisms.

To the contrary: Brocchi and other early transmutationists were keenly aware that for their science to approach others, such as the already established role-model science of physics, they had at least to consider the thorny question of mechanism.

One reason why transmutational mechanisms remained elusive, of course, is that relatively little was understood of the basic biology of organisms. By the time Darwin discovered natural selection in the late 1830s, those interested in biology knew that organisms tend to resemble their parents, and that there is heritable variation in populations of every species. Beyond that, no one had the faintest idea why—though it must be said that with the simple addition of the principle of population-size regulation borrowed from Thomas Malthus, Darwin had all he needed to formulate what still stands as the essential statistical law of evolution: natural selection.

But self-conscious awareness of the primitive state of biological knowledge is not the whole story that explains the dearth of speculations on the origin of species. Martin Rudwick (2008) has remarked that "it was widely accepted among savants, even in Britain, that some kind of natural process, as yet unknown, must be responsible for the origin of new species" (480).

Rudwick was writing of the 1830s, but his comment applies, with somewhat lesser force, to the two or even three decades preceding the 1830s. Rudwick was alluding to the hush-hush nature of the question of the origin of species, saying that smart money knew that there must be a natural causal explanation for the existence and origin of species, but with no one daring to address the issue publicly. Very simply, there was a general reluctance to discuss the origin of species in natural causal terms throughout the 1820s and beyond. That attitude prevailed even in Darwin's time and was the main reason Darwin kept his thoughts to himself for so long. I say this even though I am aware that it has recently become fashionable again to blame Darwin's delay in publishing not on the frightening prospect of public outrage, but on his methodical attempt to keep his scientific ducks in a row in order to present as convincing a case as possible for transmutation. That's ridiculous. The structure (content and order of chapters) of Darwin's *Origin* of 1859 is the very same as his first extended essay on evolution—"Pencil Sketch" of 1842.

If people shied away from discussing transmutation, they found the discussion of extinction much less daunting. Cuvier led the charge with his claim that extinction is real, and that there were multiple extinction events, not just one (the "biblical deluge") that accounted for the apparent loss of ancient species, including the large Tertiary mammals. The biblical deluge remained a hot topic of controversy in geology/paleontology at least into the early 1830s. For example, in the early pages of Darwin's *Beagle* voyage Geological Diary, Darwin himself was constantly mooting (and rejecting) the idea of a single biblical flood as he studies the sediments of southern Argentina.

The early transmutationists considered four basic causes of extinction: external, physical, natural causes (either staggered, as Lyell thought, or concentrated in multi-taxon spasms, as Cuvier proposed); and—as virtually everyone considered as a possibility, or as Lamarck proposed—that species suffered a sort of "evolutionary pseudo-extinction" simply by evolving directly into their ancestors (an idea taken seriously as recently as in the mid-twentieth-century writings of the paleontologist George Gaylord Simpson). Or extinction is a false signal: early naturalists were mindful that the world had yet to be fully explored, and new species were being discovered all the time. Perhaps no species had truly become extinct, but are instead still alive, awaiting discovery. Darwin wrote in Geological Diary that he had heard reports when first arriving in southern South

America that a giant mammal—perhaps the missing and presumed extinct *Megatherium*—might still be alive and well in the South American interior.

And then there was the fourth possibility. This one was the brainchild of Brocchi (1814), as part of his core "beautiful speculations": "I thought I had enough inductions to venture to say that it is an established law that species die like individuals, and that they are bound to make their appearance in the world for a fixed span of time" (Dominici 2010:591).

Hence "Brocchi's analogy": specifically, with respect to extinction, Brocchi saw little evidence for external, physical causes for the extinction of species, though he admitted that such could "cut short" the life of species before their time. Instead, he suggested that species might have innate longevities, much as individuals do. Again, he posed this as a natural cause, implicitly suggesting that, however much God may be watching over an individual, and ultimately be responsible for their births and deaths, nonetheless the births and deaths of humans (and of course all other individual organisms) have natural causes.

And neither individuals nor species live forever. Absent compelling evidence of physical, external causes, Brocchi simply suggested that species, like individuals, must have internal longevities.

And as Rudwick has said, by implication, at least, the births of species also have natural causes. This, too, is analogous with the situation in individuals. But that's it: even less speculation as to causes for species births than Brocchi offers for species extinction.

But the key here is the search for natural causal explanations of natural phenomena. The clear implication is that species have births through natural causes every bit as much as they suffer extinction through natural causes. And this is the way early naturalists often skirted the topic of the origin of species per se: talking about extinction—the deaths of species—and suggesting that extinction is the converse of the problem of species' origins.

We need now to show that Brocchi's work was known and discussed in Great Britain from about 1816 to 1830. Beyond that point, Darwin picked up and developed these themes on the natural births and deaths of species on his own.

BROCCHI AND LAMARCK IN GREAT BRITAIN, 1816–1830

Tying the issue of the advent of the species of the modern fauna to the search for natural causal explanations is relatively straightforward, at least in the British literature of 1816 to 1830. Undoubtedly much more exists to be uncovered by future historians. Yet there is enough known at this point to link both Jean-Baptiste Lamarck's and Giambattista Brocchi's disparate transmutational views to Charles Darwin, who, even before he became a medical student in Edinburgh in 1825 (at the age of sixteen), had apparently already read his grandfather Erasmus Darwin's *Zoonomia*, replete with his evolutionary views.

As copies of Brocchi's (1814) two-volume monograph began to show up in Paris, London, Edinburgh, and other European seats of learning of the still-young scientific enterprises of geology and paleontology, one came to the hands of Leonard Horner, a Scottish geologist and the future father-in-law of Charles Lyell. Horner appears to have been fluent in Italian, and had visited family members who were living in Italy. So it was only natural that Horner would take an interest in Brocchi's work—and in 1816, he published a twenty-four-page review of it in the *Edinburgh Review*.

At the outset of the review, Horner expresses his overall admiration for Brocchi's work, including Brocchi's plates illustrating his fossils, which Horner (1816), in the very last sentence, proclaims to be "more beautifully executed than any thing of the kind we have ever seen before." Of the work in general, Horner wrote that "this appears to us to be a work of very great value and merit." He summarizes in very general terms Brocchi's contents, saying that the "chief object is to describe the fossil shells that are found in the clay, and gravel, of which the hills that skirt the base of the Apennines are composed, and to compare them with their prototypes now existing, either in the adjoining or more distant seas" (156).

But then, after a short summary of the organization and contents of Brocchi's monograph, Horner broaches his main criticism of the work, as he launches into a short lecture on the nature and philosophy of science. Basically Horner says that "facts" should be kept strictly separated from theory, and that "although there is not, we think, any reason to suspect that the facts have been in the slightest degree distorted, for the purposes of adapting them to some favourite system, we should have been glad to have had, in this Introduction, the descriptions, and the author's reasonings upon them, less mixed up together" (157).

So Horner is saying that Brocchi has a "system," or at least some conclusions too freely interspersed with the objective results to please Horner's taste. But Horner hastens to add: "We are by no means of opinion, that the geologist ought to confine himself to a bare narration of facts, and that he ought to abstain from all theoretical explanations upon them. . . . It must be admitted that theory is the ultimate object of all geological researches" (157).

He goes on to cite Brocchi to the same effect, in which Brocchi says that those who oppose hypotheses in geology are generally "ignorant of the use of them." So Horner is basically in Brocchi's corner, though he concludes this short disquisition of the philosophy of science by saying, for the rest of the review, that "we shall confine ourselves principally to the matters of fact. To enter into a consideration of the author's theoretical opinions, would extend our remarks beyond our limits, unless we were to omit what we have no doubt will be more generally interesting to our readers" (158)

So Brocchi indeed has a system, or at least some general conclusions, what Brocchi himself refers to at one point as "beautiful speculations." But Horner is not going to bother his readers with them, although he does tell his readers that they are there.

And true to his word, most of the rest of Horner's review of Brocchi's monograph applies to Italian geology in Brocchi's discussion, augmented to some degree by the results and observations of others.

Only near the end of this review does Horner turn to what he said at the outset was the main purpose of Brocchi's work: the description of the fossil shells flanking the Apennines, along with a comparison with their "prototypes now existing" in the present seas. Horner reports that Brocchi's fossil shells "are not scattered confusedly through the different beds, but often appear to be distributed in families and in distinct species"—meaning that many mollusks of the same distinct species (and even families) seem to co-occur. On the other hand, the species are usually not restricted to the same type of sedimentary rock, and they are not always present in otherwise identical lithologies.

Horner then reports that "the fossil shells of the sub-Apennines may be divided into two general classes, the one comprehending the shells that are still found in the sea, the other comprehending those whose prototypes are unknown" (174). Two pages later, he reports:

> In the catalogue which Lamarck has given of the fossil shells that have been found in the neighbourhood of Paris, there are about five hundred species; and it is wonderful how few of them resemble those found in the Sub-Apennine Hills, and how many genera there are among them, wholly unknown in Italy. But the most remarkable difference in the fossil shells of the two countries, is in those of which the prototypes are unknown. These greatly predominate in France, and, with a few exceptions, are wholly different from those which exist in Italy. (176)

Never mind that Horner is determined to designate the "still living" members of a species as the "prototype," rather than calling the fossil specimens the prototypes of the living. Perhaps there is some creationist perspective at work here. But nevertheless, Horner makes it clear that members of the fossil faunas in both France and Italy have living counterparts in the seas in their still-adjoining neighborhoods. Though this, as Horner promised, is a far cry from recounting Brocchi's deeper conclusions, he is nonetheless reporting an analysis based on Brocchi's (and Lamarck's) comparison of their respective fossils and the still-extant species of the modern biota in Italy and in France: analytic conclusions based on empirical evidence that unmistakably points to the appearance of elements of the modern fauna at some point in the geological past. Indeed, the direct comparison of Lamarck's and Brocchi's results is in itself significant, given what the world already knew about Lamarck's conclusions. Horner's statement that the search for theory to explain phenomena in general terms is legitimate and important, and his clear pronouncement that Brocchi in fact has published such a theory, apparently served to draw attention to the potential importance of Brocchi's work, even though Horner himself was loath to describe Brocchi's conclusions in any detail.

ROBERT JAMESON, ROBERT GRANT . . . AND CHARLES ROBERT DARWIN

Charles Darwin (1809–1882) grew up helping his father tabulate the flowering peonies in the family garden, making collections of rocks, roaming the fields, and shooting the wildlife. As Darwin ([1876] 1950) recalled in *Autobiography*, his father once told him that "you care for nothing but shooting, dogs and rat-catching, and you will be a disgrace to yourself and all your family" (17). Despite growing up in a well-to-do household, Darwin's father, Robert Waring Darwin, was determined that his son have an education befitting his family background—and more than that, have a profession.

Robert Waring Darwin was a physician, as was his famous father, Erasmus Darwin, before him, as well as many of the male collateral kin. Edinburgh was well established as *the* medical school of choice, and Robert decided to send young Charles there to follow in the family footsteps. This was in 1825, when Charles was only sixteen. Charles's older brother Erasmus ("Ras"—who trained in, but never practiced, medicine) was already enrolled at Edinburgh, softening the blow of leaving home and training for a profession that alternately bored and frightened him (though he did show some promise as he went with his father on some of his rounds back at home).

When first looking at Darwin's pre-*Beagle* education in detail, I was surprised to learn that Edinburgh was a hotbed of advanced, modern, even radical thinking. Radical scientific ideas, particularly the growing debate and open advocacy of various forms of transmutation in continental Europe, were definitely part of the intellectual currency in Edinburgh in the 1820s.

I knew about the medical school of course. But I had no idea that it housed, along with its lecture halls and operating theaters, the best natural history museum at the time in Great Britain. I had never heard of the *Edinburgh New Philosophical Journal*—a goldmine of scientific observations, results, and discussions of theory. I hadn't known, either, of the Plinian Society and other discussion groups, formal and informal, that regularly met to discuss the latest ideas. Nor had I ever heard that at least one of the lectures in one of Darwin's courses was entitled "On the Origin of the Animal Species."

I knew none of these things about the medical school—even when I glimpsed its shadowy presence one winter's night when I alighted in Edinburgh from a train from London. I was there to look over some of the cornets in the musical instrument collection overseen by Arnold Myers, renowned brass musical instrument historian. I was living outside my paleontological/evolutionary skin on that quick trip. I had no idea whatsoever that I was crossing Darwin's footsteps as I entered Reid Concert Hall to see the horns in the basement, paying no heed whatsoever to the remains of a building next door that had been so important in the emergence of evolutionary theory.

In retrospect, I wonder why I was so surprised to discover that Edinburgh had been such a hotspot of radical biological thinking. After all, I had known since the 1960s that Edinburgh had been home to James Hutton (1726–1797), considered the founder of modern geology. Hutton showed conclusively that molten rocks that later cooled and hardened (igneous rocks, such as lavas) can be injected into, or lie on top of, sedimentary rocks. His was the linchpin demonstration that not all rocks were formed from chemical precipitates: the theory called "Neptunism," which up until then had been the prevailing geological theory.

But it was Hutton's demonstration of the enormity of geological time, and his conviction that processes of erosion, deposition, volcanoes and the like were always at work in the geological past, that was really important to me as a young paleontologist. Hutton taught us that the history of the earth can be interpreted in a rational way by considering the world around us today, and by paying close attention to what the outcroppings of rock have to tell us if we only examine them carefully. This aspect of Hutton's work was the direct forerunner to Charles Lyell's principle of uniformity: the processes we see operating today are the very same that shaped the events in the long history of the earth.

Hutton was a farmer, geologist, and also a physician, initially taking courses at the Edinburgh medical school before completing that phase of his studies on the Continent. My first trip to Edinburgh was in 1973, the year after "Punctuated Equilibria" (Eldredge and Gould 1972a) had appeared. I was there to do some field work collecting trilobites with Euan N. K. Clarkson, whose work on trilobite eyes in the 1960s had shown me the way to understanding the evolution of my own species in

North America. That, in turn, was the work that led me to the notion of punctuated equilibria.

But this was my first trip to Europe, and of course we saw the sights in Edinburgh. One stop was Holyrood Palace, Scottish residency of the British royal family. And just outside, in Holyrood Park, loomed the Salisbury Crags and the hill known as "Arthur's Seat," where Hutton made his observations that led him to demolish Neptunism, as the evidence there is unequivocal that igneous rocks are *not* formed as chemical precipitates from cold oceanic waters. Hiking up the stretch to Arthur's Seat, with its magnificent view of Edinburgh (and the Firth of Forth to the north) was in itself exhilarating. But even more exciting was examining the evidence first-hand of the baking of sedimentary rocks by a molten mass of lava that had subsequently cooled to stone. I saw the evidence, knowing what role it played in the history of my science, and was thrilled.

The "deep time" revealed in Hutton's work, and his direct stimulus to the later work of Charles Lyell, which in turn had had such an effect on the young Darwin, was an utterly necessary precondition for the serious contemplation of the history of life on earth. Small wonder, then, that Edinburgh's intellectual environment also fostered equivalent, radical ideas in zoology, as they had a generation earlier in geology.

But the story of Darwin's mentors in medical school, and what he himself absorbed while he was there, was far less known in the 1960s than the saga of Hutton (and his "Boswell," John Playfair). I was ignorant of all of this, but should not have been particularly surprised when I finally learned, in the first decade of this century, of the foment around transmutational ideas that seethed in Edinburgh. The creative intellectual ferment of Edinburgh in science in the late eighteenth century and first decades of the nineteenth had carried on in biological topics well into the 1830s.

Robert Grant (1793–1874), himself an Edinburgh-trained physician, was an out-and-out champion of Jean-Baptiste Lamarck. Almost solely through Darwin's word, Grant has become easily the most famous of what historian James Secord has called "the Edinburgh Lamarckians." His connection to Darwin was extremely important.

But it was Robert Jameson, originally known to me only as an important early geologist who was stodgily one of the last defenders of

Neptunism, who was the real central mover-and-shaker of the new wave of thinking. It was Jameson who founded and ran the natural history museum. It was Jameson who founded, edited, and wrote for the *Edinburgh New Philosophical Journal*, almost undoubtedly the author of the first (and most probably the second) of two anonymous essays exploring transmutation, as well as, presumably, a précis of the work of the transmutational ideas of the Frenchman Étienne Geoffroy Saint-Hilaire, also Giambattista Brocchi's death notice, Georges Cuvier's eulogy of Jean-Baptiste Lamarck, and, I suspect, many more goodies yet to be unearthed by modern scholars.

Darwin ([1876] 1950) himself writes that "the Plinian Society was encouraged, and, I believe, founded by Professor Jameson" (23)—the latter statement disputed by some historians. The Plinian Society was a social group where students would gather to discuss their research and opinions on the hottest topics of the day in natural history. And it was indeed Jameson, in the fifth edition (1827) of the translation of Cuvier's *Essay on the Theory of the Earth*, who added a discussion to the section on illustrations, couching his thoughts on transmutation (not labeled as such) in terms of the advent of the modern fauna.

Darwin wrote a list of "books that I have read thro since my return to Edinburgh," presumably meaning the start of his second term in medical school in 1826. Ras had departed for London, and Darwin had turned to a deeper involvement with his fellow students, and with the invertebrate zoologist Robert Grant. The list of books is eclectic, but includes three or four items of special interest: his grandfather Erasmus's *Zoonomia*, Cuvier's *Theory of the Earth* (Darwin had a personal copy of Jameson's fifth edition, which appeared early in 1827), "several numbers in the New Edinb: Philos Journal" (Jameson's journal), and seven "Pamphlets by Drs. Grant & Brewster on Nature History"—some of the last undoubtedly including Grant's papers in Jameson's journal, one of which was a paper by Grant with Darwin's own (unattributed) original observations on the motile "ova" (larvae, actually) of the bryozoan *Flustra*. Grant had trained Darwin in field collecting and microscopic study of marine invertebrates.

If it weren't for the second volume of Lyell's *Principles of Geology* (1832), I might never have heard of Brocchi, at least from my own readings, despite the fact that his ideas are all over the place in books and articles of the 1820s and 1830s, and were demonstrably the basis of Darwin's deliberate testing of Lamarckian versus Brocchian versions of transmutation while on the *Beagle*. And while Grant seems to have been performing the still widely practiced petty pilferage wrought by mentors mining their students' work for their own gain and glory when he ripped Darwin's data off, it is Darwin's own behavior that has arrested my attention in this regard.

As James Secord has said, basically the only thing anyone seems to know about Jameson comes from Darwin's *Autobiography*, written for his family's eyes only when Darwin was sixty-seven years old, about fifty years after he was at Edinburgh. I think the portraits that Darwin paints of both Grant and Jameson are fraught with a desire to retain an aura of his own originality, and are, at least in Jameson's case, deceptive and off-putting.

Of Jameson, Darwin ([1876] 1950) writes: "During my second year at Edinburgh I attended Jameson's lectures on Geology and Zoology, but they were incredibly dull. The sole effect they produced on me was the determination never as long as I lived to read a book on Geology, or in any way to study the science" (23).

We'll meet Darwin's denial that someone had had any "effect" on him again, as he says the very same thing, not only about Robert Grant, but also about his own grandfather. It is like a mantra, this "no effect on me" repeated claim, yet it is always followed by a mitigating, second-thought statement. Here, in the next sentence on Jameson, is Darwin's "on the other hand" second thought: "Yet I feel sure that I was prepared for a philosophical treatment of the subject" (23).

Darwin embellishes his disdain for Professor Jameson, writing in *Autobiography*:

> Equally striking is the fact that I, though now only sixty-seven years old, heard the Professor, in a field lecture at Salisbury Craigs, discoursing on a trap-dyke, with amygdaloidal margins and the strata indurated on each side, with volcanic rocks all around us, say that it was a fissure filled with sediment from above, adding with a sneer that there were men who maintained that it had been injected from beneath in a

molten condition. When I think of this lecture, I do not wonder that I determined never to attend to Geology. (23)

Darwin was a hip young student who knew that the Salisbury Crags (I love the Scottish "Craigs") were a huge piece of Hutton's evidence. Darwin's probably anachronistically advanced terminology in this passage shows that he, even as a young student, thought that Professor Jameson was hopelessly out of date.

And I cannot help but observe that our visit to Salisbury Crags/ Arthur's Seat in Holyrood Park in 1973 was the first, wholly inadvertent, place where my path crossed the footprints, not only of James Hutton, but also those of Charles Darwin. I love that.

As to Jameson not inspiring him, never mind that he had been reading the fifth edition of Jameson's translation of Cuvier's *Theory of the Earth*, replete with Jameson's own illustrations. Never mind that one or more lectures (the number isn't clear) in Jameson's course had been entitled "On the Origins of the Animal Species." Methinks the sixty-seven-year-old Darwin doth protest too much.

Apparently it was the anthropologist Loren Eisley who first drew attention to the short, unsigned essay "Observations of the Nature and Importance of Geology" (1826), in the first volume of the *Edinburgh New Philosophical Journal*. That it was in fact written by Jameson, the journal's editor, as James Secord (1991) has persuasively argued, is bolstered by a lingering, wistful hint of Neptunism, that school of geological thought that held that the rocks of the earth's crust all originated as precipitates in the primordial ocean. Near the end of this essay, the author writes "whether granite be a production of fire or water." Jameson was known to have been one of the very last defenders of a doctrine that most geologists had abandoned by the mid-1820s.

"Observations on the Nature and Importance of Geology" ([Jameson] 1826) is a strong statement on transmutation. It mixes distinct Brocchian and Lamarckian elements, though Lamarck is the only one of the two mentioned by name (perhaps the reason why many historians have supposed that Robert Grant was the author). It delivers a thumbnail sketch of the economic (including agricultural) and even moral significance of geology, constituting an encyclopedic survey in capsule form of the values of this branch of science in fact echoed in the structure of Grant's "Essay on the Study of the Animal Kingdom" ([1828] 1829), discussed shortly. There is no direct record that "Observations on the Nature and Importance of Geology" was in one of the "numbers" (issues) that Darwin read in this journal, though some historians have found it difficult to believe that he hadn't in fact read it. Among the notable, isolated gems in this essay are the word "evolved" in a totally unambiguous context: Lamarck "maintains, that all other animals [other than "infusorians"], by the operation of external circumstances, are evolved from these [worms] in a double series, and in a gradual manner" ([Jameson] 1826:297). He also uses the catch-phrase "origin of species" when he writes, "Geology does not inform us merely of the origin of animal species, but also of their destruction" (297).

Jameson (the presumed author of "Observations on the Nature and Importance of Geology") makes the very interesting point that the physical side of the history of the earth seems very different from the history of life. There is no discernible progress from simple to complex in the rocks of the earth's crust. If anything, to Jameson, the oldest rocks seem to be "the most compound."

In contrast, there is a gradation from simple to complex in both the spectrum of the anatomies of currently living species (in the two separate sets of them distinguished by Lamarck) and the fossil record of the history of life: We "meet with the more perfect classes of animals, only in the more recent beds of rocks, and the most perfect, those closely allied to our own species, only in the most recent; beneath them occur granivorous, before carnivorous, animals; and human remains, are found only in alluvial soil, in calcareous tuff, and in limestone conglomerates" (297). As we shall soon see, Jameson was not reluctant to include humans in his views on the advent of the modern fauna.

And the "sagacious" Lamarck "has expressed himself in the most unambiguous manner" to the effect "that all other animals, by the operation of external circumstances, are evolved from these [worms] . . . in a gradual manner." Thus "the scale of gradation, according to which he arranges the animal kingdom, is, at the same time, the history of their origin; and the discovery of this truly natural method, the most important problem of the natural philosopher" (297).

Thus Jameson (if indeed he wrote these words) is strongly in favor of the search for natural causes underlying natural phenomena. The rest of this paragraph is worth quoting in full:

> Although it should not be forgotten, that this meritorious philosopher [Lamarck], more in conformity with his own hypothesis than is permitted in the province of physical science, has resigned himself to the influence of imagination and attempted explanations, which, from the present state of our knowledge, we are incapable of giving, we nevertheless feel ourselves drawn towards it, and these notions of the progressive formation of the organic world, must be found more worthy of its first Great Author than the limited conceptions we commonly entertain. (297)

This presumably includes such "limited conceptions" as the notion that each species is separately created by God, and that the unproven "explanations" probably referring to Lamarck's mechanisms is, I think, borne out by yet a third anonymous contribution (in the same journal and almost certainly again by Jameson), summarizing the experimental results of Étienne Geoffroy Saint-Hilaire, discussed shortly.

There follows a remarkable page examining theories of extinction, headed by the already quoted comment that geology sheds light not only on "the origin of animal species, but also [on] their destruction." First the author says that, of the "vast number of animal remains" in the fossil record, "but few belong to species now living, and these only, in the most recent rock-formations." The vast majority of species once alive are long since gone. And he says that one explanation for apparent extinction is that many living species have yet to be discovered.

The writer then asks whether "this destruction . . . ha[s] been the result of violent accidents, and destructive revolutions on the earth"? We have independent evidence that Jameson did not accept Cuvier's ideas on "revolutions" (mass extinctions), from his own sections accompanying the translation of Cuvier's text. He goes on:

> Or does it not rather indicate a great law of nature, which cannot be discovered by reason of its remote antiquity? Within the narrow

> circle of vision in which the organic world manifests itself to our observation, we observe individuals only going to destruction, and in opposition to that, great preparations made for the preservation of the species. But if all living perish, may no point of duration have been fixed for the species; or do we not rather, in these signs of a former world, discover a proof, that, from a change in the media in which organic creatures lived, and from powerful causes operating upon them, their power of propagation may be weakened, and at length become perfectly extinct? Is the continual decrease, then, which we observe among some species, a consequence of the various modes of destruction they experience from the hand of man, or may it not rather be produced by natural circumstances, and be a sign of the approaching old age of the species? (298)

So here is Brocchi, with individuals and species analogized, and the interesting addition of individuals "going to destruction," but making "great preparations" for the survival of the species as a whole—by "generation" (reproduction). Like Brocchi himself, the discourse is limited to the causal explanation of the extinction of species, in the context, though, of the assumption that natural causal explanations for the origins of species will ultimately be found. And though the phrase "old age" resonates in pure Brocchian fashion in the last words of the paragraph's final sentence, mention is made also of the possibility that the "powers of propagation" may be weakened from external, environmental causes.

But then the author of this essay turns to Lamarck's idea that many species become extinct simply by slowly evolving themselves out of existence: extinction of a species is a side-effect of its having been transformed into a descendant species. The writer begins with the admission that "the distinction of species is undoubtedly one of the foundations of natural history; and her character is the propagation of similar forms"—meaning that species are discrete, and that the individuals within species "propagate" similar offspring:

> But . . . are these forms as immutable as some distinguished naturalists maintain; or do not our domestic animals and our cultivated or artificial plants prove the contrary? If these, by change of situation, of climate,

> or nourishment, and by every other circumstance that operates upon them, can change their relations, it is probable that many fossil species to which no originals can be found, may not be extinct, but have gradually passed into others. (298)

So here, right after Brocchi, is pure Lamarck: the reason why so few living species can be found as fossils is at least in part because those fossil species kept on slowly being transformed into the species now alive in the modern world. Note that "originals" here means not the ancient species, but the living ones—just as Horner (1816) used the word "prototypes" in the review of Brocchi. And it is interesting, too, that the author of this essay invokes domestic animals as evidence for mutability.

But why is there little direct evidence of such slow, steady, gradual change from an ancestral species into another, including living species from recently "extinct" species whose remains are found in the youngest sediments? The author of this piece does not invoke a poor fossil record (though he does concede earlier in the essay that the record is incompletely known and often the fossils that have been found are poorly preserved). Rather he invokes, in this 1826 essay, the vast reaches of geological time. He maintains that change is very slow. True, he remarks, "We indeed observe that the Ibis, which was worshipped in ancient Egypt, and preserved as a mummy, is still the same in modern Egypt; but what are the few thousand years to which the mummy refers, in comparison with the age of the world, as its history is related by geology" (299).

Napoleon's foray into Egypt ignited a fierce discussion among the Parisian natural philosophers. Cuvier twitted his pro-transmutational colleagues Lamarck and Geoffroy on this lack of change between mummified and modern sacred ibises, and the author here is simply agreeing with the transmutationists.

One possible, albeit slight, argument that Darwin indeed read this essay was that, though he examined plankton hauls and individual specimens of bryozoans, octopi, corals, nudibranch mollusks, and fossil marine invertebrates, almost none of his observations on invertebrates can be firmly linked with his early speculations on transmutation. Rather, it was the living and fossil species of vertebrates (chiefly birds and mammals—though one snake also stands out) that provided the patterns that brought Darwin to favor, not only transmutation, but a distinctly Brocchian, rather than a Lamarckian, brand of transmutation. I always thought that Darwin focused on vertebrates over invertebrates because he was surer of the biogeographic affinities of vertebrates: edentate mammals, for example, were known to be characteristic of South America. He wanted to be sure that he was looking at the patterns of stability and change in time and space of species that could not also be found—and evolving—elsewhere.

Thus it is striking to read these words in "Observátions on the Nature and Importance of Geology" ([Jameson] 1826):

> As there are among dicotelydons, that is, among the most perfect plants, no species, which are at the same time indigenous to the hot climates of the old and new worlds, so both halves of the globe in the same zone possess mammiferous animals, birds, reptiles, and insects peculiar to each. Species common to both are found only among inferior gradations of organization, and species of a higher order are found only in those high northern latitudes, where the continents were undoubtedly at one time conjoined. (299)

Darwin's obsession with endemism while on the *Beagle*, and his consequent focus on higher animals, may have been inspired in part by reading these words.

This anonymously penned essay, as far as I am aware, is the first post-Lamarckian, post-Brocchian work to examine, and even to develop a bit further, the transmutational debate in terms of the search for natural causal explanations, patterns of transmutation (and biogeography), and putative causes underlying the origins and extinctions of species. Matched in scope only by Grant's lecture of 1828 (published in 1829), this short paper (1826) in the *Edinburgh New Philosophical Journal* is a landmark in the history of evolutionary biology. It is impossible for me to imagine that Darwin somehow managed to miss it.

The fifth edition of Jameson's translation of Cuvier's *Essay on the Theory of the Earth. With Geological Illustrations by Professor Jameson* was published in 1827. As we have already seen, this book appears on the list of works Darwin "has read thro" since "my return to Edinburgh." Though Jameson was no fan of Cuvier's vision of multiple catastrophes wiping out most species in earlier faunas, replaced by newly created species as geological time wears on, he nonetheless admired the man and his work, manifestly enough to prepare successive editions in which presumably the main additions were in his own "geological illustrations." And Jameson added a lot to the geological illustrations for this new, fifth, and final edition of Cuvier.

At first glance, the reader looking for signs of transmutational thinking in Jameson's geological illustrations is apt to be a bit disappointed. Nothing like the clear, ringing declarations of "Observations on the Nature and Importance of Geology" (1826) present themselves in nearly as much depth, scope, and intensity—with the exception of two short passages. The first occurs in the preface to the fifth edition, hinting at transmutation, but also firmly linking Jameson to the anonymous essay "Observations on the Nature and Importance of Geology":

> Can it be maintained of Geology, which discloses to us the history of the first origin of organic beings, and traces their gradual development from the monade to man himself,—which enumerates and describes the changes that plants, animals, and minerals—the atmosphere, and the waters of the globe—have undergone from the earliest geological periods up to our own time, and which even instructs us in the earliest history of the human species—that it offers no gratification to the philosopher? Can even those who estimate the value of science, not by intellectual desires, but by practical advantages, deny the importance of Geology, certainly one of the foundations of agriculture, and which enables us to search out materials for numberless important economical purposes? (Jameson 1827b:vi–vii)

The first sentence is a pretty stark declaration of a natural causal process underlying the changes from "monades" to "man," while the second sentence links Jameson firmly with "Observations on the Nature and Importance of Geology."

The second passage on transmutation is a rather remarkable statement in the section "On the Universal Deluge." Here we read:

> These [physical geological] operations at the earth's surface generally appear to have produced its present figure, and to have designed it for the habitation of numerous organic beings; This appears as early as a suitable element occurred; first in water, then in land animals; and, like the formation of rocks, we observe produced a regular succession of organic formations, the later always descending from the earlier, down to the present inhabitants of the earth, and to the last created being who was to exercise dominion over them. (431)

Again, pretty stark, as Jameson transforms a generalized statement of the geological succession from simple to complex into a declaration of transmutation—with the simple yet unmistakable phrase "the later always descending from the earlier, down to the present inhabitants of the earth." That the advent of the modern fauna is a key to his thinking is thereby made especially plain, though Jameson pulls his punches a trifle when he refers to human beings as having been "created."

The next sentence is also arresting, if for no other reason than it also seems to confirm the suspicion that Jameson was indeed the author of "Observations on the Nature and Importance of Geology." Jameson writes: "But here occurs this important distinction: the organic world, with youthful vigour, renews itself daily, and decomposes its materials only to reunite them by fresh combinations in uninterrupted succession; while the powers of the inorganic world appear almost extinguished" (431).

This is the very same contrast between the dynamism and progressive changes in life through time contrasted with the immanence, sameness, and non-progressive aura of the physical history of the earth. Jameson does allude, several sentences earlier, to the effect that "new elements" (meaning water, first, then dry land) had on life, where the inorganic world seems to set the stage for changes in the succession of life. Despite this slight inconsistency, the distinction between the inorganic and organic histories of the earth, mentioned in these two successive publications, strongly suggests that Jameson was the author of both.

As far as Jameson's text per se is concerned, though, that's it for transmutational themes in *Essay on the Theory of the Earth. With Geological Illustrations*. But as it turns out, that is by no means the only material of direct transmutational import to be found in these pages. Jameson (Jameson 1827b:547–550), in fact, saved the best, literally, for last: just before we read "The End," there are two remarkable tables ("tabular views"), the first taking up three pages, the second only one (figure 1.3). They amount to the first quantitative analysis of the fossil record since Brocchi published his observations on the percentages of living species of mollusks known from the fossil record.

The first table is "The Genera of Fossil Mammifera, Cetacea, Aves, Reptilia and Insecta,—exhibiting their Geognostical Number and Distribution." The genera are listed down the left-hand column. Then there are six vertical columns: genera that are found (1) living only; (2) living and in the fossil state; (3) fossil only; (4) in the Strata anterior to the Chalk; (5) in the Strata of the Chalk; and (6) in the Strata posterior to the Chalk. Then there are two more columns: Number of Species (1) in the Living State (he makes no entries here) and (2) in the Fossil State; Jameson concludes the table with a column for miscellaneous observations.

The "table of genera" makes it obvious at a glance that many genera of the tabulated groups occur in the fossil as well as the living "state," and that most of these occur in Tertiary sediments (that is, "posterior to the Chalk," which is Cretaceous in age). There is in general a clear pattern of progressive modernization of the fauna as one approaches the "Recent." That this pattern is a function of descent (species within genera) is implicit, and certainly borne out by the brief passage on descent in the text itself.

The second table lists the genera that are found in higher taxa ("Classes, Orders, or Families") in the same columnar organization of geological occurrence as in the first table. For the univalved and bivalved mollusks, as well as the mammals and Cetacea (whales—which are of course mammals), most of the genera found as fossils are, again, Tertiary in age, though the "lower" orders of animals display less of a pattern of progressive modernization of the fauna as one ascends the stratigraphic column.

547

TABULAR VIEW

OF

The GENERA of FOSSIL MAMMIFERA, CETACEA, AVES, REPTILIA, and INSECTA,—exhibiting their Geognostical Number and Distribution.

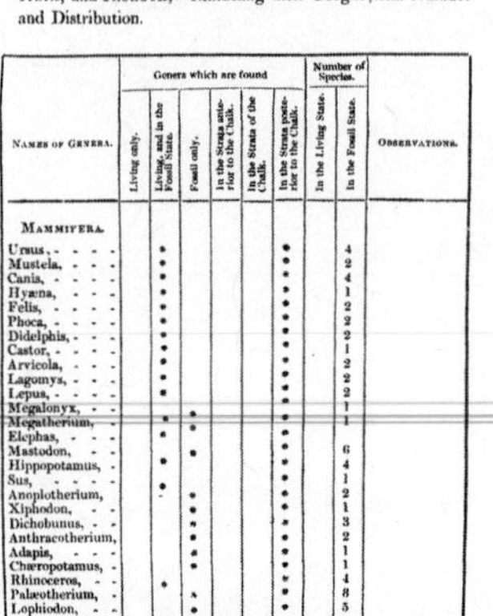

NAMES OF GENERA.	Genera which are found: Living only.	Living, and in the Fossil State.	Fossil only.	In the Strata anterior to the Chalk.	In the Strata of the Chalk.	In the Strata posterior to the Chalk.	Number of Species: In the Living State.	In the Fossil State.	OBSERVATIONS.
MAMMIFERA.									
Ursus, - - - -		•				•		4	
Mustela, - - -		•				•		2	
Canis, - - - -		•				•		4	
Hyæna, - - -		•				•		1	
Felis, - - - -		•				•		2	
Phoca, - - - -		•				•		2	
Didelphis, - - -		•				•		2	
Castor, - - - -		•				•		1	
Arvicola, - - -		•				•		2	
Lagomys, - - -		•				•		2	
Lepus, - - - -		•				•		2	
Megalonyx, - -			•			•		1	
Megatherium, -		•	•			•		1	
Elephas, - - -		•				•			
Mastodon, - -			•			•		6	
Hippopotamus, -		•				•		4	
Sus, - - - -		•				•		1	
Anoplotherium,			•			•		2	
Xiphodon, - -			•			•		1	
Dichobunus, - -			•			•		3	
Anthracotherium,			•			•		2	
Adapis, - - -			•			•		1	
Chæropotamus, -			•			•		1	
Rhinoceros, - -		•				•		4	
Palæotherium, -			•			•		8	
Lophiodon, - -			•			•		5	

548

TABLE—continued.

NAMES OF GENERA.	Genera which are found: In a Living State only.	Living and Fossil.	In a Fossil State only.	In the Strata anterior to the Chalk.	In the Strata of the Chalk.	In the Strata posterior to the Chalk.	Number of Species: In the Living State.	In the Fossil State.	OBSERVATIONS.
MAMMIFERA.									
Tapirus, - - -		•				•		1	
Elasmotherium, -			•			•		1	
Equus, - - - -		•				•		1	
Mus, - - - -		•				•		1	
Cervus, - - -		•				•		5	
Bos, - - - -		•				•		4	
Myoxus, - - -		•				•		2	
CETACEA.									
Manatus, - - -		•				•		1	It is extremely difficult to make out the genera of the Birds, whose remains occur in a fossil state, and there are more of them than those mentioned.
Delphinus, - -		•				•		4	
Balæna, - - -		•				•		3	
AVES.									
Sturnus, - - -		•				•		1	
Pelecanus, - -		•				•		1	
Charadrius, - -						•		1	
REPTILIA.									
Testudo, - - -		•				•		6	
Crocodilus, - -		•			•			6	
Plesiosaurus, - -			•					1	
Ichthyosaurus, -			•	•				4	
Pterodactylus, -			•	•		•		3	
Rana, - - - -		•				•		1	
Mosasaurus, - -			•					1	
Salamandra, - -		•			•	•		1	

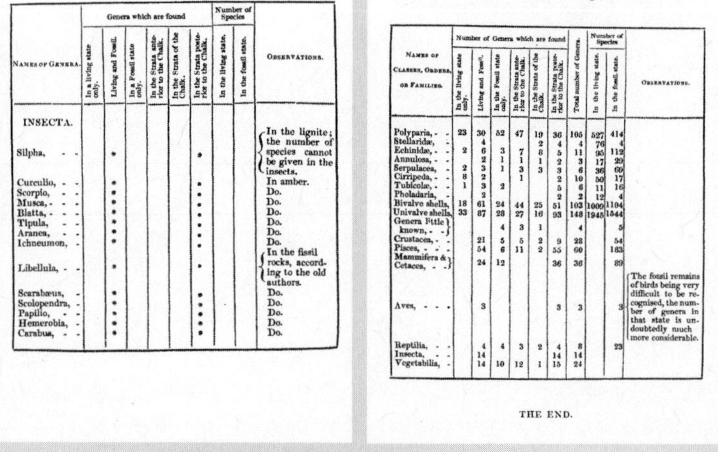

549

TABLE—continued.

NAMES OF GENERA.	Genera which are found: In a living state only.	Living and Fossil.	In a Fossil state only.	In the Strata anterior to the Chalk.	In the Strata of the Chalk.	In the Strata posterior to the Chalk.	Number of Species: In the living state.	In the fossil state.	OBSERVATIONS.
INSECTA.									
Silpha, - -		•				•			In the lignite; the number of species cannot be given in the insects.
Curculio, - -		•				•			In amber.
Scorpio, - -		•				•			Do.
Musca, - - -		•				•			Do.
Blatta, - - -		•				•			Do.
Tipula, - -		•				•			Do.
Aranea, - -		•				•			Do.
Ichneumon, -		•				•			Do.
Libellula, - -		•				•			In the fissil rocks, according to the old authors.
Scarabæus, -		•				•			Do.
Scolopendra, -		•				•			Do.
Papilio, - -		•				•			Do.
Hemerobia, -		•				•			Do.
Carabus, - -		•				•			Do.

550

TABULAR VIEW

OF

The CLASSES, ORDERS, or FAMILIES, of ANIMALS, occurring in a Living and Fossil State, with their Geognostical Distribution.

NAMES OF CLASSES, ORDERS, OR FAMILIES.	Number of Genera which are found: In the living state only.	Living and Fossil.	In the Fossil state only.	In the Strata anterior to the Chalk.	In the Strata of the Chalk.	In the Strata posterior to the Chalk.	Total number of Genera.	Number of Species: In the living state.	In the fossil state.	OBSERVATIONS.
Polyparia, - -	23	30	52	47	19	36	105	527	414	
Stellaridæ, -		4			2	4	4	76	4	
Echinidæ, - -	2	6	3	7	8	5	11	95	112	
Annulosa, - -		2	1	1	1	2	3	17	29	
Serpulacea, -	2	3	1	3	3	3	6	36	69	
Cirripeda, - -	8	2		1		2	10	50	17	
Tubicolæ, - -	1	3	2			5	6	11	16	
Pholadaria, -		2				2	2	12	4	
Bivalve shells,	18	61	24	44	25	51	103	1009	1104	
Univalve shells,	33	87	28	27	16	93	148	1945	1544	
Genera little known, - -			4	3	1		4		5	
Crustacea, - -		21	5	5	2	9	28		54	
Pisces, - - -		54	6	11	2	55	60		183	
Mammifera & Cetacea, - -		24	12			36	36		89	
Aves, - - -		3				3	3		3	The fossil remains of birds being very difficult to be recognised, the number of genera in that state is undoubtedly much more considerable.
Reptilia, - -		4	4	3	2	4	8		23	
Insecta, - -		14				14	14			
Vegetabilia, -		14	10	12	1	15	24			

THE END.

FIGURE 1.3 Robert Jameson, "The Genera of Fossil Mammifera, Cetacea, Aves, Reptilia and Insecta" and "The Classes, Orders, or Families, of Animals, occurring in a Living and Fossil State." (From *Essay on the Theory of the Earth. With Geological Illustrations by Professor Jameson* [1827b:547–550])

Jameson does not discuss these results in the text, or even allude to them, as far as I can find. But someone, again almost undoubtedly Jameson, does in fact utilize these data in the second anonymous transmutational essay, again in the *New Edinburgh Philosophical Journal*, published in the same year as Jameson's fifth edition of Cuvier: 1827. Its title: "Of the Changes Which Life Has Experienced on the Globe."

The very first paragraph contains, as far as I have been able to ascertain, the first explicit usage of the imagery of new species replacing older extinct ones: the author writes "new animals and vegetables have assumed the place of those that have been destroyed and whose ancient existence is only revealed to us by their fossil remains" (Jameson 1827c:298). This imagery soon took center stage as the pattern to be explained by natural (secondary) causes in the British literature.

And the advent of the modern fauna is also well represented here: "As we approach nearer to the present time, we find in all places remains more and more resembling those of the plants and animals which now live in the same country" (299). Note how endemism is again a part of the mix when it comes to comparing fossil species with those of the "Recent."

Going on to discuss changes in biogeographic distribution, the author of this little piece also says that physical changes in fact contribute to extinctions: "The beings, which were unable to resist the influence of these various causes were destroyed and disappeared from the earth, with the circumstances for which they were created; new species appeared with new conditions of existence" (300). And then, in what must have been a direct reference to Jameson's tables, he writes:

> But, in examining the series of fossils that are found buried in the strata of the globe, there is nowhere perceived a distinct line of demarcation between the different terms of that series, so as to prove that life has been once or oftener totally renewed on the earth. On the contrary, we discover in it a proof of the successive and gradual change which we have pointed out. Certain primitive types have indeed completely disappeared, but they are found existing at various epochs, and their remains are blended with those of more modern types; along with new species of types still existing, we find some of anterior epochs; certain

> genera that yet obtain are common to all the terms of the series; and toward the end of the series, we find the remains of some of our present species along with ancient types and extinct species. (300)

It is very much as if the author of "Of the Changes Which Life Has Experienced on the Globe" had Jameson's tables in front of him when he wrote this passage. And that he sees in these data irrefutable proof that Cuvier's ideas on multiple mass extinctions and subsequent wholesale reinvention of life are wrong is clear also from his conviction that his "theory, which has been founded on all the facts that have been established, cannot but prevail over the systems hitherto proposed."

Then there is the more celebrated and better-known Robert Grant. Darwin's ([1876] 1950) passage on Grant in *Autobiography* is well known: "He [Grant] one day, when we were walking together, burst forth in high admiration of Lamarck and his views on evolution. I listened in silent astonishment, and as far as I can judge without any effect on my mind. I had previously read the *Zoonomia* of my grandfather, in which similar views are maintained, but without producing any effect on me" (21).

Just as he did for Jameson, Darwin uses this almost mantra-like repetition of "no effect," as he seeks to minimize the influence of both his grandfather's and his mentor Grant's evolutionary views. But he does go on to say in the very next sentence: "Nevertheless it is probable that hearing rather early in life such views maintained and praised may have favoured my upholding them under a different form in my *Origin of Species*" (21).

In other words, Darwin finds he can't deny some kind of influence on him in his grandfather's work (indeed, in his grandfather's previous stature as England's most outstanding, even notorious, evolutionist) and in the words and research program of his mentor Robert Grant. Though he was a medical doctor, Grant had been trained in invertebrate zoology in Paris, although not by Lamarck. That his work is distinctly transmutational lies in the fact that he was searching for connections between the plant and animal kingdoms. Indeed the very word "bryozoan" means "moss animals," in spite of the fact it was known before the 1820s that these minute colonial organisms are clearly animals; Grant also used the word "zoophyte" ("animal plant") regularly.

But the hallmark of Grant's Lamarck-infused research was the search for anatomical intermediacy between different groups of species, genera, or even higher taxa: for example, in the paper naming the new genus *Cliona*, Grant (1826) claimed that the combination of characters present in *Cliona* provided the missing connection between *Alcyonium* (a cnidarian) and sponges.

Darwin "often" went to the shoreline of the Firth of Forth on the northern outskirts of Edinburgh with Grant. They were collecting marine invertebrates, and Darwin's own research was on the bryozoan *Flustra*. Grant taught Darwin how to use a light microscope, and Darwin made some novel observations on what they thought was the "ova" (in reality, the larvae) of these bryozoans. Darwin discussed these results at a meeting of the Plinian Society. Writing in *Autobiography*, Darwin ([1876] 1950) said: "The papers which were read to our little society were not printed, so that I had not had the satisfaction of seeing my paper in print; but I believe Dr. Grant noticed my small discovery in his excellent memoir on Flustra" (22). In other words, Grant ripped Darwin off, publishing Darwin's observations without crediting his young disciple. And I think it likely that it was on one of these excursions to the Firth of Forth that Grant "burst forth" in his paean of praise of Lamarck.

The Firth of Forth means something to me, as it ranks now to have been the second time that I crossed paths with Charles Darwin. I was to start chasing him down in earnest only in the first decade of the twenty-first century, after I had become involved in developing the exhibition *Darwin, Discovering the Tree of Life*—and thereby becoming so hooked that I had to pursue him in Great Britain and in South America.

We went out to the Firth of Forth with our host Euan Clarkson on a pleasant weekend excursion to collect Carboniferous fossil shrimp along the shoreline. We were in a mixed group of amateurs and professionals. Among us was the by-then-elderly Robert Watson-Watt, who invented radar during World War II. Watson-Watt sported a three-piece heavy mustard-colored tweed suit and spats—rather formal attire for such a shoreline experience, especially as the day was rather warm for those seaside Edinburgh climes. We were lucky indeed to have met Watson-Watt, as he died shortly thereafter at the age of eighty-one.

But once again, I had no idea at the time that Darwin had been there in the course of his training as a field zoologist, and under the guidance of an avowed adherent of Lamarck, whose own research focused on deep historical connections between such fundamentally disparate groups as plants and animals.

We ended that 1973 journey with a visit to London. Westminster Abbey was a don't miss, and I found myself gazing at the slab covering Isaac Newton, the patron spirit of the entire enterprise of modern science, and especially of the search for natural causes for natural phenomena. But then I looked down, and found I was standing above Charles Darwin. Instantly there were tears in my eyes.

I had no idea Darwin was buried in Westminster. Many years later, in conjunction with doing that Darwin exhibition, I found out that Darwin had planned to be buried in the church cemetery in Downe Village, less than one mile from his house. But politics intervened. Darwin, as the Newton of evolutionary biology, could not have landed up in a more fitting place.

In any case, in *Autobiography*, Darwin dismisses Grant as having done nothing "more in science" after taking a position at the University of London in 1828—evidently not taking Grant's inaugural address entitled "Essay on the Study of the Animal Kingdom" (1829) seriously. In fact, Grant's essay is so beautifully fraught with Brocchi and Lamarck (though mentioning neither by name) that it must stand as one of the clearest and strongest exponents of a transmutational viewpoint published in England, with the exception of the earlier work of Darwin's own grandfather, of course. Darwin does laconically acknowledge Grant's notice of Darwin's "small discovery in his excellent memoir on Flustra." Whether his dismissal of Grant as a serious player in evolutionary thinking arose over that plagiarism flap, or simply from a desire to de-emphasize Grant's importance on the development of Darwin's own thinking, is impossible to judge.

Grant's "Essay on the Study of the Animal Kingdom" (1829), the publication of the introductory lecture at the University of London (1828), is an understated, yet eye-opening gem in the early history of overtly published transmutational thinking in Great Britain. True, as Grant discusses the "succession" of animal species, he uses the word "created," and goes out of his way in one particular passage to allude to the "Author of Nature." The fiery, explosive words effusive in their praise of Lamarck, which one might have expected given Darwin's report of Grant's "outburst," are basically missing here: this was, after all, a public lecture in the intellectually more staid environment of London. Grant was a very early faculty appointee in Great Britain, and he evidently chose his words carefully.

What we find, instead of a ringingly explicit proclamation of Lamarckian transmutation, is a measured description of the range of subject matter of zoology, including especially comparative anatomy and comparative physiology, but also biogeography, behavior, and even the human economic and philosophical interests attached to the animal world.

But the origins, nature of the histories, and extinctions of species are also here, as Grant (1829) writes:

> This science [zoology] enquires into the origin and duration of entire species, and the causes which operate towards their increase or their gradual extinction; the laws which regulate their distribution, and the changes they undergo by the influence of climate, domestication, and other external circumstances. . . . It investigates the characters and relations of the extinct races of animals, the remains of which are found every where imbedded in the crust of the earth; and thus enables us to read, in their imperfect and mutilated remains, the history of the former inhabitants of this globe. It points out to us the order followed in the successive creation of animals, as discovered by their fossil remains, by their degree of organization, and by their relations to the strata of the earth, and unfolds the nature of those remarkable revolutions that have repeatedly taken place in the Animal Kingdom, in consequence of sudden or gradual changes in the condition of the surface of this globe. (6–7)

Grant's rhetoric here is fraught with the search for "laws" and "causes" explicitly linked with everything on the history of species: their histories, the changes they undergo, and their eventual extinctions—everything but their origins. Yet we know from Darwin himself (and of course other evidence, including Grant's empirical work) that he was in fact a transmutationist. One can only assume that on the very question of actual origins of species, Grant was doing the conventional thing by treading lightly; and indeed it is in the following paragraph that Grant does the obligatory genuflect to the "Author of Nature," "to exalt our conceptions of the infinite wisdom, power and goodness of the great Author of Nature, as displayed in his minutest works."

In other words, had Darwin not assured us of Grant's fiercely held support of Lamarckian transmutation, it still might have been possible—though just barely so—to read Grant's words cited earlier as not especially transmutational, if not downright creationist.

Nor is the passage purely Lamarckian in flavor, though the allusion to the "changes" that species undergo in the course of their histories is definitely reminiscent of Lamarck's predicted pattern of the constant flux that species undergo through time. Rather, the passage has all the elements of the sort of hybrid vision of transmutation that was first developed in Jameson's writings (assuming the two anonymous essays that had appeared shortly before were his), melding the births and deaths of species with evidence of continuity and change both within and among species.

Grant also alludes to the "constant warfare between the species"—ideas generally associated with the Frenchman de Candolle. From physiology to ecology, Grant is manifestly interested in dynamic, natural processes.

But it is in the passage that Grant indeed "bursts forth"—and what he writes is a brilliant synthesis of Brocchian and Lamarckian viewpoints: Brocchi's analogy between the births and deaths of individuals and species is clearly here (though once again restricted to their deaths), and Grant (1829) tells us why Lamarck's expectation of consistently smooth intergradations between species is in fact not observed, as he writes:

> In this vast host of living beings, which all start into existence, vanish, and are renewed, in swift succession, like the shadows of the clouds in a summer's day, each species has its peculiar form, structure, properties, and habits, adapted to its situation, which serve to distinguish it from every other species; and each individual has its destined purpose in the

> economy of nature. Individuals appear and disappear in rapid succession upon the earth, and entire species of animals have their limited duration, which is but a moment, compared with the antiquity of the globe. (11–12)

This is pure Brocchi. Grant is saying, despite his Lamarckian propensities, that species are discrete entities, and are *adapted* to their own "situation." That means that every individual, as a representative of its species, plays a "destined" role "in the economy of nature." Sticking with individuals, he then says that they "appear and disappear" in "rapid succession," and jumps back up to the level of species: species themselves have their own "limited duration."

Like Brocchi, Grant shies away from an explicit assessment of the analogy between the births of individuals and of species. Nor does he make clear the cause of the short durations (in the scale of geological time) of species. But the basic analogy between individuals and species is abundantly clear in this passage, and Brocchi's fundamental "beautiful speculation" is unmistakably in print, in English, three years before Darwin set sail on the *Beagle*.

Grant (1829) continues:

> Numberless species, and even entire *genera* and tribes of animals, the links which once connected the existing races, have long since begun and finished their career, and their former existence can now only be recognized by their skeletons, embalmed in the soft superficial strata of the earth, or by their casts preserved in the more solid rocks. . . . Almost every stratum [contain fossils that] the Zoologist must learn to decipher the history of the species, and discover their relations to the existing races. (11–12)

And there it is: all species (and "races") were once connected. The reason we don't find the Lamarckian "smear" of intergradation is that extinction has long since claimed those intermediates. It is the job of the "zoologist" (that is, paleontologist or, nowadays, "paleozoologist") to discover these ancient remains and to "decipher" their relations with one another—*and with the existing fauna*. And though these words call to mind Grant's well-publicized admiration of Lamarck, it must also be said that Brocchi would have had no objection to them either: in Brocchi's lineages of ancestral and descendant species, the degree of anatomical differences between species was usually rather modest.

These two cited passages are in reality parts of a single, rather remarkable, modern-sounding paragraph. Use of the word "adapted," the phrase "economy of nature," the notion that species are indeed discrete entities with origins and eventual deaths (like individuals), and that all species and "races" were once interconnected, the skein only being broken by extinction, is as concise a characterization of the evolutionary history of life anyone (especially a paleontologist like myself) could ever hope to see. And these words were written in 1828—a culmination and integration, I think, of the transmutational ideas being developed and bandied about by those "Edinburgh Lamarckians"—especially Jameson and Grant.

There are further passages in which Grant reiterates the need to compare fossil species with living species, and by "pointing out the extensive and terrible catastrophes to which the Animal Kingdom has often been subjected," thereby understanding a "cause of the many apparent interruptions in the chain of existing species": his allegiance to Lamarck once again showing through, though once again Brocchi would not have objected. And while we have come to think of Grant strictly as a marine invertebrate zoologist (and there is indeed little evidence that he actually collected and analyzed fossils, or indeed delved into any other geological subject, on his own), nonetheless it was *de rigeur* for anyone braving the quest for a non-miraculous explanation of the origins of species in terms of material, natural (secondary) causes to address the fossil record and the connections between extinct and modern species simply in order to solve that cardinal question: How did modern species originate? Grant was no exception.

Giambattista Brocchi died in Egypt in 1826, and a brief notice of his passing was published—again anonymously, but again almost certainly written by Robert Jameson (1827a)—in the *New Edinburgh Philosophical Journal*. Brocchi is memorialized as "so well-known by his numerous works on geology and conchology" (383). Though nothing of Brocchi's analogy and "beautiful speculations" is mentioned, the death notice is important simply in the fact that it is only the second time we know of that Brocchi's name is mentioned in a British publication. That Jameson thought so highly of him is clear from the fact of the death notice itself.

One last, again anonymous, and once again undoubtedly Jameson-produced essay concludes this short review of the early transmutational literature centered in Edinburgh, mostly in Jameson's own journal. The essay, published in 1829, is entitled "Of the Continuity of the Animal Kingdom by Means of Generation, from the First Ages of the World, to the Present Times" (subtitled "On the Relations of Organic Structure and Parentage That May Exist Between the Animals of the Historic Ages and Those at Present Living, and the Antediluvian and Extinct Species"). This is a short (3½-page) précis of a forthcoming "series of memoirs" by Étienne Geoffroy Saint-Hilaire "of which the first only has been read."

The essay proclaims Lamarck to have been a genius ahead of his time—and also proclaims that Geoffroy's experimental work proves that Lamarck was correct in asserting that transmutation occurs as a reaction to change in external circumstances. It is difficult to judge whether some passages are meant strictly to summarize Geoffroy's views, or whether they do so, but with the anonymous author's wholehearted agreement and approval.

Summarizing Geoffroy's basic transmutational interpretation of the history of life, the author (Jameson [1829]) writes:

> M. Geoffroy St. Hilaire believes in an uninterrupted succession of the animal kingdom, effected by means of generation, from the earliest ages of the world up to the present day. The ancient animals, indeed, whose remains have been preserved in the fossil state, are all, or at least almost all, different from those which now exist on the surface of the globe. But this is not a reason for thinking that they could not have been the ancestors of these latter. In the *first* place, the extinct species are united with the living species by the closest analogy. All have without difficulty entered into the prescribed limits of our great classifications; all, as being formed of analogous organs, *seem* to be nothing but modifications of the same being, of what is now called the vertebrate animal. (153)

This summary of Geoffroy's views suggests that he (that is, Geoffroy) basically agreed with Lamarck that comparatively few species known as fossils are still to be found in the modern fauna; given the emphasis on vertebrates in Geoffroy's work, this is hardly surprising. Yet Geoffroy used "mastodon" phylogeny to show that a progression of fossil and living species may belong to the same genera, as species of some genera of fossil elephants are still alive in the modern fauna. Jameson himself had (in 1827) already listed nearly thirty vertebrate genera known to be "living and in the fossil state."

The anonymous author proceeds with his précis of Geoffroy's views—this time pertaining to his older colleague Lamarck: "M. Geoffroy St Hilaire cited, as the performance of an author who outstripped the age in which he lived, the work in which M. de Lamarck treats *Of the Influence of circumstances upon the actions and habitudes of living bodies*, and, reciprocally, *on the influence of the actions and habitudes of living bodies upon the modification of their parts*" (154).

In what now seems to be as much the opinion of the anonymous author as it is of Geoffroy, we read that though Lamarck's facts are not all trustworthy, his conclusions are undoubtedly true ("right for the wrong reason"): "The particular facts on which M. Lamarck rests his grand idea, are far from being perfectly correct. Perhaps there is not even one of them that is not blemished by some inaccuracy; yet the conclusion which he draws from them is true—such is the power of genius on foreseeing the great truths of nature." But "to establish M. Geoffroy's opinions in a solid manner, the important point is to demonstrate that the differences of atmospheric constitutions may have been sufficiently great and powerful to bring the different species and genera, from the types they originally presented, to what we now see them to be. Now, of this the author thinks no doubt can be entertained" (154).

First, consider the effects of introducing species into new geographic areas—new circumstance, new "atmospheric conditions." Using data published, again in Jameson's journal, by "Dr. Roulin with respect to the animals transported from Europe to America," we see "the modifications which the species may still undergo, in consequence of a mere transportation from one latitude to another" (154).

But something more is required: experimental evidence on the production of monstrosities. We read that Geoffroy performed experiments on chickens, kept secret until now in a hostile political atmosphere, "to determine the power of external causes in modifying the development of living beings" (155).

And here, in what must be the anonymous author's own conclusions on the significance of Geoffroy's experimental work, we read: "The experiments here alluded to are conclusive. M. Geoffrey St Hilaire, by varying the phenomena of heat, dryness, and motion, not only produced monstrosities at pleasure, but even produced a given species of monstrosity, by means of a particular precaution" (155).

And the denouement: "And let it not be objected that the monstrous species thus produced in an artificial manner, were incapable of being reproduced and perpetuated. Nature, aided by time, which he had not at his disposal, acting by more numerous and gentler modifications, could have done what will always be impossible in the most judiciously conducted experiments" (155).

That Jameson (whether as author or simply as editor) is publishing here in 1829 such a ringing endorsement of the truth of Lamarck's "grand idea" of transmutation, is in itself ringing testimony that transmutation was finally out of the closet as a subject to be discussed openly by the late 1820s—albeit still in anonymous authorship.

There is no evidence that Darwin ever read this short précis of Geoffroy's work. By then he was long gone to Cambridge to take an undergraduate degree, en route to becoming a clergyman in the Church of England—though of course the *Edinburgh New Philosophical Journal* was available far beyond Edinburgh's city limits. But there is plenty of evidence that Darwin's interest in species and geology (despite what he later had had to say about Jameson's lectures) if anything intensified while he was in Cambridge from 1828 to 1831.

CHARLES DARWIN IN CAMBRIDGE, 1828–1831

While Charles Darwin, acquiescing to his father's desire to see him gain a profession after his "failure" at medical school, went to Cambridge for a general undergraduate degree with the nominal thought of eventually training for the clergy, he appears to have focused more on natural history than philosophy, literature, or religion. He continued his well-established habits of collecting specimens (mostly beetles) and participating in discussions of a scientific nature with many of the leading figures of the emerging professional science he met in Cambridge, among whom were John Stevens Henslow, Adam Sedgwick, William Whewell, and John Herschel. Sedgwick by then was already one of Britain's most highly skilled and respected geologists. Whewell, a philosopher, had coined the very term "scientist," and although he forever opposed ideas of transmutation, was very well read on this and all natural history subjects. Herschel, son of the famous astronomer William Herschel, was himself a scientist, but also a philosopher and a great thinker. He was an eloquent proponent of the search for natural causes for natural phenomena, which he felt was the central task of science, and yet certainly not antithetical to the observance of religion.

But it was John Stevens Henslow who was the main figure for Darwin. Henslow served for Darwin the same mentoring role as Grant did in Darwin's second year at Edinburgh, after his older brother Erasmus had departed for London. Darwin became known as "the man who walks with Henslow," frequently dining as well at the Henslow table.

There was one great difference between Darwin's elders, teachers, and mentors at Edinburgh, and those he later met at Cambridge: in Edinburgh they had medical degrees. At Cambridge, most, including scientists such as Henslow and Sedgwick, were ordained clergymen. Little is known of the discussions with these men and others that Darwin was fortunate enough to take part in. But it seems a very safe bet that the radical ideas (including most certainly transmutation) that were so freely discussed, and often strongly supported, in Edinburgh were treated with great circumspection, if brought up at all, in those days in Cambridge.

Henslow had begun as a crystallographer, but had then accepted a professorship in botany, and, becoming a quick and enthusiastic study, founded the Cambridge Herbarium and the Botanical Gardens, and began giving lectures in botany. Darwin took his course three times—every year that he was there.

Of course there is no way of knowing what Henslow and Darwin discussed during all their meetings; a syllabus of Henslow's botany course survives. But the greatest insight into what Henslow may well have taught Darwin resides in the 10,172 individual plants mounted by Henslow, his staff, and students, including Darwin, on 3,654 sheets preserved in the Cambridge Herbarium. These have been analyzed recently by a team consisting of Darwin historian David Kohn, botanists Gina Murrell and John Parker, and information specialist Mark Whitehorn (2005).

Henslow was clearly out to determine the limits of variation within stable species. That he saw species as inherently stable despite their internal variation has generally been assumed to reflect his creationist position—though transmutationists such as Robert Jameson and Robert Grant especially were aware of within-species variation as well. Just because someone saw species as inherently stable (Giambattista Brocchi, for example; Jameson, for another) did not automatically imply the person was a creationist. Whewell saw species as stable, and was an adamant creationist. But it was only Jean-Baptiste Lamarck (and later, apparently, Étienne Geoffroy Saint-Hilaire, and arguably Jameson in the end) among the transmutationists who saw species as non-stable, inherently variable, and changing entities though time and space. To see species as stable entities was not to automatically be someone who rejected transmutation.

Henslow's Herbarium sheets show that he was equally aware of within-population species variation and geographic variation. There can be no doubt that this emphasis on variation played a big role in Darwin's thinking, though probably more so in the years after he discovered natural selection (in 1838) than in his earliest transmutational forays, in which he stressed the births and deaths of species, paying relatively little heed to patterns of within-species variation. But there is absolutely no doubt that Darwin's hands-on experiences collecting British plants in the field and laying them out on sheets to document variation deepened his experiences in empirical field and observational research, extending his earlier experiences with marine invertebrates with Grant, as well as his own work with beetles. What he privately made of the variation Henslow taught him to see is of course unknown.

Adam Sedgwick, who was to write one of the more scathing denunciations of the *Origin of Species* when Darwin at long last published his transmutational views in 1859, was also, ironically, instrumental in training the young man to make geological observations in the field, which would be critical to the emergence of Darwin's initial ruminations on transmutation. Darwin was right to worry about what his older colleagues, including his teachers, would think of his transmutational ideas: the major reason, I persist in thinking, that Darwin was so reluctant to publish his evolutionary ideas.

In roughly a week, Sedgwick taught Darwin the rudiments of measuring the angles of bedding planes in sedimentary rocks, a critical first step in arranging the sporadically outcropping rocks of a region in ascending stratigraphic (hence chronological) order, thence paving the way for an interpretation of the sequence of events in the physical history of a region in the remote geological past. Darwin ([1832] 2008b) was later to write Henslow from the *Beagle*, asking him to convey his thanks to Professor Sedgwick for this absolutely invaluable training: "Tell Prof: Sedgwick he does not know how much I am indebted to him for the Welch [*sic*] expedition" (129). Apparently, Sedgwick trusted Darwin's observational skills so highly that the two entered into spirited discussions, often disagreeing on their interpretations of the geological history of that part of Wales they were visiting: a sign of great respect from the learned Professor Sedgwick.

As critically important as these experiences with the Reverends Henslow and Sedgwick undoubtedly were, there was one more circumstance that must have played an important role in shaping the course of Darwin's intellectual life: he read Herschel's *A Preliminary Discourse on the Study of Natural Philosophy* (1830). Considered by some to be the first modern statement of the nature and practice of science, the book amazed me. Following Herschel's plea not to regard science (344; though Whewell had not yet coined the term "scientist") as antithetical to religion, but simply as a rational way of understanding and explaining natural phenomena, Herschel then gives a thoroughly modern-sounding (to my non-philosopher's ear, in any case) account of how science is done: basically what has come to be called the "hypothetico-deductive method" (figure 1.4)

FIGURE 1.4 John Herschel. (By permission, Library of Congress, Prints and Photographs Division)

Herschel presents scientific laws as invariant and eternal. Speaking of the combining of chemical elements, he says that "no chemist can doubt that it is *already fixed* what they [the elements] will do when the case does occur. They will obey certain laws, of which we know nothing at present, but which must *be* already fixed, or they could not be laws. It is not by habit, or by trial and failure, that they will learn what to do." Herschel saw laws starkly: If A, then *B*. No room for trial and error—the attitude that led him (it was rumored to Darwin) to pronounce natural selection as the "Law of Higgledy-Piggledy" once the *Origin of Species* appeared in 1859. Given the importance of Herschel as one of the main inspirations to pursue a career in science, this must have been a cruel and crushing blow.

The remainder of Herschel's book is an account of how natural philosophy sees the world, as of 1830. I was, in my naïveté, stunned to read "that, in one second of time, light travels over 192,000 miles"—surprisingly close to the modern figure. But it is what Herschel has to say on the history of life, and the patterns of resemblance linking up the species of the modern biota, that are most germane to understanding the influence Herschel had on the young Darwin. For though, in the early pages, Herschel says that "to ascend to the origin of things, and speculate on the creation, is not the business of the natural philosopher" (that is, science cannot know the ways of God), he does nonetheless boldly state the modern, scientific understanding of the modernization of the fauna right up to the origin of present-day species. This passage, on the fossil record, clearly embraces a vision of continuity of descent within lineages, and replacement of old species by new, à la Brocchi, Jameson, and Grant:

> These remains [fossils] are occasionally brought to light; and their examination has afforded indubitable evidence of the former existence of a state of animated nature widely different from what now obtains on the globe, and of a period anterior to that in which it has been the habitation of man, or rather, indeed, of a series of periods, of unknown duration, in which both land and sea teemed with forms of animal and vegetable life, which have successively disappeared and given place to others, and these again to new races approximating gradually more and more nearly to those which now inhabit them, and at length comprehending species which have their counterparts existing. (283)

It couldn't be more clear, really, that Herschel (like everyone else, omitting citation) was repeating the conclusions, and even the language, of the radical thinking in Edinburgh. As to the modern fauna, considered alone, Herschel (1830) writes that "the great progress which has been made in comparative anatomy has enabled us to trace a graduated scale of organization almost through the whole chain of animal being; a scale not without its intervals, but which every successive discovery of animals heretofore unknown has tended to fill up" (290). A tip of the hat, perhaps, to Lamarck, though no one, including Brocchi, would have disagreed.

And before turning to Darwin's epochal experimental transmutational work on the *Beagle*, let us not forget that famous line from

Herschel's letter to Charles Lyell written in February 1836, not to be seen by Darwin until the letter was published in 1838 (though he may have discussed the same points when he met Herschel in Cape Town in June 1836). This letter was written long after Darwin had not only sailed, but (as we shall soon see) had long since developed his own transmutational ideas (no later than February 1835—but in my opinion several years earlier). Obviously, then, Herschel's words to Lyell could have had "no effect" on the actual development of Darwin's views—beyond, that is, cheering him on when he read the words and noted them in Notebook E, and later resurrected them in the second sentence of the *Origin of Species*. But the passage is instructive, as it leaves little doubt that Herschel was almost certainly thinking that natural rather than miraculous processes underlie the origin of species when he wrote the passage, cited earlier, in *Preliminary Discourse on Natural Philosophy* (1830)—the passage Darwin must have read.

In the more complete quote, Herschel writes to Lyell:

> Of course I allude to that mystery of mysteries, the replacement of extinct species by others. Many will doubtless think your speculations too bold, but it is well to face the difficulty at once. For my own part, I cannot but think it an inadequate conception of the Creator, to assume it as granted that his combinations are exhausted upon any one of the theatres of his former exercise, though in this, as in all his other works, we are led, by all analogy, to suppose that he operates through a series of intermediate causes, and that in consequence the origin of fresh species, could it ever come under our cognizance, would be found to be a natural in contradistinction to a miraculous process—although we perceive no indications of any process actually in progress which is likely to issue in such a result." (quoted in Babbage 1838:225–227; see also Kohn 1987:413n.59-2)

Only Lamarck had offered a natural causal process (and by 1829 Jameson, citing Geoffroy, had agreed), and that was not for the replacement of species by "fresh" species, but through the steady modification of one species transforming it into an apparent descendant. But for those, beginning with Brocchi, who saw the problem as replacement of extinct, stable species by descendant "fresh" species, no one had yet proposed a mechanism—and were not to do so until Darwin, safely home from the *Beagle* voyage, did so in 1837, in Notebook B. Contrary to what one might imagine, that causal, theoretical explanation was *not* natural selection. It was, instead, geographic ("allopatric") speciation.

Time now to look at how Darwin did that.

2

DARWIN AND THE *BEAGLE*

Experimenting with Transmutation, 1831–1836

There can be no question that Charles Darwin was fully apprised of the transmutational debate taking shape around him as a student at Edinburgh and Cambridge. He would have been aware of the search for natural causal explanations of the modern fauna through the births and deaths of species, seen as inherently real and stable entities, analogized with the births and deaths of individuals: "Brocchian transmutation," whether or not he knew Giambattista Brocchi's name in that connection. And of course he was also apprised of the Lamarckian construct of continual transformation of the anatomies of organisms. He knew that Jean-Baptiste Lamarck posited that all species slowly and inexorably grade into one another, not only through time, but geographically as well. As we shall see in this chapter, Darwin began his journey looking at the evidence for both positions while performing his duties as an unpaid naturalist on the *Beagle*.

Relatively few historians appear to have read Darwin's Geological Diary (or Geological Notes) thoroughly, at least from the explicit standpoint of tracing the growth of Darwin's own transmutational thinking. Yet the published debate in Great Britain on transmutation centered on the work of Lamarck and Brocchi, both of whom focused heavily on the fossil record (especially, though not exclusively, Tertiary marine mollusks). In hindsight, Darwin's own experiences with fossils was precisely where one would expect to find the most obvious "gold" in terms of Darwin's own empirically based interpretations on the subject. Coupled with Darwin's Zoological Notes and items from his correspondence, the course of development of Darwin's transmutational thinking emerges quite clearly—despite the obvious caution Darwin took while writing his notes. Indeed, Peter Grant, whose modern work with Rosemary Grant on the morphology, genetics, and evolution of some species of "Darwin's finches" of the Galápagos is justly celebrated, recently remarked (privately) that any biologist can see by merely reading Darwin's Zoological Notes what that young man had been thinking. The same is if anything more true when a paleontologist reads the Geological Diary.

Darwin was indeed circumspect in his notes—an early start to his reluctance to go public with his evolutionary ideas that lasted until the fateful letter and manuscript from Alfred Russel Wallace in mid-1858 forced him into action. The deeply religious Captain Robert FitzRoy had the authority to inspect all writings aboard the *Beagle*, reason enough for Darwin to be discreet in his notes. For example: Darwin's essay "February 1835"—his first, albeit short, discursive exploration of transmutation—is far more cryptic than his restatement of the gist of the essay in his famous, explicitly transmutational passages in the Red Notebook, not written until Darwin was safely home a little more than two years later.

Prevailing historical opinion over the last few decades is that Darwin did not embrace transmutation until after he arrived home in England and heard the opinions of naturalists, especially Richard Owen on fossils and John Gould on birds, on the identification of materials he had collected over the years (Sulloway 2009). But after completing work on the American Museum exhibition *Darwin: Discovering the Tree of Life* in late 2005 I was fairly sure that Darwin sailed home already thoroughly convinced that life had evolved through natural causes. I told my colleague, Darwin historian David Kohn, whose work had been critical to the success of the exhibition, that I had formed an image of this young man chomping on a cigar, with one foot propped on the bow of the *Beagle* as it finally set a course for home in the fall of 1836, thinking that he had the keys to a successful scientific career in his pocket: his knowledge of South American geology, and especially the evidence to put the very concept of evolution on a firm scientific footing. David was skeptical and, telling me in effect to put my money where my mouth was, suggested we go to the Rare Book Collection of the Cambridge Library to read Darwin's Geological Notes.

David Kohn was well known in Cambridge circles. He was an editor on the Darwin Project team that began the task of publishing Darwin's letters, a work still very much in progress. He transcribed and annotated Darwin's Transmutation Notebooks, perhaps the single most important documents written by the young Charles Darwin. And he has remained a Darwin scholar throughout his career.

Through Kohn, I met Randal Keynes, a great-great-grandson of Darwin's, whose book on Darwin's family life was originally published as *Annie's Box* in 2001. It was with Keynes's input that Kohn developed the wish list of the "crown jewels" of Darwiniana: letters, manuscripts, and artifacts crucial to a successful museum exhibition.

Keynes's parents lived near the massive Cambridge University Library. His father, Richard, had been a zoologist on the Cambridge faculty and had begun transcribing some of his great-grandfather's early works from the *Beagle* days, including Darwin's Diary (1988) and Zoological Notes (2000). They had us over to lunch one day, and in addition to a delightful meal of smoked salmon and white wine, and riveting conversation, we also saw many of the Darwiniana goodies that Richard Keynes personally owned. Among the highlights were a number of paintings by Conrad Martens from the early days of the *Beagle* voyage.

Nora Barlow's house was just up the street and around the corner. She was one of Darwin's granddaughters, and one of the earliest to begin publishing the treasure trove of letters and manuscripts that Darwin left behind. We were surrounded by a richly Darwinian world.

After lunch, we went to a windowless inner sanctum of the library, there to be treated to a spectacular array of Darwiniana. We had to wear white gloves and check our pens at the door. The flow of manuscripts, letters, and books was overwhelming. The next day we went on to Darwin's house in Downe Village, where the Field Notebooks and the famous Red Notebook were stored. We were able to borrow the latter, along with Darwin's geological hammer and a number of other objects. A visit to the Natural History Museum in London netted some of Darwin's fossils, including the huge *Toxodon* skull. All told, our exhibition *Darwin* undoubtedly contained the richest and most diverse array of genuine Darwinian objects ever assembled.

That earlier trip was a scouting expedition. Only after the exhibition was up and running in the fall of 2005 could Kohn and I contemplate going back to Cambridge to read Darwin's Geological Notes. We arrived back in Cambridge in early February 2006.

The rare book room in the Cambridge Library houses the Darwin Correspondence Project, and includes a rich holding of Darwin's notebooks and manuscripts. Kohn's friends in the library made everything accessible. I'll never forget Kohn saying to me at one point, "We have to look at Lyell." By which he meant Charles Lyell's three-volume *Principles of Geology* (1830–1833). Out they promptly came, and I was stunned to see they were Darwin's own personal copies from the *Beagle* days. I refused to touch them. They seemed like sacred texts. In any case, I had always preferred reprinted versions over the originals—nowadays augmented by the rapid rise of texts on the internet.

The Geological Diary (or Geological Notes) were on sheets of watermarked paper. Kohn would pick one up and hold it toward the light streaming into the room through the large windows. He would find the watermark and tell me the range of dates of manufacture of those sheets. He was also familiar with the color of the pen or pencil Darwin used in successive periods through his life.

It took me a while to feel comfortable reading Darwin's handwriting, but I finally got the hang of it. Little tricks include knowing the shorthand symbol for the word "ye," itself a shortened (not to mention antiquated) version of "the." Kohn, of course, could read Darwin's writing with ease—though there were times when Darwin's scrawl was more like a scribble. Darwin apparently seldom wrote anything while the *Beagle* was under way. Presumably his chronic seasickness played a role in that, but the mere rolling and pitching of that little ship would have made writing difficult in any case.

Midway through our study of the Geological Notes, Kohn turned to me and said, "You know, I don't think anyone with your particular background and expertise has ever seriously looked at these notes before." While I don't think that is quite true, I was further emboldened to read them carefully and thoroughly. I have continued to reread them for the past eight years.

What we discovered in those two quiet but intellectually eventful weeks prompted even more detailed study of all of Darwin's notes, letters, and manuscripts from the *Beagle* days. This is what I have found out.

EARLY DAYS ON THE *BEAGLE*, DECEMBER 1831–SEPTEMBER 1832

HMS *Beagle* left England in late December 1831. Twenty-two-year-old Charles Darwin, as primed as he was ever going to be for what would turn out to be a nearly five-year circumnavigation of the globe, was naturally glad that the waiting and false starts were finally over (figure 2.1).

FIGURE 2.1 The young Charles Darwin, shortly after his arrival home from the *Beagle* voyage. (By permission, Bridgeman Art Library)

He was especially looking forward to his first glimpse of tropical rainforests that he knew he would see at the very first port of call of Tenerife in the Canary Islands, having read of them in Humboldt's *Narrative*. But a cholera outbreak in England meant a twelve-day quarantine period for the *Beagle* that Captain Robert FitzRoy was unwilling to endure. The basic mission of the voyage was to survey the Atlantic and Pacific coasts of southern South America, and the captain was eager to get on with the job.

January–February 1832

Darwin hardly fared better in his quest to see tropical vegetation at the next stop, St. Jago in the Cape Verde Islands, yielding only a hint of the splendors of the tropical forests Darwin was yearning to see. Yet here, on his very first stop, Darwin established a pattern typical of his work for the remainder of the voyage: he simply took advantage of what presented itself to him. On St. Jago, it was the geology that arrested him, and led him to his very first in-the-field empirical observations, that in turn led to his initial exploration of a natural causal explanation of the births and deaths of species.

Exploring the beach on Quail Island, Darwin encountered an outcrop of fossiliferous sedimentary rock. Darwin prized some fossils from the rock, and compared them with the dead shells of the living invertebrate fauna cast up on the beach adjacent to the outcrops. He wrote in Geological Diary: "To what a remote age does this in all probability call us back & yet we find the shells themselves & their habits the same as exist in the present sea" (DAR 32.1:34 St. Jago: 20).

It was the very first stop of the trip, and here was Darwin already comparing fossil shells to the recent invertebrate fauna. In this regard, Darwin was extremely fortunate, for his initial paleontological experiences were with formations ringing continental margins, young enough that their included fossils immediately invited comparison with the modern fauna. Later, as the trip wore on, Darwin collected Paleozoic brachiopods and other invertebrates on the Falkland Islands. Then still later, he found Mesozoic ammonoids and other mollusks in the Andes.

He had little to say about these older fossils except to denote their age. As Giambattista Brocchi did before him, Darwin arguably saw little significance in these much older fossils, since they shed no light on the origin of the species of the modern biota. They were interesting solely from the point of view of age determination of the rocks in which they were found, and their putative correlations with European fossils and sediments.

These Quail Island fossils—so young they retained traces of color and gave off an organic stink when struck by Darwin's hammer—were nonetheless of "some remote age," covered as the outcrop was by a cooled, hardened lava flow. That he judged the species to be the "same" in terms of their anatomy and their "habits" (best rendered as "ecology" I think here) at the very least shows that Darwin was thoroughly comfortable with ascribing a relatively impressive (albeit not known) old age to existing species. That Darwin was to be able so matter-of-factly, so confidently, to point to the prior geological existence of the very species that remain alive today implies, I think, his familiarity with the phenomenon through the work of his predecessors—at the very least his own professor Robert Jameson. And of course the main point of interest is simply his matter-of-fact willingness to look at fossils and compare them with members of the living fauna. All this occurred at the very first outcrop from which Darwin collected fossils on the *Beagle* journey.

As a result of his geological experiences in the Cape Verdes, certainly including his paleontological discoveries and comparisons, Darwin ([1876] 1950:41) wrote in *Autobiography* of his resolve to come home from the voyage to write a book on the geology of South America.

February–July 1832

Darwin finally got his chance to experience a tropical rainforest when the *Beagle* came to Salvador, Bahia, Brazil, in late February 1832. Darwin was enthralled, famously remarking in Diary on February 28, admixed with descriptions of those delights, and mentions of Humboldt's illuminating commentaries and descriptions, that "the mind is a chaos of delight" (Keynes 1988:42). Darwin's Bahian experiences in what is now known as the Atlantic Tropical Rainforest (around Salvador, where remnants still persist, as well as later that year in the environs of Rio de Janeiro) arguably ultimately emerge in the "entangled bank" metaphor of the very last paragraph of the *Origin of Species* published twenty-seven years later. The whole point of riots of vegetation, and diverse ecosystems in general, is the diversity of species involved—a very different sort of biological organization from the pure genealogical systems that are the traditional focus of evolutionary biology.

Yet, with one exception, there is relatively little in Darwin's Brazilian rainforest experiences that seems to bear on the development of his transmutational ideas. Collecting and observing in the Rio de Janeiro region, Darwin came upon some terrestrial planarians (flatworms). He would have been familiar with marine flatworms from his work with Robert Grant at Edinburgh. But these were terrestrial, and were rather unexpected. Darwin's letters to his Cambridge mentor John Stevens Henslow are, after Zoological Notes and Geological Diary (and far more than the more personal Diary), the most important sources of insights into his scientific thinking while on the *Beagle*. In a letter to Henslow begun on July 23, Darwin ([1832] 2008c) wrote: "Amongst the lower animals, nothing has so much interested me as finding 2 species of elegantly coloured true Planariæ, inhabiting the dry forest! The false relation they bear to Snails is the most extraordinary thing of the kind I have ever seen.—In the same genus (or more truly family) some of the marine species possess an organization so marvelous.—that I can scarcely credit my eyesight."

Darwin's identification of his specimens as terrestrial planarians, and his obvious delight in not being fooled by the "false relation they bear to snails," is not in itself inherently transmutational. Ever since Carolus Linnaeus developed his system of classification in the mid-nineteenth century (and probably before), the early biologists and paleontologists routinely spoke of "natural kinds"—or as Darwin tended to refer to them in the *Beagle* notes and letters, "allied forms." But the passage does show that Darwin was very aware of groups of natural affinity, as opposed to collections of unrelated species that form diverse ecosystems that also delighted him. Darwin's keen interest in natural kinds—groups of "allied species"—would soon prove to be one of the keys to the development of his transmutational thinking as the voyage wore on.

The planarian passage in the letter to Henslow began a desultory correspondence that stretched over a few years, with letters crossing in the mail. Henslow said he doubted that his brother-in-law Leonard Jenyns (who, like Henslow himself, been invited to join the *Beagle* in the slot Darwin eventually accepted) would agree with his identifications. Henslow later wrote that Jenyns did in fact disagree. Darwin was already gaining confidence and allowing himself to disagree with his mentors.

The rest of Darwin's Brazilian experiences in mid-1832 also proved to be formative and invaluable in the development of his thinking, if not exactly about transmutation. His famous encounters with slavery stand out in particular. He also bounded up Corcovado Mountain in Rio de Janeiro twice, to measure its elevation barometrically, as well as to take in the view and demonstrate his physical prowess.

Whether or not Darwin actually encountered tree sloths in the Brazilian rainforest is not known—though they were to play a role in his thinking later that year. And he had narrowly missed seeing the Pleistocene rocks, with their rich marine invertebrate fauna, that crop out just north of All Saints Bay, where Salvador lies, and run all the way up the Brazilian coast to Pernambuco.

Ironically, it was these very rocks and fossils just north of Salvador, Bahia, that I encountered as a nineteen-year-old fledgling anthropology student in 1963 that in fact inspired me to take up paleontology. The context, of course, was very different: by the time I got there, evolution had long since been established as a scientific generality on a par with gravity. Yet somehow I was led to feel that fossils were the real way to study evolution. Evidently, and unbeknownst not only to me but to virtually all historians and evolutionary biologists, Darwin felt much the same way when he was there 131 years earlier. But I utterly failed to see any remnants of the Atlantic Tropical Rainforest in 1963, the sight of which had so transfixed the young Darwin. Instead, the geology he saw in Brazil was igneous and metamorphic—with no fossils, and of no special interest to him, as he felt they had already been well studied.

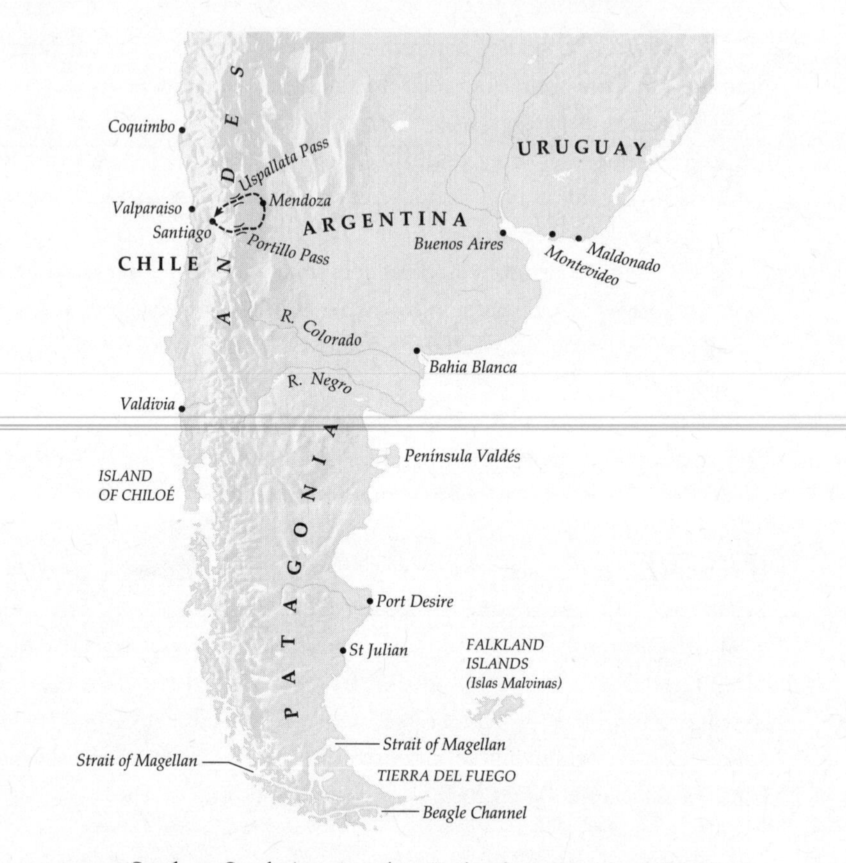

FIGURE 2.2 Southern South America, showing locales important to Darwin on the *Beagle* voyage, 1832–1835. (Artwork by Network Graphics)

All that was to change when he encountered the rocks and fossils at Bahia Blanca, Argentina, in September 1832 (figure 2.2).

BAHIA BLANCA, SEPTEMBER–NOVEMBER 1832

The Fossils—and Shades of Giambattista Brocchi

For an evolutionary-imbued paleontologist to visit the beach outcrops of Bahia Blanca these days is to generate feelings tantamount to those that religious pilgrims to Santiago de Campostela in northern Spain must have had centuries ago on reaching their destination. Even the presumably quite secular mid-twentieth-century paleontologist George Gaylord Simpson evidently felt so, when he visited the cliff at Monte Hermoso (now known as Farola Monte Hermoso, to distinguish it from the village of Monte Hermoso miles farther along the beach to the east). Simpson had had a grand paleontological venture to Patagonia in 1930 at the age of twenty-eight, writing of his adventures both political and paleontological in his first book, *Attending Marvels* (1934). Surprisingly, Simpson didn't mention Charles Darwin's earlier visit in this book. But twenty-five years later, Simpson (1984) paid a visit to Monte Hermoso and had the following to say about it:

> In 1887, fifty-four years after Darwin, the great Argentinean paleontologist Florentino Ameghino sat at Monte Hermoso and mused about it. . . . In 1955, one hundred twenty-two years after Darwin, and sixty-eight after Ameghino, I also sat and mused at Monte Hermoso. . . . The paleontological Monte Hermoso is a high, wave-cut scarp, and at its foot there is a wide platform . . . exposed at low tide and containing bits of fossil mammals. Here I had a feeling of awe, or even of piety, as it is a place made holy, in a proper sense, by the fact that these two truly great men had stopped and worked and mused here long before me. (26)

And yet Simpson was convinced that Darwin's paleontological experiences in South America had little to do with formulating his evolutionary theory—Darwin himself to the contrary notwithstanding. Darwin (1859) wrote in the first, topical sentence of the *Origin of Species*: "When on board H.M.S '*Beagle*,' as naturalist, I was much struck with certain facts in the distribution of the inhabitants of South America, and in the geological relations of the present to the past inhabitants of that continent" (1).

Simpson himself had written the introduction to an edition of Darwin's *Autobiography*. Simpson was a sentient reader, and could of course hardly have missed Darwin's ([1876] 1950) statement in *Autobiography*:

> During the voyage of the *Beagle* I had been deeply impressed by discovering in the Pampean formation great fossil animals covered with armour like that of the existing armadillos; secondly, by the manner in which closely allied animals replace one another in proceeding southwards over the Continent; and thirdly, by the South American character of most of the productions of the Galapagos archipelago, and more especially by the manner in which they differ slightly on each island of the group; none of the islands appearing to be very old geologically. (52–53)

Yet near the end of his life, Simpson (1984) steadfastly denied that Darwin's experiences with fossils—whether while in the field in South America, or later, when he got Richard Owen's determinations of the identities of his fossil samples—had anything much to do with the genesis of Darwin's evolutionary convictions:

> Taken only by themselves and with no more knowledge of fossils or of their stratigraphic succession, Darwin's collections of fossil mammals could not and did not lead directly to evolutionary conclusions. With the possible but improbable exception of the little tucutucu-like rodent, none of these fossils could reasonably be considered as ancestral, in an evolutionary sense, to any living species. . . . Now, too, there was *Glyptodon*, anatomically somewhat like the living armadillos. (36)

To be fair to Simpson, Darwin's Geological Diary lay unpublished in the Cambridge Library (it still is, though it is beginning to appear online at long last), and even the Red Notebook had yet to be discovered and published by the time Simpson wrote these words. His demurral on the "tucutucu-like" rodent fossil that Darwin recovered in the cliff at Monte Hermoso is presumably predicated on Darwin's somewhat suggestive mention of it in *Journal of Researches* (or *Voyage of the Beagle* [1839, 1845]).

But Simpson's treatment of Darwin, based on what little he knew, is equally unfair. Simpson criticizes Darwin for being basically wrong about his identifications. Darwin never thought that the extinct glyptodonts from Bahia Blanca (and elsewhere later) could possibly have been directly ancestral to modern armadillos. But the larger, and vastly more important, point is not whether Darwin's (or later Owen's) identifications were correct: instead, what is manifestly important is what Darwin thought was true. That is, after all, how people reach conclusions.

Simpson mentions the "tucutucu" from Monte Hermoso; Darwin calls it that in the *Voyage of the Beagle*, after returning home and finding that is what Owen decided the rodent fossils were. While in the field, though (and all the way throughout the remainder of the journey on the *Beagle*) Darwin thought he had found a fossil cavy closely related to the modern Patagonian cavy, which is the third largest rodent species still living (after the South American capybara and the somewhat smaller northern hemisphere beaver).

Darwin's "cavy" is probably the most important single fossil species in the history of evolutionary biology.

Like Simpson and so many others, I have trod the beach at Monte Hermoso. Like George Gaylord Simpson, Florentino Ameghino, Charles Darwin, and no doubt many others, I have sat and mused at this wonderful place. Yet it is not simply the ghosts of great thinkers that engender this feeling of awe and reverence in me, so much as it is the certain knowledge that this place was as important in the development of Darwin's evolutionary thinking as the Galápagos Islands. It would take nearly another three years for him to reach the Galápagos. Indeed, had Darwin not gone to Bahia Blanca first, it is unlikely that the Galápagos Islands would have exerted the powerful effect on him that they did.

When Darwin showed up in Bahia Blanca, there were two fossiliferous outcrops along the northern stretch of beach, east of what eventually grew into the small city of Bahia Blanca itself. They were Punta Alta, a low escarpment (*barranca*), and the Monte Hermoso outcrop farther along to the east a few miles away. In the 1890s, the beach outcrop at Punta Alta was obliterated to make way for an artificial inlet to dock ships of the Argentinean navy; it remains a naval base to the present day.

During a trip to track Darwin around southern South America, our host at Bahia Blanca, vertebrate paleontologist Teresa Manera, took us to Farola Monte Hermoso. Dr. Manera is at the Departamento de Geología, Universidad del Sur, Bahia Blanca, and the Museo Carlo Darwin in Punta Alta.

The blackened bones of small mammals were not hard to spot. Even a leg bone of a false camel (litoptern) macraucheniid was being carefully excavated while we were there. And for an extra treat, Dr. Manera pointed to a place just up the beach at Playa del Barco where a rock unit was exposed at the water's edge. Part of it was a conglomeratic mix of bits and pieces of rock and fossil shells. Though we saw no bones, Argentinean geologists and paleontologists believe this small exposure is the same formation that Darwin had sampled at the now long-gone Punta Alta farther west along the coastline.

In contrast to the Punta Alta situation, the outcrop at what is now called Farola Monte Hermoso has survived nearly intact, though the colony of burrowing social parrots, apparently not there in 1832 (at least, Darwin does not mention them), has altered the cliff face to some degree. So has the inexorable process of storm-wave erosion, and the relatively minor effects of continued fossil collecting.

Darwin was worried from the start that his assumption that the outcrops at Punta Alta and Monte Hermoso were the same age might be wrong. After all, though there were fossils (like glyptodonts, the giant armadillo relatives) common to both places, the sediments themselves, and much of their included fossil faunal content, were different. Geologically, the sediments at Punta Alta, some of which contained coarse pebbles along with the shells of marine invertebrates and disarticulated and worn bones of fossil mammals, contrasted strongly with the fine-grained reddish silts comprising the rocks at Monte Hermoso. And the fossils at Monte Hermoso were black bones, with no marine invertebrates to be found.

It turns out that Darwin was justifiably worried that the rocks and included fossils at Punta Alta and Monte Hermoso might not be of the same age: it turns out that the Punta Alta is Pleistocene in age (that is, within the last 1.65 million years), while Monte Hermoso is considerably older: Lower Pliocene, maybe in part even Upper Miocene, roughly, say, five million years old.

Yet Darwin saw instantly that at both of these two places he was dealing with fossils that were recognizably South American in character, connected in some way with the modern fauna. When he was safely back in London five years later, Darwin asked the mammalogist George Waterhouse to make a list of the known genera and species of living edentate mammals of the world. The mammalian list was written in what has been identified (by David Kohn) as Waterhouse's hand, on the back of a business card listing the meetings of the Zoological Society of London for 1837, with "Sept. 7" underlined in green ink (DAR 39:133). Darwin must have buttonholed Waterhouse at the meeting to pick his brain.

Darwin was simply checking to be sure that the edentates were largely, if not wholly, restricted to the New World. Edentate mammals include the various species of anteaters, sloths, and armadillos of the modern (mostly South) American fauna. Waterhouse listed nineteen species of South American edentates but also included five Old World species (for example, African and Asian pangolins) no longer thought to be closely related to South American sloths, anteaters, and armadillos. These South American taxa are now included in their own separate mammalian order, Xenarthra.

Darwin was very lucky to have found relatively young fossil mammals. His teacher Robert Jameson had emphasized the importance of focusing on higher plants and animals. Especially in the middle and low latitudes, the higher plants and animals tend to differ on the different continents: different species, and often entirely different higher taxa. In contrast, lower groups, such as invertebrates and more primitive plants (along with microbes, or "infusorians"), as well as higher groups (for example, mammals, birds, tree species, and the like) living in the higher latitudes, tend to occur in both the Old and New Worlds (attributed in "Observations on the Nature and Importance of Geology" [(Jameson) 1826] to the fact that the continents had once been conjoined).

Why is this important? More especially, why was this important to Darwin? Georges Cuvier and other natural historians had suggested that many instances of new faunas and floras replacing old, extinct ones were a matter of replacement by biogeographic migration. This is indubitably true. But successive replacement of taxa that have never been found, dead or alive, in any place other than a particular region helps narrow the focus: if attention is paid solely to patterns of births and deaths of species of such endemic species, the chances are maximized that the births occurred in more or less the same place as the deaths.

And Darwin could not be sure of the biogeography of typically far-ranging invertebrate species, especially marine mollusks. Of these, only the extinct "Patagonian oyster" (there are at least three distinct species of extra-large oysters in the extensive marine Miocene deposits along the Atlantic coastline of Patagonia) struck Darwin as probably endemic to the region. His Zoological Notes desultorily reflect his incessant queries of gauchos, Indians, fishermen, seafaring captains, and the like, as to whether they had ever seen a living Patagonian oyster. None had.

Darwin was obsessed with endemism while collecting, thinking, and writing about the fossils at Bahia Blanca in September and October 1832. This was before he received volume 2 of Charles Lyell's *Principles of Geology* in late November 1832, when the *Beagle* arrived in Montevideo, Uruguay. Volume 2 was Lyell's anti-Lamarckian diatribe. In that volume, Lyell extensively discusses the distributions of animals and plants. Captain FitzRoy had presented volume 1 of the *Principles* to Darwin before the *Beagle* set sail in December 1831—a book focusing on physical ("inorganic") geological phenomena (and one of the volumes we saw in Cambridge). Thus Darwin either came to a realization of the importance of endemism when contemplating the births and deaths of species on his own, or he had at least a glimmer of it from his days as a student back in Great Britain. In any case, he did not get it from reading Lyell. Darwin had already written of the Bahia Blancan fossils in his Geological Notes, his Diary, and in a letter to Henslow, mailed in late November from Montevideo at the time he received Lyell's volume 2, but with the discussion of the fossils in this letter composed earlier in the month.

In a nutshell, the fossils Darwin found at the two localities at Bahia Blanca fall into three categories regarding the development of his transmutational thinking. The invertebrates at Punta Alta struck him as belonging to the very same species as are still to be found living in the Bay today—the same conclusion, in other words, that he made regarding the molluscan and other invertebrate fossils in the Cape Verde Islands earlier in the year. But the bones of the large ground sloths and the giant armadillo-like glyptodonts (though he waffled a bit on the identity of these latter fossils) seemed to belong entirely to extinct species nonetheless belonging to endemic groups of edentates.

And third there was the cavy. Darwin thought that the fossil cavy was an extinct, smaller species belonging to the same genus as the recent

Patagonian cavy. For the most part, he called them "agoutis" in his notes and letters; again, incorrectly by later taxonomic standards. But it is his thinking on the implications of these identifications that matter in the development of his evolutionary thought. From the beginning of his notes on these fossils, Darwin consistently expressed his belief that these rodents are endemic to South America. And in that supposition, he was entirely correct.

Here's what Darwin (1832–1836a) had to say about the fossil invertebrates from Punta Alta in Geological Notes, written in September 1832:

> The cemented gravel . . . contains numerous organic remains: 1s. shells there are so numerous as in places, especially the upper bed, almost to compose it; they appear to me to be exactly the same species which now exist on the beach: And it is to be *especially* remarked the proportional numbers of each species are about the same; the most abundant in both cases are Crepidula Voluta. . . . & Venus; the rarest. Pecten Fissurella &c.— 2d: 2 Coralls. An encrusting Flustra & an Ostrea. These both appear identical with what now exist:— 3d a piece of wood converted into calcareous matter. (DAR 32:52)

A few pages later in his notes, Darwin writes: "That they [the formations from which Darwin collected his fossils at Bahia Blanca] are comparatively modern is certain from the similarity of shells & the features of the country not being greatly different.— Some geologists have been surprised that the extinction of land-animals, has not occurred, without destroying the inhabitants of the sea; this would seem to be a case in point." (DAR 32:59)

That these invertebrates (plus the nondescript piece of wood) were in beds that were "cemented" and contained extinct mammals implied to Darwin that they had some age to them. Yet the marine invertebrate species seemed to all remain alive and well and still living right there along the shoreline. Thus not all species living at any one time are destined to become extinct at the same time, a lesson hammered home by Jameson's tables, and a theme that fits in with the earlier writings of both Jean-Baptiste Lamarck and Giambattista Brocchi—and as opposed in particular to Cuvier's catastrophic view of extinction.

Of the fossil mammals, Darwin's notes on the fossils of Punta Alta continue:

> 4th: the number of fragments of bones of quadrupeds is exceedingly great: I think I could clearly trace 5 or 6 sorts.— the head of one very large animal (with singular anterior cavity) has 4 large square, hollow molar teeth; perhaps it may be the Megalonyx: the lower jaw & one molar tooth of some smaller animal. I conjecture one of Edentata & perhaps allied to the Armadilloes: the molar tooth of some large (a) animal. (Rodentia?) 741–744: bones of some smaller quadruped. like deer: These bones are generally white, soft & friable; Hence it is difficult to cut out of the gravel perfect specimens. (DAR 32:52–53)

Megalonyx is a genus of North American ground sloth, originally named by President Thomas Jefferson. The "large molar tooth" that Darwin thinks might belong to some form of rodent is indeed superficially very rodent-like, though it turns out to have belonged to a species of an extinct and then as-yet-unknown group of endemic South American mammals, the Notoungulata. Owen later described and named the genus *Toxodon*, based on the Punta Alta specimens, and others (including the large skull we saw in London and borrowed for our exhibition) that Darwin collected elsewhere in Patagonia later on in the voyage.

Darwin continues his notes:

> At Punta Alta the only organic remain I found in the Tosca (excepting mere particles of shells) was a most singular one: it consisted in an extent of about 3 feet by <4> <<2>> [Darwin originally wrote "4" but crossed it out and inserted "2"] covered with thick osseous polygonal plate; forming together a tessellated work: it resembles the case of Armadillo on a grand scale: these plates were double or an interval of few inches between them.—(a) At present the case of the <<dead>> Armadillos are oftener found separate from the body. than connected with any part.— In this case the envelope of the great animal would easily be carried by the water. & by the pressure of Tosca would be doubled up as described.— It is stated. that "recent observations" show

> the Megatherium had such an envelope; it certainly is probable it belongs to some of the animals the bones of which are so abundant in the gravel.— In connection with the Megatherium I may mention a curious fact.— It is a common report. <<in all this part of S America>> that there exists in Paraguay an animal larger than a bullock, & which goes by the name of "gran bestia." (b) The Commandante at the Fort. states that he many years ago saw a young one, when in Paraguay; that it had great claws & snout. like Tapir. (He added also that it is carnivorous; having only seen a young one this must be conjectural.) Now these are the very words <<with>> which Cuvier describes the probable figure of the Megatherium; the fossil bones of which are well known to come from Buenos Ayres & Paraguay.— If no credit is given to the actual existence of the "gran bestia." we must suppose it is either traditional or that it is <an> a report arising from the occurrence of very perfect skeletons.— the resemblance is too striking to be attributed to mere chance. (DAR 32:53–54)

Darwin's initial impulse turns out to have been correct. These large, tessellated curved sheets comprised of "osseous polygonal plates," often found as disarticulated fragments, indeed came from various species of extinct glyptodonts, collateral kin of modern armadillos. And there had been a recent report suggesting that these bony plates were instead embedded into the skin of the back of species of the giant ground sloth *Megatherium*. So Darwin was understandably unsure of himself, variably referring to these osseous polygonal plates that he occasionally encountered throughout his Patagonian travels to *Megatherium* or to extinct armadillos.

Though Darwin himself crossed out the passage on the possible existence of a living, breathing "gran bestia," I left this passage in as it calls to mind the thought, common then, that unexplored terrains might still have living representatives of species otherwise only known as fossils. The author of "Observations on the Nature and Importance of Geology" ([Jameson] 1826) posed this as a possible explanation for "extinction," and Thomas Jefferson himself thought that many of the extinct fossil mammals he described while in the White House might be still alive and well in the unexplored reaches west of the Mississippi. Indeed, there are still reports coming from remote areas of South America's Pantanal (wet grasslands), home to capybaras, anacondas, tapirs, giant anteaters, and, some still claim—of giant ground sloths, which seem to have achieved a mythic, Yeti-like status.

Yet Darwin, on the whole, throughout the voyage preferred to think that these plates belonged to an extinct form of giant armadillo. Writing to Henslow from Montevideo, Darwin ([1832] 2008d) says: "In the same formation I found a large surface of the osseous polygonal plates. . . . Immediately I saw them I thought they must belong to an enormous Armadillo, living species of which genus are so abundant here."

This passage of course reveals his complete ease with comparing his fossil finds with the living South American fauna. Indeed, this letter is perhaps the most important single letter that Darwin wrote to Henslow (or anyone else, from a purely scientific point of view) on the entire *Beagle* voyage.

But then there is the matter of the cavy at Monte Hermoso. Darwin's (1832–1836a) notes continue:

> I will now describe the section of <<120 ft. high:>> cliff at Monte Hermoso. situated at the entrance of B Blanca: & like Punta Alta forming a head—land, which the sea is continually washing away: The bones in their nature were singularly different from those at P Alta; in the one case <<they>> had been immediately enveloped in the Tosca, in other exposed to the action of the water: Here bones were very hard & of great specific gravity, their surfaces polished & blackened externally; in the smaller ones they, from this cause resembled jet.— I could perceive traces of 4 or 5 distinct animals: two of which certainly belonged to the Rodentia. One must have been allied to the Agouti; the tarsi & Metatarsi belong to an animal less than the present. common inhabitant, Cavia patagonica. (b) The Agoutis are all proper to S. America; & none have hitherto been found in a fossil state:— To conclude with the organic remains I have shown that some of the bones probably belong to the Edentata. & that the osseous plates are supposed to belong to the Megatherium. (DAR 32:57–59)

The letter (b) refers to a footnote on the reverse side of the page—one of the last entries that Darwin made in the Geological Notes of 1832 on the Bahia Blanca fossils: "It is interesting to observe that this tribe of animals [the Agoutis], which is now peculiar to S. America, should in this epoch when the Megatherium flourished, also be present—showing that with the extinction of one genus, that of others did not follow" (DAR 32.1.71).

Taken by itself, this footnote seems innocuous enough—seeming to be about differential extinction of genera. Darwin's passage quoted earlier, contrasting the extinction of *Megatherium* with the persistence of the fossil invertebrates, immediately follows the passage on the "agouti." That the footnote extends the observation to differential survival among mammalian genera themselves—a stronger point—might indicate the footnote had been written at some point after Darwin had completed the Bahia Blanca Geological Notes of 1832.

Yet I feel certain this passage has meaning that transcends the point about differential extinction of mammalian genera. For it is clear from Jameson's tables and the discussion of the meaning of such data in "Of the Changes Which Life Has Experienced on the Globe" (1827c) (again, almost surely written by Jameson, though published anonymously) that the survival of genera implies the persistence of component species—*or the replacement of extinct species by others still alive in the modern fauna*. The context of such discussions reviewed in chapter 1 is blatantly transmutational.

And Darwin insists from the very beginning that the fossil "agouti" and the living Patagonian cavy ("mara") are distinct, separate species of the same endemic genus *Cavia*. The author of "Of the Changes Which Life Has Experienced on the Globe" ([Jameson] 1827c) would consider this to be a documented case of transmutation—of replacement of an extinct species by a descendant in true Brocchian transmutational form. Whether Darwin himself thought as much at that point in 1832 (as he certainly did later) is impossible to say.

But clearly he was thinking about the possibilities. And given all the connections between Darwin and the Edinburgh literature on transmutation (not to mention the authors of those papers!), it is hard to resist the thought that he himself felt he was looking at an actual example of transmutation with the fossil and recent cavy species.

Darwin shipped off the precious Bahian Blancan fossils, among other specimens, to Henslow in late November 1832 from Montevideo. As he ([1832) 2008d) wrote in a separate letter to Henslow: "If it interests you to unpack them, I shall be *very curious* to hear something about them:— *Care must be taken*, in this case, not to confuse the tallies [identifying labels]. They are mingled with marine shells, which appear to me identical with what now exist."

Darwin returned to Bahia Blanca in August 1833. Reading his diary, recounting the horrors of the war between General Rosas (who had given him a safe conduct pass) and the native American population, it is difficult to see how Darwin could still keep his mind on natural history. But he did, though Geological Notes reveal a disappointed young scientist, more disheartened then than perhaps any time before or since on his five-year adventure. In an "Appendix" to the notes on Bahia Blanca, Darwin writes "Theories revisited"—in itself a somewhat foreboding portent. The Punta Alta beds cannot be the same age as the great Tosca beds, Darwin writes. Nowadays, "Tosca" in Argentina simply means a calcareous rock type. But Darwin, taking up the common usage in the 1830s, saw it as a great continuous formation widespread in Patagonia. He had indeed interpreted the beds of both Punta Alta and Monte Hermoso to belong, at least in part, to the Tosca formation. Having subsequently seen the Tosca all over Patagonia, Darwin was sure the Bahia Blancan rocks did not match up. And he also couldn't say how old the Monte Hermoso rocks really were.

In what looks like a lighter pen (yet not a lighter heart), at the end of the "Appendix" of 1833, in Geological Notes, Darwin writes, rather poignantly: "The chief thing proved by this place, is an upheaval posterior to the great Tosca plain; & the coevality [meaning "of the same time"] of certain animals at M. Hermoso; and the extreme lateness of existence of animal (1405).—the Punta Alta bones can prove nothing; nor indeed the M. Hermoso with reference to other formations—" (DAR 32:74).

Pretty grim. Darwin was rethinking his earlier conclusions on Bahia Blancan geology in light of his later travels, and quite possibly in light of having read more of Charles Lyell. But he was, by his own admission, concocting theories, and here, in 1833, he was worried that the bones of Punta Alta, by themselves "prove nothing." All of which means that he had been doing a lot of thinking—and now was entertaining a lot of doubts. Only natural in an unseasoned scientist who was quickly maturing into a seasoned observer and analyst of his own observations.

But the nature and significance of the rocks and fossils of the two localities at Bahia Blanca continued to stick in his mind. And his thoughts on their meaning seemed to brighten up as the trip wore on. In March 1834, on his second trip to the Falkland Islands, Darwin had occasion once again to write to Henslow. And once again, he brought up the importance of the "tallies" associated with the fossil specimens from Bahia Blanca:

> I have been alarmed by the expression cleaning all the bones, as I am afraid the printed numbers will be lost: for the reason I am so anxious they should not be, is that a part were found in a gravel with recent shells, but others in a very different bed.—Now with these latter were the bones of an Agouti, a genus of animals I believe now peculiar to America & and it would be curious to prove some one of the same genus coexist with the Megatherium, such & *many other* points entirely depend on the numbers being carefully preserved. ([1834] 2008:259–263)

Less than a year after his crisis of confidence in his own theories, Darwin was still holding out hope that the cavy of Monte Hermoso was alive when *Megatherium* was alive. The point, arguably, was not simply that the genus *Megatherium* became extinct before *Cavia* did, but that a related species of *Cavia* still lives. Why else would Darwin care? By trip's end, in what was evidently the first page of the post-voyage writings in Red Notebook, Darwin (1836–1837) wrote "I really should think probably that B. Blanca & M. Hermoso contemp" (113; see also Herbert 1980:57). By then he had satisfied himself (and I think probably before he wrote "February 1835" and even earlier) that the rocks and fossils of Punta Alta and Monte Hermoso were indeed contemporaneous.

Despite finding additional Tertiary mammalian fossils here and there during his remaining nearly two years in Patagonia, Darwin never had a paleontological experience again that came remotely close to the richness and diversity of materials he had experienced at Bahia Blanca. Nor did his further fossil adventures generate the kind of speculations recorded in Geological Notes and, to a lesser degree, in a November 1832 letter to Henslow based on his work at Bahia Blanca. That his geological and paleontological observations at Bahia Blanca continued to percolate in his thoughts is revealed in a letter to Henslow from the Falklands ([1834] 2008), and soon thereafter in his two surviving essays from the *Beagle* voyage: "Reflection on Reading my Geological Notes" (apparently written in 1834) and "February 1835," housed together in the "Earthquake Portfolio" within Geological Notes at Cambridge.

Darwin did unearth the bones of what he thought was a mastodon at Port St. Julian in early 1834. The specimen was later identified as an extinct camel by Richard Owen and named *Macrauchenia*. Darwin used this specimen in the essay "February 1835" to form some conclusions about South American climate and geology. Later, in the second half of the Red Notebook, written in approximately March 1837, in his most overtly and explicitly transmutational passage to date, he substitutes the extinct camel/living guanaco for the cavy example of replacement of an extinct species by a living descendant species. Owen had told him the fossil cavy was really the remains of a different (yet also endemic) rodent, the tucutucu. Ironically, *Macrauchenia* turns out not to have been a camel at all (whereas guanacos are indeed true camels), but a very camel-like member of still another endemic South American group, the Litopterna. But again, it is what these earliest scientists thought at the moment that really matters in understanding their interpretations.

The Living Fauna—and Shades of Lamarck and Robert Grant

Darwin's Zoological Notes (and, in a less formal but equally interesting manner, the more personal Diary entries for the same period) for Bahia Blanca in 1832 abound with descriptions of many of the most familiar, and ultimately to Darwin, most significant, elements of the living Patagonian fauna. He talks about the anatomy, behavior, and often even the flavor of the local "ostriches" (greater rhea), guanacos, cavies ("agoutis"), pumas, armadillos, and many other species—comparing (as we have seen) the living armadillos with the extinct glyptodonts, and even including a brief comparison of the fossil cavy from Monte Hermoso with the living Patagonian cavy ("agouti") in the Diary entry for October 19: "The Captain landed for half an hour at Monte Hermoso (or *Starvation* point as we call it) to take observations.—I went with him & had the good luck to obtain some well preserved bones of two or three sorts of Gnawing animals.—One of them must have much resembled the Agouti but it is smaller" ([1831–1836] 1988:110)

But Darwin made observations on two other species of the living fauna around Bahia Blanca without any paleontological comparisons—observations with surprisingly frank transmutational overtones. The fossil cavy (discovered within a short half hour's casual visit to the outcrop!) and its putative relationship with the modern Patagonian cavy was Darwin's main example of replacement of an extinct species by its living presumed congeneric descendant while on the *Beagle*. Darwin's most striking example (in more ways than one—a species of highly venomous snake, now known as *Bothrops alternatus*) of the sort of anatomical intermediacy he was taught by Robert Grant to look for in his marine invertebrate zoological research also came from his experiences at Bahia Blanca in 1832. Grant had not only taught Darwin the rudiments of field collecting and microscopic examination of the often tiny colonial animals like bryozoans, but had apparently done so in the spirit of the search for connections between major groups of animals such as sea anemones, corals (coelenterates), and sponges in the transmutational spirit of Jean-Baptiste Lamarck.

Indeed, in one of the four papers he published in Robert Jameson's *New Edinburgh Philosophical Journal* in 1826, Grant describes the new sponge genus *Cliona*, and makes the claim that the combination of properties observed in this new taxon provides the connection between sponges and the coelenterate *Alcyonium*. According to historian Janet Browne (1995), Grant had given copies of the 1826 papers to Darwin, and Darwin had read them "almost as if collecting all possible points of view about transmutation and secular science in general" (85). And we already know by Darwin's own autobiographical admission that he was well aware of Grant's admiration of the transmutational ideas of Lamarck.

Darwin seems to have been excited by the specimens of the venomous snake he observed and captured at Bahia Blanca. Struggling at first with its identification, he soon realized that it belonged to the general group of what are now called pit vipers. And he saw in his specimen a species that seemed to him to lie anatomically between Old World adders and true, New World rattlesnakes, especially in the structure of the tip of the tail, and its use as a warning device. As Darwin wrote in the Diary for October 8, 1832: "I also caught a large snake, which at the time I knew to be venomous; but now I find it equals in its poisonous qualities the Rattle snake. In its structure it is very curious, and marks the passage between the common venomous and the rattle-snakes. Its tail is terminated by a hard oval point, & which, I observe, it vibrates as those possessed with a more perfect organ are known to do so" (Keynes 1988:109).

In Zoological Notes, Darwin wrote "how beautifully does this snake both in structure & *habits* connect Crotalus & Vipera" (Keynes 2000:91). And he says nearly the same thing in that same November 1832 letter to Henslow, though omitting the word "structure": "a new Trigonocephalus beautifully connecting in its habits Crotalus and Viperus."

One measure of what I perceive to be Darwin's excitement over this zoological discovery is that he wrote about it in three different places: in Zoological Notes, naturally enough; in a letter to Henslow posted in late November from Montevideo; and in his own personal Diary. The last entry is the most graphic, with its use of the phrases "marks the passage" between two well-known groups of poisonous snakes, and the just as striking "more perfect organ"—referring of course to the more elaborate structure of the rattle of a rattlesnake over the simple hard cask found on the tip of the tail of what Darwin identified as *Trigonocephalus*.

Darwin rarely included any detailed scientific observations in his personal Diary. Yet here he is, still quite early in the voyage, enthusing over just the same sort of observations he knew Grant, and no doubt others, were looking for as evidence of Lamarckian transmutation. Indeed, he seems to be more circumspect in a letter to Henslow than he was in this remarkable passage in Diary, although the intermediate nature of his snake's tail tip morphology, linking as it does adders with rattlesnakes, is clear enough in all three accounts.

Perhaps less striking, but I think intriguing, is yet another example of a living species encountered at Bahia Blanca that seemed to Darwin to combine the characteristics of two different groups in a curious mélange: another form of anatomical intermediacy. Writing to Henslow, Darwin ([1832] 2008d) described "a poor specimen of a bird, which to my unornithological eyes, appears to be a happy mixture of a lark pidgeon & snipe. . . . Mr Mac Leay himself never imagined such an inosculating creature." William Sharp MacLeay's "Quinarian System," published a decade earlier, held that groups of animals and plants typically have five subordinate groups, arranged in "circles of affinity." Each circle is contiguous ("inosculant") with two others. As we have seen, it was commonplace for systematists to speak of "affinity" or "allied forms," as Darwin himself was already doing. Such terminology was used without any necessary overt connotation of transmutation, though by the 1820s, the "Edinburgh Lamarckians" were indeed openly linking degrees of affinity with propinquity of transmutational descent. Whatever Darwin's intent, the passage is important for the later (and somewhat confusing) use of the term "inosculate" in Darwin's writing in conjunction with geographic replacement patterns of living species, especially the two species of rhea, all the way at least to his frankly transmutational passages in the second half of the Red Notebook. "Inosculating" here clearly means "mixture of characters of two different groups." For the remainder of his notes throughout the voyage and beyond, "inosculate" is used strictly for instances in which closely related distinct species sharing attributes (the very evidence of their close relatedness) meet (geographically), yet do *not* mix or blend into one another in the region where their ranges overlap.

And so we must ask if these observations, especially on the morphology and behavioral use of the tip of the tail of *Trigonocephalus*, did indeed convey transmutational meaning to Darwin as he wrote them? If so, why would he say such things to Henslow? And how could he feel comfortable including them so graphically in his personal Diary? Except for a page-long retelling of the story in *Journal of Researches* (1839:114), written upon arrival home, and at a point when even the most conservative Darwin scholars admit Darwin was a confirmed and convinced transmutationist, Darwin never mentions the snake again.

But the fact that, in the *Journal of Researches*, he repeats his observations on the intermediacy of that snake after he was a confirmed transmutationist suggests that Darwin was thinking along transmutational lines when he wrote the original passages in his notes and letter in late 1832.

For whatever reason, Darwin never again wrote such a Lamarckian suggestion of transmutation as the *Trigonocephalus* passage in the Diary. On the one hand, even Zoological Notes fails to reveal another case of "anatomical intermediacy" between two well-known groups. Whether, in a trend carried out until he finally published the *Origin* in 1859, Darwin grew more cautious as his thoughts progressed; or examples similar to the *Bothrops* (that is, *Trigonocephalus)* intermediacy between adders and rattlesnakes failed to materialize again; or whether he simply started to focus on geographic replacement patterns among closely allied living South American species (which he began to do soon after leaving Bahia Blanca), dropping a focus on Lamarckian transitions, is difficult to say.

Historian Sandra Herbert has pointed out that Darwin had available in the *Beagle* library at least some of the volumes of Edward Griffith's encyclopedic *The Animal Kingdom* (1827), an extended and modified English edition of Cuvier's earlier work. Apparently among the volumes Darwin had on hand was volume 2 on mammals. Herbert (1995) writes: "In Volume 2, three ideas are broached with general relevance to the species question: universal descent, development, and the notion of secondary causation for the origin of new varieties" (28).

And indeed there is much of transmutational interest in Griffith's volume 2. Just as an example, consider the following passage on the African wild dog (or Cape Hunting Dog):

> After the Caninae, or at least as a distinct section of the race, and before the Hyenas, must be placed a newly-discovered or described animal, partaking of several points of both these genera, and consequently intermediate between them; the number and character of its teeth corresponding with those of Dogs, would place it in that subgenus of the "Animal Kingdom," in which, as may be observed, dentition is selected as the most influential distinctive character.

> This, and such like intermediate animals, appear to claim the particular attention of the zoologist, from which results remain to be deduced—they form the connecting links, which, as it were, chain organization together: they seem to multiply the extent and enlarge the influence of secondary causes in the great work of creation, and stand decidedly opposed to a host of other facts which display the impassable barriers interposed by nature between the several creatures and their respective races. (375)

As we have seen in chapter 1 and briefly recapitulated here, emphasis on morphological intergradations in a pro-Lamarckian transmutational context were, if not exactly common, present in the work of Grant, Jameson, and others at just about the same time, and would have been known to Darwin before he left on the *Beagle* voyage. Whatever the effect of Griffith's volume on Darwin while he was on the *Beagle*, Herbert (1995) does point out that it is one of the three references he cites in the earliest essay written while on board: "Reflection on Reading My Geological Notes" (1834), discussed at the chronologically appropriate place later in this chapter.

Thus I think it fair to suggest that Darwin was indeed comparing the relative merits of a Lamarckian transformational versus a Brocchian wholesale species replacement approach to transmutation beginning at Bahia Blanca and perhaps continuing throughout the voyage. Yet as we have seen, especially in the two anonymously published essays ([Jameson] 1826, 1827c), from Étienne Geoffroy Saint-Hilaire's work reported anonymously ([Jameson] 1829), in Grant's inaugural address of 1828, and in Herschel's (1830) book, it was commonplace for those seeking a secular, natural causal explanation for the origin of the species of the modern world to cite evidence of both a Lamarckian-style anatomical intermediacy nature alongside Brocchian-imbued passages on the replacement of extinct species by living species. Both categories of example were deemed admissible; no conflict was seen between them in a general sense, even though Lamarck's and Brocchi's projected patterns of change through time were themselves so deeply different.

Again, it is difficult to say which possibility best describes Darwin's approach to transmutation while on the *Beagle*. But it is a simple fact that, after posting a letter to Henslow late in November 1832 from Montevideo, Darwin ([1832] 2008d) wrote no more of anatomical intermediacy or of species (like his inosculating "lark-pidgeon") that seemed to him to combine the characters of several distinct groups. From then on, right through the Galápagos, it was all replacement of closely allied species (or sometimes "varieties") seen as discrete, stable, albeit closely related entities. Darwin began to emphasize the non-blending aspects of the constant, consistent anatomical differences between closely related species (or "varieties") that replaced one another over the Patagonian plains, or on separate islands within archipelagos of the South American Atlantic and Pacific coasts, or across the Andes: virtually everywhere he went.

LYELL'S *PRINCIPLES OF GEOLOGY*, VOLUME 2

Darwin received his copy of the second volume of Charles Lyell's *Principles of Geology* in Montevideo, shortly after his first sojourn at Bahia Blanca (figure 2.3). This time on the flyleaf, Darwin wrote simply "Charles Darwin M: Video Novemr. 1832." Because Darwin's ([1832] 2008d) letter to Henslow summarizing his Bahia Blanca findings was sent from Montevideo and was dated November 24, 1832; because Lyell's second volume was so fraught with births, deaths, and successions of species; and because Lyell's second volume also explicitly discusses Giambattista Brocchi's analogy, especially his ideas on innate species longevities as the explanation for species extinction (indeed, this may have been the first time Darwin actually encountered Brocchi's name in conjunction with these ideas), one must wonder if it was not until after Darwin received and possibly quickly read Lyell's second volume that he wrote up his suggestive notes, including his thoughts to Henslow.

Fortunately, the details of Darwin's correspondence resolve this matter: Darwin wrote up all his discoveries, as well as his thoughts on the extinction and succession of endemic, congeneric species, and his thoughts on *Trigonocephalus* and other members of the living biota observed at Bahia Blanca, before he received Lyell's tome. Darwin was

FIGURE 2.3 Charles Lyell. (By permission, Library of Congress, Prints and Photographs Division)

anxiously awaiting a box of books he had asked his brother Erasmus to assemble. When did the box arrive?

The text of the letter to Henslow in itself provides the first clue: it is actually dated "[ca. 26 Oct–]24 November" and begins by saying that the *Beagle* had arrived at Montevideo on October 24 (though apparently the *Beagle* log reveals they could not have arrived until late on October 25). The *Beagle* then sailed to Buenos Aires from Montevideo on October 30, staying for over a week, until sailing back to Montevideo and arriving on November 14. In the letter to Henslow, after recounting his Bahia Blancan scientific discoveries, Darwin picks up by saying, near the end, that "we have been in Buenos Aires for a week.—Novr. 24th." and proceeds to finish the letter. No mention of books—and, indeed, the only time Darwin was in Montevideo in November was from November 14 to 24.

The second clue is in a letter that Darwin ([1832] 2008a) wrote to his sister Caroline, dated October 24 to November 24. He writes about his joy in receiving letters from home: "November 14th .—M. Video.—I have just been again delighted with an unexpected stock of letters . . . from Erasmus (Aug.) 18th. . . . As it is a special favor, thank dear old Erasmus for writing to me & doing all my various commissions—I am sorry the books turn out so expensive & not to be procured. . . . No schoolboy ever opened a box of plumcake so eagerly as I shall mine, but it is a pleasure which shall not come for the next 9 months."

But fortunately, Darwin only had to wait another week and a half to receive his treasure trove. Continuing his letter to sister Caroline: "Hurrah, (Nov 24th): have just received the box of valuable [*sic*] thanks everybody who has had a finger in it, & Erasmus for packing them up so well."

It seems a safe bet, then, that Darwin made his observations, and drew his conclusions on extinctions and species replacements, stressing endemism and closely related taxa, before he received Lyell's anti-transmutational, anti-Lamarckian diatribe. This is important, as it has been sometimes supposed that Darwin's first real introduction to transmutational issues came from Lyell's second volume of the *Principles*, rather than his familiarity with the subject through his teachers and contemporary scientific literature while a student at Edinburgh and even at Cambridge in the 1820s.

Moreover, there is a passage in Lyell's second volume that alludes directly to the potential future discovery of the "laws which govern this part of our terrestrial system" (that is, on the births and deaths of species) in the context of paleontological study of the advent of the species of the modern biota—a passage that some have taken as the impetus for Darwin's early transmutationally imbued paleontological work. That he did not receive Lyell's second volume until after he had done his work and thought his thoughts at Bahia Blanca in 1832 shows otherwise.

Yet how eagerly Darwin must have devoured Lyell's second volume! Though Darwin had applied Lyell's approach to geological thinking, modifying his initial thoughts on the formation of the uplifted Patagonian steppes, as well as the sequence of events forming the Patagonian strata (including those at Punta Alta and Monte Hermoso) by the time he returned a year later to Bahia Blanca, most of that inspiration must have come from Lyell's first volume on physical geology. This second volume is focused on the organic realm.

With chapter titles such as "Changes of the Organic World-Reality of Species," "Theory of the Transmutation of Species Untenable," and the like, the book is usually, and correctly, seen as an anti-transmutational screed. Lyell was trained as a barrister, and much of the second volume reads like a lawyer's brief, laying out the case against transmutation, especially, if not quite solely, the ideas of Jean-Baptiste Lamarck. Giambattista Brocchi is virtually the only other naturalist mentioned by name. And though Lyell professes much respect and admiration for him, Lyell dismisses Brocchi's analogy, and especially the idea that species have innate longevities.

Lyell makes short work of Lamarckian transmutation, taking pot shots at Lamarck even as he presents his interesting and accurate summary of Lamarck's ideas in chapter 1. He then launches into an out-and-out refutation of Lamarck in chapter 2. Lyell criticizes both the empirical claims of progressive modification (the slow transformation of one species into another), as well as Lamarck's suggested mechanism for transmutation. By then, of course, Lamarck's views on the spatial and temporal transformation of one species into another were widely taken to be untenable, both by anti-evolutionists such as the naturalist Georges Cuvier and the philosopher William Whewell, as well as by pro-transmutationists like Robert Jameson, who thought Lamarck to be correct in the main claim that life has evolved, but not especially in terms of the predicted patterns of intergradations among species.

In other words, Lyell picked what was by then an easy target to raise against transmutation. True, Lamarck's ideas were the most explicitly transmutational and easily the most famous. But Lyell manages to essentially gloss over the ideas of his predecessors and contemporaries who had made it clear that the origin and extinction of species—their births and deaths—can and must be understood in terms of natural (secondary) causes. Thus when I first reread Lyell's second volume a few years back, I found myself saddened by what I thought to be the prevarications of a bright man who really ought, by then, to have known better. I had been trained to think of Lyell as essentially the father of modern geology, a hero who put our profession, including paleontology, on a thoroughly modern, rational, scientific footing. There may be something to historian James Secord's suggestion in his introduction to a 1997 edition of Lyell's three volumes that Lyell was trying to placate the clergyman who sat on the examining board governing an appointment in geology to the faculty of King's College—an appointment Lyell is said to have coveted. Yet it remains true that Lyell remained reluctant to endorse the arguments of his by-then long-term friend Charles Darwin in favor of evolution, until the *Origin of Species* was finally published in 1859.

Yet a closer reading reveals that Lyell was a brilliant, original, *ecological* thinker. Lyell saw faunas and floras—entire biotas—as interlinked combinations of animals and plants, of herbivores and carnivores. In modern terms, Lyell saw the connections between species primarily in terms of energy flow in ecological settings. He was aware that biotas differ in different regions of the globe. He was also aware, of course, of the Linnaean hierarchy: that species are parts of genera and genera are arranged together in "natural groups," such as families, and the like. But his gaze is directed far more at the ecological side than at data that speak to patterns of propinquity of descent.

Lyell is best known in paleontological history for using the general stratigraphic sequence of Tertiary marine invertebrate (mainly molluscan) fossils to divide Tertiary time into the divisions Eocene, Miocene, and Older and Newer Pliocene. The general approach was to define and identify these "time-rock" subdivisions on the basis of percentage of identity of fossil species with those of the modern biota—an approach clearly adumbrated, especially by Brocchi's work.

Thus Lyell was comfortable acknowledging, analyzing, and utilizing the palpable "younging" of an entire fauna, even while tiptoeing around the gorilla in the room: the strong suspicion held by many, starting with Lamarck and Brocchi, that that "younging" was achieved through the births of new species from old, forming actual genealogical lineages of descent. Surprisingly little of what we would today call "phylogenetic" thinking, in which a succession of closely related species is followed up the stratigraphic column up to and including the appearance of modern species, appears in Lyell's *Principles*.

That Darwin thought so too is nicely revealed in some correspondence with Lyell and Richard Owen in December 1859, soon after the *Origin* was published. Darwin (1859a) writes to Lyell complaining that Owen is trying to rip him off in claiming priority to what Darwin thought was clearly his own concept of the law of succession. Darwin published the law of succession in the first edition of the *Journal of Researches* (1839:209–210). This was after he had already, in the Red and Transmutation Notebooks, begun to write candidly of his evolutionary ideas, including natural selection. To Darwin, at least, the law of succession (that is, of allied species through time) was a way to talk about the patterns of replacement of species and the relation of recently extinct species to modern species, without admitting he was a committed transmutationist.

Darwin (1859b) had evidently told Owen that others, too, had made much the same point—citing the work of William Clift, who was Owen's father-in-law and predecessor at the Royal College of Surgeons. Darwin goes back to Lyell and finds the reference to the work of Clift, who noted in 1831 that Australian fossils reveal older, extinct forms of marsupial mammalian life prior to the marsupials of the living fauna—an observation fraught with endemism and continuity of natural kinds in a given region. And Darwin (1859a) says to Lyell: "You will also find that you were greatly struck with the fact itself, which I had quite forgotten." Darwin never did associate Lyell very strongly with thoughts on the succession of species within natural groups. Reading volume 2 of Lyell's *Principles* strongly bears this point out. Lyell habitually steered clear of exploring any such species-by-species replacement patterns in the fossil record.

Lyell ([1830–1833] 1997) begins volume 2 with four questions:

> to inquire, first, whether species have a real and permanent existence in nature; or whether they are capable, as some naturalists pretend, of being indefinitely modified in the course of a long series of generations? Secondly, whether, if species have a real existence, the individuals composing them have been derived originally from similar stocks, or

> each from one only, the descendants of which have spread themselves gradually from a particular point over the habitable lands and waters? Thirdly, how far the duration of each species of animal and plant is limited by dependence on certain fluctuating and temporary conditions in the state of the animate and inanimate world? Fourthly, whether there be proofs of the successive extermination of species in the ordinary course of nature, and whether there be any reason for conjecturing that new plants and animals are created from time to time to take their place? (183)

As to the first question, Lyell, unsurprisingly, firmly believes that species are real and fixed entities. Elsewhere in the text, Lyell writes: "It appears that species have a real existence in nature, and that each was endowed, at the time of its creation, with the attributes and organization by which it is now distinguished." His view is indistinguishable from the creationist philosopher William Whewell's—especially as he does not (save the teasing hint discussed later) openly discuss the possibility that new species are born of others through natural causes. But it is also indistinguishable from the views of Robert Grant expressed in that remarkable, near-poetical paragraph in the inaugural address of 1828. In short, Lyell's opinion on the reality and stability of species differs from that of only one person: Jean-Baptiste Lamarck.

As to the second question, Lyell prefers to think of newly created species beginning at a geographic point with a few original individuals derived from a single stock, but admits that they may be created from a number of similar stocks over a wider area. Lyell's rhetoric in stating this question clearly implies that he knows full well that new species are often found to bear distinct resemblances to other, pre-existing species. Yet the point seems to be wholly rhetorical, as the pattern of introduction of new species that Lyell discusses later in the text is predominantly couched in ecological, rather than in "phylogenetic," terms.

The third question—on the longevity of species being determined predominantly by physical questions—lies very close to Lyell's own heart. Lyell is convinced that extinction of species is a real phenomenon: species have deaths that, though in a vague sense are foreordained by, or at least with foreknowledge of, the Creator, are nonetheless caused by physical environmental change. Lyell is eloquent on this subject (the physical causes underlying extinction) and these discussions are where he shines the most.

Lyell is a steady-stater. He sees relative stability in the number of species inhabiting the planet through geological time. He estimates that, on average, one species of plant or animal becomes extinct each year, and, likewise, another is created to take its place. But to Lyell, this is a numbers game first: as the balance in species level diversity is maintained, not by one molluscan species dying out in some tropical sea, to be replaced immediately by another, closely similar species. No indeed, the species that becomes extinct in a given year may be a species of pine on the mountainous slopes in one part of the world, while the new species might be some form of aquatic animal living somewhere else. Moreover, here Lyell's ecological perspective comes to the fore, again pushing aside lineage thinking when it comes to the replacement of extinct by new species: Lyell makes it clear that, for example, in order for a new species of carnivore to be created, there must first be one or more suitable prey species in place. The dynamics of ecosystem interspecific processes—in this case, who eats whom—determines what and when species can be "called into creation."

And thus the fourth question—whether or not there are successive extinctions followed by successive creations of new species "from time to time to take their place"—is also answered. Only in one passage does Lyell get close to suggesting that "take their place" has the same meaning as encountered in the earlier Edinburgh literature: similar species, derived from others, within a lineage, often seen in terms of localized, endemic faunas and floras. For the most part, it is indeed a numbers game: "takes their place" seems to mean the birth of a new species in an ecosystem far removed from where the death of another species occurred.

The exception to this lies in the passage in which Lyell comes closest to linking his narrative with the contributions from the 1820s, and comes near the end of chapter 11, "Extinction and Creation of Species." It follows a discussion of estimated intervals of time needed to elapse before a large mammalian species restricted to a particular region might be expected to become extinct, assuming the "background" rate of extinction of one per year. Thus Lyell concludes, assuming the same rate for the births of new species, similarly large periods of time must elapse before it could be possible to "authenticate the first appearance of one of the larger plants or animals."

Then Lyell (1830–1833] 1997) speaks of the possibility that naturalists over the course of a few centuries might "accumulate positive data, from which an insight into the laws which govern this part of the terrestrial system [births and deaths of species] may be derived." But then Lyell says that we have gone as far as we can at present go, save for looking at the fossil record:

> Geological monuments alone are capable of leading us on to the discovery of ulterior truths. To these, therefore, we must now appeal, carefully examining the strata of recent formations wherein the remains of *living* species, both animal and vegetable, are known to occur. . . . From these sources we may learn which of the species, now our contemporaries, have survived the greatest revolutions of the earth's surface; which of them have co-existed with the greatest number of animals and plants now extinct, and which of them have made their appearance only when the animate world had nearly attained its present condition.
>
> From such data we may be enabled to infer whether species have been called into existence in succession or all at one period; whether singly, or by groups simultaneously; whether antiquity of man be as high as that of the inferior beings which now share the planet with him, or whether the human species is one of the most recent of the whole. (297–298; see also Brinkman 2009)

Yet even here, in this passage in which Lyell comes closest to acknowledging the key issue of the advent of the modern fauna, his words are far more of a ringing cry for "more data" than of an explicit recipe for adducing natural causal explanations for the births of species. It is hard to read much sincerity in John Herschel's (1836) letter ostensibly congratulating Lyell by saying that no doubt some will think his speculations on the "mystery of mysteries" "too bold," before saying that of course the job is to look for and specify the intermediate causes that must be there in the births of species. Lyell, in my opinion, is just talking around the problem. He is not even allowing that one species might actually give rise to another—for instance, within the same genus.

In contrast, Lyell was certain that the rock record does yield a secondary causal explanation for the deaths of species: environmental change, which a biota might absorb for some time before species begin to go extinct. This is precisely why he opposes Giambattista Brocchi's ideas. In two places (one each in volumes 2 and 3), Lyell reports Brocchi's conclusion that fully half of the Tertiary fossils collected and examined from the Subapennine Tertiary beds in Italy are still alive and well in the modern fauna (Lyell says that that number is just about right).

Lyell did the world a service by at least linking Brocchi to his empirical work, and to some of his ideas on the deaths of species. As we shall see, Darwin sides with Brocchi on this very point in the essay entitled "February 1835." And there are other points in Lyell's second volume that surface in Darwin's thinking—points best deferred to when they crop up in Darwin's writing, especially in the critical essay "February 1835."

DARWIN AND THE GEOGRAPHIC REPLACEMENT OF CLOSELY ALLIED SPECIES IN PATAGONIA, 1833–1834

By the time Charles Darwin set sail, there had been a thirty-year history, beginning with Jean-Baptiste Lamarck in 1801, of searching for secondary causal explanations of the advent of the modern biota, mostly at the outset utilizing the Tertiary molluscan fossil record. By the time the 1820s rolled around, at least in Edinburgh, the problem was seen as the replacement of extinct species by new, congeneric descendants, up to and including the modern fauna. So it comes as no surprise that, as soon as the data presented themselves, Darwin was comparing fossils on St. Jago and at Bahia Blanca with the living fauna, looking at *temporal* patterns of the replacement of species.

Yet as far as I am aware, Darwin was the first naturalist on an extensive expedition to record his thoughts on the temporal succession of species, and also the replacement of extinct species by elements of the modern fauna, at the very moment he was collecting specimens in the field. Lamarck and Giambattista Brocchi may have done so, but there is no direct evidence to that effect. Robert Jameson and Robert Grant, and for that matter John Herschel, in contrast, are not remembered for their paleontological field work: indeed, only Jameson among them was a geologist.

And the thirty-year history of discussions of temporal replacement patterns of species within lineages does not explain, at least entirely, what happened next in Darwin's research life on the *Beagle*: Darwin began looking for examples of the *geographic* replacement of living species. This, as he alluded to in the opening of the *Origin*, and as he said more explicitly in the 1950 edition of *Autobiography*, was one of the three sets of phenomena that had brought him to evolution in the first place; first mentioning the fossils, and before he singles out the Galápagos fauna as his third point, he wrote: "secondly, by the manner in which closely allied animals replace one another in proceeding southwards over the Continent" (1).

With the geographic replacement of species, Darwin was once again forging a new approach. He was looking for geographically disjunct species ("allopatric" or "parapatric"), congeneric endemic species, and writing up his thoughts as he encountered examples. And of the early transmutational forerunners to Darwin, only Lamarck ventured an opinion on the relationships of closely allied species geographically: Lamarck ventured the speculation that were species to be sampled carefully and completely over their geographic ranges, many species thought to be distinct would be shown to grade imperceptibly into one another. Lamarck's critics, including his admirers like Jameson, tended to tread lightly over the subject, though they clearly did not buy Lamarck's argument.

Then, too, there were Darwin's experiences as a fledgling student of botany under John Stevens Henslow at Cambridge. Darwin learned about patterns of within- and among-population variation. And he assembled his own herbarial sheets demonstrating geographic variation within species. Henslow, though not a transmutationist, had taught Darwin that species, though fixed, nonetheless often tended to vary considerably geographically, as well as within local populations.

Thus Darwin seems to have been testing Lamarck's speculations on the geographic blending of species versus Henslow's view of geographic variation in the context of the fixity of species. But he was adding a search for patterns of geographic replacement of "closely allied" congeneric species endemic to South America. He sought a parallel phenomenon in space that mirrored patterns of replacement in time.

Exactly when Darwin began formally equating the two patterns—the replacement of one closely allied, endemic species by another in space and in time—is hard to say. By the time he arrived home and completed Red Notebook in early 1837, exploring and rewriting the transmutational thoughts that had been more cryptically expressed in the earlier *Beagle* essay "February 1835," Darwin explicitly drew a parallel between a Patagonian example of fossil replacement by a modern species and the rheas' example of geographic replacement recounted later.

It is also clear that Darwin was accustomed to interpreting a series of geological events recorded in a vertical column of rocks—sediments interbedded with lava flows, for example—and relating them to a geographic landscape of marine basins and nearby volcanoes. As Darwin ([1835] 2008b) wrote to Henslow in July 1835 from Lima, Peru, recounting his observations when he traversed the Andes to and from Santiago, Chile, and Mendoza, Argentina:

> The Cordilleras of the Andes so worthy of the admiration from the grandeur of their dimensions, to rise in dignity when it is considered that since the period of Ammonites, they have formed a marked feature in the Geography of the Globe.— The geology of these mountains pleased me in one respect; when reading Lyell, it had always struck me that if the crust of the world goes on changing in a Circle, there ought to be somewhere found formations which having the *age* of the great European secondary beds, should possess the *structure* of Tertiary rocks, or those formed amidst Islands & in limited basins. Now the alternations of Lava & coarse sediment, which form the upper parts of the Andes, correspond exactly to what would accumulate under such circumstances.

In other words, if Lyell was right that the same processes shape geological history throughout geological time, there should be examples of volcanoes dotting the landscape around shallow marine sedimentary basins in the Mesozoic, just as had been described by Lyell and others from the younger Tertiary beds of Europe. And Darwin is telling Henslow in this passage that he found the evidence for this landscape in the Mesozoic rocks of the Andes, as determined by the presence of Jurassic ammonoids instead of Tertiary clams and snails in the sedimentary beds.

Physical geological history is far less controversial than the evolutionary history of organisms. But it is highly likely that Darwin began equating the processes of geographic and stratigraphic replacement of species while still on the *Beagle*, whether or not his turn to the geographic pattern of replacement of species beginning in 1833 in Patagonia was done explicitly with the analogy between space and time already taking shape in his thoughts.

In any case, after an eventful, if not evolutionarily significant, cruise down to Tierra del Fuego and back, Darwin found himself in Maldonado, Uruguay in May 1833. With little of geological interest to be seen, Darwin decided to collect specimens of as many species of the local birds as possible. He wrote that he was satisfied that he had collected virtually every species of birds present around Maldonado. Whatever his original intent for doing so was (he was, after all, expected to come back with all sorts of natural history specimens), this collection served as the template for comparisons with the bird species he observed and collected as he explored the Argentinean pampas and south into Patagonia proper over the next two years. Indeed, they provided a baseline for comparisons with birds collected beyond Argentina when the *Beagle* finally arrived in Chilean waters in 1835.

Thus, turning from Darwin's Geological Notes to Zoological Notes, after Maldonado the incidence of phrases like "takes the place of" or, simply, "replaces" becomes fairly common as Darwin's chronological list of living species he encounters accrues. An early example is the "ovenbirds" of the genus *Furnarius*. Ovenbirds are relatively easy to observe (and shoot), commonly found on open ground. Darwin begins his Maldonado notes with his observations on the morphology and behavior of the common ovenbird *Furnarius rufus*. Thereafter, many of the specimens attributable to *Furnarius* he observed or collected elsewhere are compared with this initial experience. Furnariid species are either considered "the same," "different," or, for some species, to be "replacing" this or other species.

But Darwin's best example by far of the geographic replacement of discrete species is the greater rhea (or common rhea, or simply "ostrich"; Darwin refers to this species as simply Avestruz in places in his notes [Keynes 2000:101–102, 188–189]; the local name is *ñandú*). The conventional demarcation between the pampas to the north and the Patagonian plains to the south is the Rio Negro. To the south, ñandús are replaced by the lesser rhea (local name *choique*), known also to Darwin as the "Avestruz petise," and, for a time thereafter, as "Darwin's rhea." Greater rheas are slightly larger and grayer than the southern species, and there are other morphological and behavioral differences as well.

Darwin saw, collected, and ate ñandús as early as his first visit to Bahia Blanca, where they still exist, in 1832. But it was through his lifelong penchant for pumping information from virtually everyone he encountered who seemed knowledgeable (including, later in life, his barber, who knew a lot about domestic dogs) that Darwin first heard reports of a second species of rhea. Visiting the Rio Negro in August 1833, he began to hear from local gauchos about the rare occurrence of another, slightly smaller, species of rhea that some of them had encountered in the region. He later heard that, south of the Rio Negro, the only rhea to be seen was this smaller species: the "Avestruz Petise."

In December 1833, at Port Desire, the ship's artist, Conrad Martens, shot a rhea that was consumed for Christmas dinner (along with the guanaco Darwin had bagged for the table). Darwin writes in his notes that he had forgotten the possible existence of the Avestruz petise, thinking he was eating an immature ñandú, when all of a sudden he jumped up and raced to the galley to retrieve the remains of the bird he now suspected was actually a choique. He managed to salvage a wing, the head and neck, the legs, and many of the larger feathers. He later noted the differences in egg coloration, feathering on the legs, and the shape of the scales on the legs—further differences between the two species. Sometime later, Darwin wrote in his notes that "Whatever Naturalists may say, I shall be convinced from such testimony, as Indians and Gauchos, that there are two species of Rhea in S. America."

In Ornithological Notes, a summary of his separate notes on the birds he had observed and collected on the voyage, compiled sometime in mid-1836 as the *Beagle* returned to Brazil from South Africa before finally heading home, Darwin mentions a "half Indian" he met along the Straits of Magellan, who assured him "beyond doubt that the Avestruz and Avestruz Petise were distinct birds," adding parenthetically, "I may observe that Indians and such people are excellent practical naturalists." Darwin (1836b) summarizes his thoughts by saying: "In conclusion I may repeat that the Struthio rhea [common rhea] inhabits the country of La Plata as far as a little south of the Rio Negro in Lat. 41° & that the Petise takes its place in Southern Patagonia, the part about the Rio Negro being neutral territory" (80–85; see also Barlow 1963:268–273).

"Neutral territory" means that the geographic ranges of the two species overlap along sections of the Rio Negro. Darwin thinks he has two discrete species that do not blend into one another where their ranges slightly overlap. Rather they "replace" one another from north and south. But unlike the case of the extinct fossil cavy replaced by the modern Patagonian cavy, and unlike the three examples of species on archipelagos connected by a "halo" effect to species living on the mainland (the Galápagos mockingbirds and tortoises, and to a lesser extent the Falkland fox story), Darwin does not draw any further, explicit transmutational conclusion until he reaches home and completes Red Notebook in early 1837.

GEOGRAPHIC REPLACEMENT OF SPECIES EXTENDED: THE FALKLAND FOX STORY

The *Beagle* had visited the Falkland Islands first in March 1833. Among other things, Darwin encountered Paleozoic brachiopods and presumably other marine invertebrate fossils. These were the remains of species so old and so long extinct that they did not invite any direct comparisons with the living biota, whether in the Falklands or in any other place. The comparisons called to mind were of similar fossils discovered elsewhere, with the attendant implication of establishing roughly equivalent ages of the beds in which the remains were found.

But Darwin's return to the Falklands a year later was another matter entirely. Now, with his observations on the Falkland fox species, Darwin significantly extended both his data and his ideas on patterns of geographic replacement of closely allied species. Once again, Darwin cites his familiar sources:

> Gauchos & Indians from nearly all parts of Southern part of S. America have been here & all say it is not found on the Continent: an indisputable proof of its individuality as a species.—It is very curious, thus having a quadruped peculiar to so small a tract of country . . . the Sealers say this fox is not found or any other land quadruped in the other Islands, as Georgia, Sandwich, Shetland &c. &c. . . . very soon these confident animals must all be killed: [Darwin himself collected four specimens by approaching them and whacking them on the head with his geological hammer!] How little evidence will then remain of what appears to be a centre of creation. (Keynes 2000:209–210)

"Individuality" of this Falkland fox species, indeed! Darwin is so struck by finding an endemic, discrete, and isolated species of mammal on the Falklands that he makes a further note (added later) that this wolf-like fox species from the Falklands is "quite different" from the "Culpen of Chili"—and no note is made of any evidence that this Falkland species is "allied with" the canids of mainland South America. Yet remarking on the "individuality" of this species does call to mind, of course, Giambattista Brocchi's analogy between species and individuals—not to mention the entire discourse of "species as individuals" from the modern era.

There is more. Darwin added some notes to his original account of the Falkland fox, interpreted by Richard Keynes as contemporaneous with his presence in the Falklands, albeit written in a different color of ink. Keynes admits that the third of these notes (the one on the Culpen of Chile) must have been made later (as the *Beagle* was over a year away from arriving in Chile), and so the other two notes may also have been made later. It is the second note that is so important: "Out of the four specimens of the ~~eyes~~ Foxes on board, the three ~~larger ones~~ are darker & come from the East [East Falkland Island];

there is a smaller and rusty-coloured one from the West Island: Lowe states that all from this island are smaller & of this shade of color" (Keynes 2000:210).

Darwin sees a difference between the foxes of the East and West Falkland Islands, though he believes the foxes on both islands belong to a single species that he refers to as *Vulpes antarcticus*. Darwin expands on these observations in Animal Notes, the compilation of his notes on mammals compiled roughly at the same time as he produced Ornithological Notes in 1836, near the end of the journey. In Animal Notes, Darwin (1836a) says of the Falkland fox:

> Out of the four specimens brought home in the *Beagle*: three will be seen to be darker coloured, they come from the East Isd. The fourth is smaller and rusty coloured: & is from the West Isd.—Mr Lowe, who has been acquainted with these islands for twenty years, and who is an accurate observer of nature, asserts that this difference between the Foxes of the two Isds is invariable and constant. He says he has long since observed it.—An accurate comparison of these specimens. I have omitted to add that the difference was corroborated by the officers of the Adventure. (23)

Lowe was a sealing captain and the *Adventure* was the companion ship that Captain FitzRoy had purchased from Captain Lowe at his own expense during the *Beagle* voyage.

This passage is exceedingly important because it clarifies an ambiguity in the most famous sentence Darwin wrote while on the *Beagle*. When he wrote up the Galápagos mockingbirds (throwing in the Falkland fox and the Galápagos tortoises for good measure) in Ornithological Notes of 1836, he basically declares himself a transmutationist. Here, in the 1836 Animal Notes, Darwin makes it ringingly clear that the pattern he is looking for is variation (differences) *between* two closely related taxa, be they discrete species, or "varieties" or "subspecies," in contrast to the lack of variation *within* the two taxa. In particular, in terms of the characteristics by which two species (or even subspecies or varieties) differ, there is no variation in those characters within one taxon in the direction of the characters states of the other. Species (or varieties) do not grade imperceptibly into one another; rather as discrete, "individual" entities, they simply replace one another geographically.

This is as true of the rheas as of the varieties of Falkland fox Darwin saw on the two main islands. Just when in the trip Darwin began seeing this clearly is difficult to say. Certainly in terms of the precise rhetoric, it was there by the time he was writing his notes on the local mockingbirds while he was still in the Galápagos. But he was thinking along the lines of replacement of discrete congeneric species—or conspecific varieties—certainly merely by pointing the pattern out no later than 1834.

In any case, I am unaware of any forerunners of Darwin's who put the matter so starkly. Conversely, there is no legacy of this pattern of "constancy of differences" emanating from Darwin's *Origin of Species*. It is an important theme, and an empirical claim, that had been largely lost sight of until the mid-twentieth century.

Most of the larger mammals and birds that figured so heavily in Darwin's experiences and interpretive thinking while in the Atlantic side of South America still exist. It has become increasingly difficult to see most of them, though guanacos seem to be fairly common nearly everywhere beyond city limits. On the other hand, Darwin had been prophetically accurate about the Falkland fox: it is now utterly extinct.

When my wife and I were chasing down Darwin's South America in late 2008, we really wanted to see as many of the iconic living species that had inspired Darwin's transformational thoughts as we could. Among these were the two rhea species, and the mara—the "Patagonian cavy."

Writing in the "Afterword" to a reprinted edition of his first book *Attending Marvels*, George Gaylord Simpson concludes with a brief reflection on a trip he took with his wife to Antarctica in 1970. They stopped for a brief while to see what Patagonia looked like now, compared to the days in the early 1930s when Simpson had first been there. After talking about old friends they met, Simpson (1982) writes: "That was the happy part of our travel down the length of Patagonia in 1970. The unhappy part was that we drove hundreds of miles around Trelew, Comodoro and Río Gallegos on now good roads. And we did not see a single guanaco, mara (called 'liebre patagónica' by the local people), or rhea (locally called 'avestruz'). In the 1930s we were never out of sight of these and other Patagonian creatures" (310).

Near Trelew there is a wild place, the Peninsula Valdés, just north of Puerto Madryn. It is connected to the mainland by the Isthmus of Ameghino, named after Florentino Ameghino, the great paleontologist that George Simpson reflected on when he was "musing" at Bahia Blanca.

I had had the good fortune to meet Dr. Rolando González-José at a genetics/evolution meeting in southern Brazil a few years prior to this 2008 trip. Rolando ("Rolo") is a geneticist-anthropologist. And though we had much in common to discuss, he walked up to me and said he plays tenor saxophone and was a deep admirer of John Coltrane—news that he knew would capture my attention. But the conversation soon turned to our own professional interests, and I was especially intrigued to hear that Rolo is on the scientific staff of CONICET, located along the shores of Puerto Madryn in Patagonia. I told him I was dying to see those rheas especially, and he assured me that if I came to Puerto Madryn, we could see both species in a single day.

Naturally I remembered that, and when I started to plan a trip to see what Darwin had himself seen in South America, I got in touch with Rolo—and he set up a visit to Puerto Madryn as a vital stop on our itinerary. One day we caravanned out to the Peninsula Valdés, a happy mixture of students and professionals, organized of course by González-José.

The *Beagle* had never called in to Puerto Madryn or the Peninsula Valdés. Nowadays it seems to be one of the better places to see the Patagonian wildlife so important to Darwin's thinking. Though much of the area is devoted to farming, there is nonetheless a conscious, successful conservation effort also in place.

We were thrilled to see maras: Darwin's Patagonian cavy. They are indeed, superficially, like giant rabbits—but of course they are not. The best way to see them is to go to an estancia (farm/ranch) and look beyond the fields and corrals. There they are, ringing the place, ducking in and out of their burrows. They are there because the farm dogs ward foxes away—and evidently leave the maras alone.

And we ran into three groups of Darwin's rheas (choiques). It had recently been breeding season, and each group consisted of a male leading around a gaggle of little ones. This was a slice of old Patagonia, replete with an incessantly strong wind and an estancia serving roasted lamb, whose salt-infused meat was conditioned by what the sheep ate, the grass growing near the sea, rather than by a salt shaker.

There were other marvels as well, including elephant seals and penguins. Miocene rocks with a riot of shells (including the Patagonian giant oyster Darwin had so wanted to find alive; actually, there seemed to be three distinct forms of "giant oyster" in these wildly prolific beds). But the maras and choiques were the real deal. They are wild there, but of course protected as well.

One final word on Darwin's experiences with the living animals of the pampas and on south down through Patagonia to Tierra del Fuego. At least once in Geological Diary, Darwin calls attention to the exceptionally long geographic ranges of species like guanacos and other mammals. This is completely counter to his search for geographic replacement patterns. Darwin called them like he saw them—and I am convinced that the less stressed, but equally true, observation that many species simply extend over vast stretches, without any evidence that they are close relatives replacing one another, ultimately weighed heavily on Darwin's thinking in the late 1830s.

DARWIN'S ESSAYS IN THE EARTHQUAKE PORTFOLIO

In early 1834, Charles Darwin was still laboring along the Atlantic coast of South America, with occasional forays inland. Darwin wrote two important essays while on the *Beagle*, both of which are housed together in the Earthquake Portfolio among the Geological Diary and Geological Notes at Cambridge.

The earlier of these two essays, "Reflection on Reading my Geological Notes" (1834), sets forth Darwin's views on the geological history of Patagonia, including observations on the geological occurrence of some of the fossils he encountered (including that cavy—but also the giant "rodent" later described by Richard Owen as *Toxodon*, first seen at Bahia Blanca and then later in other localities). The essay was written, in the opinion of historian Sandra Herbert (1995), sometime in early 1834, after Darwin had visited Port St. Julian in January 1834: the large fossil mammal Darwin discovered there, which he at first thought to be a mastodon (and later, and also erroneously, identified by Owen as a camel, which he named *Macrauchenia)*, is mentioned in the text of "Reflection."

In this essay, in fact, Darwin remarks on the large geographical ranges of many of the mammalian species in Patagonia—that is, *not* the sorts of geographic replacement patterns such as he saw with the rheas and other genera of birds—tying the ranges in with the geological development of Patagonia. Thus this essay discusses neither the temporal nor geographic patterns of replacement he had already been thinking about, at least in the context of species-within-genera. The succession of fossil animals, and the geographic distribution of living ones, are both discussed more in a faunistic way, and ecology is a dominant theme. In one passage, Darwin admits he knows little about recent shells, but supplies yet another example of a living mollusk that he feels he has also identified as a fossil. And he reiterates his belief that the Patagonian "giant oyster" must surely be extinct ("so remarkable a shell could not escape observation") in the modern biota. But that's about it for the spatiotemporal ranges of particular species.

And yet the essay holds some observations, opinions, and conclusions that burst forth in a fully transmutational context in the second, obviously later, essay, "February 1835" (Hodge 1983). The first of these thoughts (that is, in "Reflection" [1834]) is Darwin's professed wonderment at the ability of large mammals to live in such "sterile plains," as he thought the beds of the Tosca represent. Darwin, in other words, saw no change in the physical environment between the days when the South American megafauna was alive and the semi-desert conditions of modern day Patagonia. He was to return to this theme in the very first lines of "February 1835."

And in a later, marginal note, Darwin observed that the way to correlate the Patagonian beds, with all their extinct marine mollusks, with European beds can only be done "by relative proportion of recent shells.—This rests on the supposition that species become extinct in same ratios over the whole world." Here Darwin is exposing the underlying assumptions of his newly acquired literary mentor Charles Lyell, and seeming to question them. As we have seen, it was Lyell (almost undoubtedly liberally borrowing from Giambattista Brocchi) who proposed the division of Tertiary rocks by percentages of living species of mollusks present in Tertiary sediments, a proposal Lyell made in volume 3 of *Principles of Geology* (1833). It is clear that Darwin, whenever he made this later notation in this essay of mid-1834, had already received and read Lyell's third volume.

In reality, both these issues focus on patterns and causes of extinction, which becomes quite clear when we turn to the essay "February 1835."

"February 1835" is Darwin's first evolutionary essay. That it is so, despite its cryptic prose that has confounded the relatively few historians who have attempted to dissect it for its meaning, is obvious both from what we have seen so far about Darwin's observations of patterns of spatiotemporal replacement of congeneric species between 1832 and 1834, and from the simple fact that the overtly transmutational passages written in early 1837 in the second section of the Red Notebook noncryptically repeat the gist of "February 1835."

It is tempting to speculate that "February 1835" was indeed written in February 1835, specifically right after Darwin experienced the horrifying shock of a major earthquake on February 20, 1835, in Valdivia, Chile. The *Beagle* was anchored in the bay at the mouth of the Valdivia River; most of the officers were visiting Valdivia itself, some 7½ miles or so upstream from the anchorage site. Darwin himself was lying down in the *Nothofagus* forest not far from the ship. Darwin's account in Diary of the events of that day starts with a measured description of the experience, but then begins to record his emotional reaction to the earthquake. As Darwin wrote in Diary:

> There was no difficulty in standing upright; but the motion made me giddy.—I can compare it to standing on very thin ice or to the motion of a ship in a little cross ripple.
>
> An earthquake like this at once destroys the oldest associations; the world, the very emblem of all that is solid, moves beneath our feet like a crust over a fluid; one second of time conveys to the mind a strange sense of insecurity, which hours of reflection would never create. (Keynes 1988:292)

There is no way of knowing which bank of the Valdivia River Darwin was on when the earthquake struck. We were there in 2008, in a misty rain, standing on the bank of the Bay, looking out at the approximate site of the *Beagle*'s anchorage. We had come there with Andrea Nespolo, wife of Roberto Nespolo, of the Instituto de Ciencias Ambientales y Evolutivas, Universidad Austral de Chile, Valdivia.

A little farther upstream, where the mouth of the river narrowed, we gazed out toward the opposite bank. The mast of a sunken ship lay in midstream. Beyond the ship's mast, on the far shore, was an intact stand of *Nothofagus* forest, standing in primordial splendor adjacent to a plantation of eucalyptus. The world had moved on since Darwin had been there.

The ship had been inundated by the tidal wave from the monster earthquake of 1960. Calculated to have been magnitude 9.5, that particular earthquake ranks as perhaps the strongest on record. The river bed dropped down, and the environs of Valdivia were "forever" changed—though what does "forever" mean in this earthquake-prone gem of a little city? There was great loss of life and property. After we left, there was yet another such event in 2010. This one was also extremely strong: 8.8 on the Richter scale.

There is no way of knowing the magnitude of the earthquake Darwin experienced in Valdivia. Captain FitzRoy repeatedly measured a vertical movement of mussel beds of 10 feet as the *Beagle* subsequently plied the coast.

Given this literal and figurative shock, it would be natural for Darwin to want to record his thoughts on what would of course prove in the long run to be by far his greatest contribution to scientific understanding: transmutation. Life is uncertain. If this supposition is correct, it is an early forerunner of what Darwin is known to have done nine years later, when he directed in his will that his wife, Emma, furnished with the then-immense sum of £400, should find a suitable editor to see to the publication of "Essay" (1844a). Following the earlier and much shorter "Pencil Sketch" (1842), and already rendered in publishable "fair copy" form, Darwin set the essay aside to be published "in the event of my death." Darwin was still reluctant to go public. But he wanted his views known, and the credit that would come with eventual publication of his thinking.

Thus it seems likely to me that Darwin, perhaps just after writing the entry in Diary for February 20, 1835, also launched into his first record of his transmutational thinking, the essay simply entitled "February 1835." I think it is more than likely that the earthquake scared him into leaving some record of where his thinking was at the time.

"February 1835" is very much in the cryptic style of all of Darwin's *Beagle* notes. But it is in the form of a short essay, albeit one laden with notes and asides. The bulk of the short text is given over to matters geological. But the commingling of his experiences with fossils—particularly mammalian fossils—with his thoughts on the depositional and uplift history of Patagonia, is much more in evidence than in the earlier "Reflection on Reading My Geological Notes" (1834).

In particular, much of "February 1835" is given over to causes of extinction, once again perhaps not surprising after Darwin's encounter with the terrifying earthquake. Yet, perhaps ironically in this context, Darwin is at pains to deny that physical events, whether sudden catastrophes such as the biblical (or any other) flood, or the longer-term process of climate change, were at work in the extinction of the large fossil mammals he had encountered at Bahia Blanca and elsewhere in Patagonia. (Darwin had been taking potshots at the existence of the biblical flood often in his notes, saying that if floods were responsible for some of the strata and geomorphological effects he had seen in Patagonia, there had to have been more than one.) Indeed, the essay begins with a return to the Port St. Julian mastodon:

> The position of the bones of Mastodon (?) at Port St Julian is of interest, in as much as being subsequent to the *re*modelling into steps of what at first most especially appear the grand (so called) *diluvial* covering of Patagonia.— It is almost certain that the animal existed subsequently to the shells, which now are found on this coast. I say certain because the 250 & 350 &c plains, must have been elevated into dry lands when these bones were covered up & on both these plains abundant shells are found. We hence are limited in any conjectures respecting any *great* change of climate to account for its former *subsistence* & its present extirpation. In regard to the destruction of the former large quadrupeds, the supposition of a diluvial debacle *seems* beautifully adapted to

its explanation; in this case however, if we limit ourselves to one such destructive flood, it will be better to retain it for the original spreading out of the Porphyry pebbles from the Andes. (1)

It is the next section that is so important in the history of the development of Darwin's evolutionary thinking, and his linkage to the Brocchian form of transmutation. Once again, the delicate subject of the births of species (that phrase appears twice here for the first time in Darwin's known writings) is couched in the far less controversial context of what in many ways is its antonym: extinction, that is, the deaths of species. To understand Darwin's thinking here, we must pay close attention not only to the sentences of Darwin's main text, but especially to how the thoughts expressed there are made clear in the footnote (denoted as [a]). I will also refer to two relevant passages in Lyell's volume 2 to clarify content and meaning. The sentences in the main text are as follows:

> With respect then to the *death* of species of Terrestrial Mammalia in the S. part of S. America. I am strongly inclined to reject the action of any sudden debacle.— Indeed the very numbers of the remains render it to me more probable that they are owing to a succession of deaths, after the ordinary course of nature.— As Mr Lyell (a) supposes Species may perish as well as individuals; to the arguments he adduces. I hope the Cavia of B. Blanca will be one more small instance, of at least a relation of certain genera with certain districts of the earth. This corelation to my mind renders the gradual birth & death of species more probable. (2)

The equally important note (a) reads as follows:

> (a) The following analogy I am aware is a false one; but when I consider the enormous extension of life of an individual plant, seen in the grafting of an Apple tree, & that all these thousand trees are subject to the duration of life which one bud contained. I cannot see such difficulty in believing a similar duration might be propagated with true generation.— If the existence of species is allowed, each according to its kind, we must suppose deaths to follow at different epochs, & then successive

> births must repeople the globe or the number of its inhabitants has Varied exceedingly at different periods.— A supposition in contradiction to the fitness, which the Author of Nature has now established. (2, reverse side)

Brocchi's analogy is here loud and clear: many different species do not perish all at once, but in succession "after the ordinary course of nature," because, as Lyell says, "species may perish as well as individuals." Like individuals, species die because they have innate longevities.

Darwin had had some experience in the grafting of fruit trees before—with his horticultural experiences back home in England, and most recently on the Island of Chiloe before arriving at Valdivia, where he took notes. He is saying that (as perhaps his botanist mentor John Stevens Henslow had taught him?) the "thousand" trees resulting from the grafting process all live to about the same age, having the "duration of life which one bud contained." In other words, grafting provided the evidence that there is an innate (we would of course now say "genetic") determinant to the longevity of an individual.

And Darwin immediately takes innate longevity up one analogical step, saying that he cannot see why sexual reproduction within species cannot convey the same sort of innate longevity in a species as grafting reveals to be the case in the longevities of individuals. This is, of course, pure Brocchi.

But why is the analogy "a false one"? Two passages in Lyell's text in volume 2 of the *Principles of Geology* reveal the answer. The first of these passages deals with the grafting of apple trees: in chapter 2, entitled "Transmutation of Species Untenable," Lyell ([1830–1833] 1997) writes: "The propagation of a plant by buds or grafts, and by cuttings, is obviously a mode which nature does not employ; and this multiplication, as well as that produced by roots and layers, seems merely to operate as an extension of the life of the individual, and not as a reproduction of the species, as happens by seed" (207).

So Darwin's analogy is "false" only because he knows Lyell has already said so. Yet he goes on to disagree, saying he cannot see why sexual reproduction cannot confer innate longevity in species, by analogy with what is empirically known to be the case of individual organisms.

Earlier in "February 1835," Darwin had already broken with Lyell, who saw extinction as solely caused by external, environmental change. In arguing that point in volume 2 (in the chapter on the extinction of species), Lyell mentions Brocchi by name, at first saying that he welcomes Brocchi's acknowledgment of the reality of species extinction in the first place. But he goes on to say:

> But instead of seeking a solution to this problem [causes of extinction], like some other geologists of his time, in a violent and general catastrophe, Brocchi endeavoured to imagine some regular and constant law by which species might be made to disappear from the earth gradually and in succession. The death, he suggested, of a species might depend, like that of individuals, on certain peculiarities of constitution conferred upon them at their birth. ([1830–1833] 1997:255)

Unsurprisingly, Lyell concludes that Brocchi is right about the patterns of extinction (piecemeal, species-by-species, instead of catastrophic near-simultaneous extinctions of species). But Lyell insists that climate change is sufficient to explain extinctions, there in any case being no credible evidence that species have innate longevities and simply lose steam and die off. It is to be noted that Lyell cites the relevant passages directly from Brocchi's text in Italian.

In "February 1835," Darwin in fact tries to do precisely what Lyell says in general needs to be done: specify the actual empirical patterns of the births and deaths of species. His allusion to the cavy in the main text is couched in terms of endemism—the details of the cavy example, otherwise not discussed in this essay, is his best example of a modern species replacing a closely allied (congeneric) extinct species. This "co-relation" of species/genus and geography (that is, endemism) renders "the gradual births and deaths of species more probable," a conclusion more fully explained in the latter half of the appended note (a): gradual (meaning, once again, staggered) deaths of species must result in replacement births in a similar, staggered manner, not long after the deaths of the previous species occur, or else there would be large oscillations of species diversity through time—a fact which we know not to be the case. Darwin puts this last thought in terms of the rules laid down by the author of nature; it is a conclusion of Lyell's that Darwin in this instance saw fit to agree with.

Thus Darwin was specifying the temporal patterns of species extinction and their replacement by congeneric species. His perspective was essentially unchanged from his first interpretive thoughts on the cavy and *Megatherium* at Bahia Blanca in 1832. But now, in "February 1835," Darwin adds explicit thoughts on the environment, and in particular the essential stability of environmental conditions. He is in effect eliminating a causal role for external environmental change in extinctions of megafaunal species.

Beyond endemism, Darwin does not include geography in his thoughts in "February 1835." When he writes, "If the existence of species be allowed," he embraces the reality of individuated taxa with births, histories, and deaths, including Brocchi's notion of internally governed longevities in each species.

In "February 1835," Darwin was looking for natural causal explanations for extinction. But the context makes it clear that those patterns are constrained by natural phenomena (endemism, for example). He is all but saying that new species arise from old. Two years later, he does this explicitly, in much the same language, using much the same data, in the Red Notebook (1836–1837).

One footnote to "February 1835" and Darwin's embrace of a nonenvironmental, Brocchian innate longevity for the explanation of species extinction is in order. Darwin thought his data precluded environmental change as a cause of extinction of the late Tertiary mammalian species whose bones he had been collecting for the past two and a half years. He thought his evidence showed that these often rather large animals were living throughout on much the same arid, scrubby plains of the Patagonian environment as still existed while he was there. And in that he was most probably correct.

But what Darwin never saw, nor did Florentino Ameghino or George Gaylord Simpson as they, in turn, "mused" on the shores of Bahia Blanca, was the amazing array of footprints preserved only a few miles east of the cliffs of Monte Hermoso (Bayón et al. 2011). They had been discovered in the 1980s by Teresa Manera's husband, a physician. The older of these beds is around 12,500 years old, preserving footprints of

Megatherium, glyptodonts, and many other species of the late Pleistocene South American megafauna in profusion and astonishing detail. Along with these are a few footprints in the same vicinity (including a trackway paralleling a series of *Megatherium* footprints), dated at about 7,000 years ago, of a newly arrived mammalian species: *Homo sapiens*—ourselves.

The entire megafauna is by now absent, though footprints of rheas and possibly guanaco (both still living there) are also present.

I think of this as mute testimony to the destructive force our species has had as we have spread over the globe in the last 50,000 years or so. Everywhere we showed up, many species soon became extinct, perhaps most notably the larger species. In this interpretation, the extinction of the South American megafauna was indeed both catastrophic and external.

On a second visit to Bahia Blanca, I went along on another well-attended field trip to see these famous footprints. They are amazing. Very clearly delineated tracks of a large variety of animals, small and large, crisscross on the exposed bedding planes. The mastodon footprints were surprisingly small, especially compared with the huge, deep, ponderous footprints of the bipedal *Megatherium*—largest of the giant ground sloths. I just stood there and felt that animal walking by—the most gripping sense of being in the presence of an ancient, extinct beast that I have ever had as a paleontologist. Bones and shells are great, but the footprints left by these now-dead species make you feel as though they're alive and just ambled by.

Had Darwin seen these footprint layers and realized their relative ages, would he have reached the conclusion that it was the arrival of humans that set off the extinctions of the South American Pleistocene megafauna? There is of course no way of knowing. But he was astute in the field, even as a relatively young and inexperienced geologist, when he visited Bahia Blanca in 1832 and then again in 1833. Had he seen them, and realized their implicit significance, he would most likely not have followed Brocchi in terms of the innate causes of species extinction. Indeed, late in the trip, still at sea, in the Red Notebook, Darwin (1836–1837) actually says: "In the History of S America, we cannot dive into the causes of the losses of the species of Mastodons, which ranged from Equatorial plains to S. Patagonia. To the Megatherium.—To the Horse. = One might fancy that it was so arranged from the foresight [*sic*] of the works of man" (85; see also Herbert 1980:48).

He had not seen the footprints, and in any case he abandoned this speculation in later passages in the Red Notebook. But hypothetical as this scenario is, this "thought experiment" is relevant as it underscores one fundamental facet of Darwin's thinking in "February 1835": his conviction that there are natural causes for the births and deaths of species, just as there are for the births and deaths of individuals, would have remained unaffected, no matter what the precise cause of extinction might be. The actual pattern of the births and deaths of species would remain the same. And it was the pattern that Darwin was concerned to establish.

CROSSING THE ANDES, THEN ON TO THE GALÁPAGOS

Charles Darwin's next big adventure after the earthquake and writing "February 1835" was a rather daring trek across the Andes, between Santiago, Chile, and Mendoza, Argentina. His mule train party took an outbound southerly route through the Portillo Pass, returning through the more northerly Uspallata Pass (where all modern traffic flows nowadays).

Darwin naturally spent much time on the geology exposed in the mountains—writing, as we have seen, to John Stevens Henslow in July and August 1835 about the geographic juxtaposition of volcanoes and sedimentary basins ([1835] 2008b). What Charles Lyell had described for the Tertiary of Europe was here manifest in older, Mesozoic geological sequences. And I have been astonished to read in modern accounts how meticulous Darwin's observations and how sophisticated his conclusions were. Most amazing to me, as someone trained in a geology department, was the fact that Darwin had correctly concluded that the principal central range of the Andes had been uplifted prior to the eastern range adjacent to the plains stretching from Mendoza to the Atlantic. I have seen the jumble of highly faulted volcanic, metamorphic, and sedimentary rocks while chasing Darwin across the Andes from Santiago to Mendoza and back again—and it blows my mind that this virtually untrained young man could have figured this all out.

Only beginning in post–World War II years have geologists seen this as the standard sequence of events in the incorporation of volcanic island arcs into the adjacent continental terrain. Volcanic island arcs are uplifted, eroded, and then thrust toward the continent in a series of upheavals that culminates with the uplift of the shallow, non-volcanic seaway. There is no way I could have made those observations and drawn those conclusions, had I gotten out of the car and studied those rocks, even given what I know of the basic picture that colleagues in tectonics have established in the past half-century.

We left Santiago one day to travel up the Yeso River Valley. That was Darwin's route back in 1835. But the border is now closed and the roads peter out. We saw plenty of birds, including the Thenca—the Chilean mockingbird Darwin went on to compare with the mockingbirds he saw in the Galápagos, in the company of ornithologist Michel Sallaberry of the University of Chile, Santiago. And we also bought some Jurassic ammonite fossils, beautifully preserved as jet-black specimens. These and similar fossils were what caused Darwin to exclaim to Henslow that he had evidence for a Mesozoic landscape of volcanoes and marine basins just like the ancient geography in Europe that Lyell had described.

The only way to Mendoza from Santiago now is over the present-day Portillo Pass. The following day, thanks to the intrepid goodwill and skilled driving of our host Dr. Rodrigo Medel, Departamento de Ecología, Universidad de Chile, Santiago, we took the modern route over to Mendoza. A day's drive—now choked with trucks, the modern beasts of burden. We were, though, taking the route that Darwin used when he returned to Santiago—and eventually the *Beagle*. Gorgeous scenery, of course, but I've already confessed that I found the geology confusing. We did see more Mesozoic ammonites on the flanks of the famed Aconcagua.

And we did finally see Andean condors. Two presumably male birds chasing one another, swooping and swirling within 25 feet of us, not so much for our benefit, we knew, as for the attention of the presumed female who had perched on the rocks across the road, bringing the others down from atop the 2,000-foot cliff. I can't honestly say that Andean condors played a discernible role in Darwin's evolutionary thinking. But I simply can't help mentioning seeing these three as we were once again on the chase for Charles Robert Darwin.

The trek over the Andes and back, in terms of the present focus on Darwin's evolutionary thinking, is most notable for his return to geographic themes in the distributions of species. And here he steps it up a notch, now embracing not single groups of allied species, but entire faunas, as when, in Zoological Notes, he observes, "The ornithology of the valleys on the Eastern slopes differ to a certain extent from the Pacifick sides; the resemblance is very strong in aspect & in zoology with the plains of Patagonia" (1832–1836b:278; see also Keynes 2000:249). It is worth noting here that in the first edition of *Journal of Researches* (*Voyage of the Beagle* [1839:399]) Darwin returns to this theme, writing that "he was very much struck" (that is, in the course of the actual journey) by the differences, not just in the avifauna, but also to varying degrees in the vegetation, mammals, and insects on either side of the Andes. But it is in a footnote to this passage that Darwin sends a strong hint of his by-then (that is, by 1838) well-developed ideas on transmutation, inviting us to wonder if he entertained similar thoughts as well back in March 1835 when he was making these original observations: "This is merely an illustration of the admirable laws first laid down by Mr. Lyell of the geographical distribution of animals as influenced by geological changes. The whole reasoning, of course, is founded on the assumption of the immutability of species. Otherwise the changes might be considered as superinduced by different circumstances in the two regions during a length of time" (400).

As we shall see, by the time Darwin published the *Voyage*, he had already formulated a theory of geographic (allopatric) speciation. And by mid-1838, he had discovered natural selection. The final sentence of this remarkable footnote is a de facto, and stark, adumbration of selection-mediated changes in species (in this case many species within distinguishable biotas) in geographic isolation: a theme Darwin began to develop in Transmutation Notebooks to some extent, even before he discovered natural selection in the late 1930s.

While in Patagonia on the Atlantic side of South America, Darwin had seen four species of armadillos. In Animal Notes, compiled and written sometime in 1836 before the *Beagle* reached England, Darwin discusses aspects of the morphology, behavior, and distributions of all four species. But the only place where all four can be found is at Mendoza, at the high plains just east of the Andes, in Argentina. And based on this and occurrences of other taxa, Darwin (1836a) concludes: "The sterile plains, at the foot of the Cordillera, elevated several thousand ft. above the sea, appear the probable birthplace of nearly all the animals, Birds, and perhaps even plants, of Patagonia" (12).

Darwin was looking for geographically restricted points of origin, not just for species within a genus, but the species within an entire biotic region. Again, we find a crucial pattern, if not an explicit acknowledgment, of a natural underlying causal process. But the interplay of geography with the births of species was becoming a strong theme for Darwin. And of course it was to prove the linchpin of his early evolutionary thinking when next he went, a few months after making it across the Andes and back, to the Galápagos.

In May 1835, Darwin visited the Coquimbo valley in northern Chile. There he had his sole encounter with a Tertiary sequence of sediments and fossils on the Pacific side of South America. In his field notes, Darwin wrote of the gradual blending of both mineralogical and zoological features as one examines the rocks from bottom to top. But here, he was not talking in transmutational terms; Darwin was instead claiming that the sequence of fossils gradually approaches the state of the modern invertebrate marine fauna strictly in terms of percentages of abundance of the species represented, unlike the fossil invertebrates of Punta Alta, which he thought had occurred in similar percentages to the recent fauna. Otherwise, the entire fossil fauna remained much the same throughout the sequence of sediments—by then a familiar observation.

The *Beagle* set sail for the Galápagos from a port near Lima, Peru, on September 7, 1835. Prior to departure, Darwin ([1835] 2008a) wrote to his sister Caroline that he was "very anxious for the Galapagos Islands,—I think the Geology and Zoology cannot fail to be very interesting." Even more specifically, he wrote at the same time to his cousin William Fox: "I look forward to the Galapagos, with more interest than any other part of the voyage.—They abound with active Volcanoes & I should hope contain Tertiary strata" ([1835] 2008c).

The Galápagos Islands have long since become the iconic symbol of the very idea of evolution. There is an almost romantic notion of a lush cluster of volcanic islands sitting astride the Equator, the very epitome of a tropical laboratory of evolution.

I've been to the Galápagos on three separate occasions. It is indeed a thrill to go there, as it is a spot that cemented, extended, and, I think, emboldened Darwin in his convictions on the natural causes underlying the births of species.

But it is no romantically lush tropical paradise. Sailors in the early 1830s disliked the place intensely. There was almost no water to be had. The only animals they could procure for their shipboard larders were the lumbering Galápagos giant tortoises. They ate the tortoises (Darwin and the *Beagle* crew certainly did). They stocked the islands with goats that still roam in a feral state and do much damage to the terrestrial ecosystems, though the addition of goats to the larder provisioning items probably helped the tortoises survive in their depleted numbers. Deserts have their charms, but in many ways these particular islands are harsh and uninviting. Herman Melville apparently hated them with a passion.

So it was largely the critical role the Galápagos played in the intellectual history of evolutionary theory that drew me to them. Otherwise I found myself largely agreeing with the sailors of yore.

Based on Darwin's prior conclusions on the replacement of congeneric species in time (the Bahia Blancan cavies, fossil and recent); in space (especially, though not solely, the replacement of the northern rhea [ñandú] by the smaller species [choique] at the Rio Negro and beyond to the south); and, more specifically, the occurrence of a distinct species of fox on the Falkland islands, and, in particular, the consistent differences to be seen between the individuals of that same species on the East and West Islands, Darwin had not only high hopes, but almost undoubtedly some specific predictions about what he would expect to see in the Galápagos.

Based on his prior three years of experience, Darwin must have expected that the species of the living fauna would be fundamentally South American in character: they would be "allied with" species he had already observed on the mainland. Yet some, at least, would differ in some respects from their mainland relatives. And he must have expected to be able to detect some differences between congeneric species from island to island. And lastly, he was hoping to see patterns of replacement of extinct species in the fossil record by closely allied (congeneric) species in the living fauna: "I should hope [the Galápagos] contain Tertiary strata," as Darwin ([1835] 2008c) had written to cousin Fox.

He was to be disappointed only in the lack of Tertiary sediments and fossils on the Galápagos. Otherwise he found precisely what he must have been looking for—based, again, on the entire range of his experiences on both sides of South America for the past three years. That Darwin, in any case, came to the Galápagos with a prepared mind is obvious from the details of his experiences over the preceding three years. Had the *Beagle* arrived at the Galápagos first, in 1832, with Darwin as a novice, before those experiences that had sharpened his field and analytical skills and that had brought him seriously to entertain a perspective on the notion of the births and deaths of species through natural causes already, it is doubtful that the Galápagos would have made the lasting impression it did make on him. The Galápagos turned out to provide the linchpin of his emerging ideas on transmutation. And it was the mockingbirds, not the finches, that proved decisive.

In his Galápagos field notebook, in a line to which Richard Keynes (1988) drew attention, Darwin observed, "The Thenca very tame and curious in these Islands. I certainly recognise S. America in Ornithology. Would a botanist?" (353). "Thenca" (or "Tenca") is the local name for the Chilean mockingbird, and other mockingbird species occurring on the western side of the Andes. Darwin had of course seen them—and also the several mockingbird species in Argentina known locally there as "Callandra"—as Darwin noted when he was in Maldonado over two years previously.

Thus from the outset, Darwin knew that, here on the Galápagos, he had at least one easily observed example of a South American bird sufficiently closely allied with familiar birds from the mainland that he gave them the same common "generic" name. The Galápagos simply had a mockingbird species that replaced the western South American

Thencas—much as the western South American mockingbird species replaced the species he had seen in Patagonia.

When Darwin (1832–1836b) wrote his notes up (while anchored in the Galápagos or perhaps just after departing in October 1835), he said of the mockingbirds:

> This birds [*sic*] which is so closely allied to the Thenca of Chili (Callandra of B. Ayres) is singular from existing as varieties or distinct species in the different Isds – I have four specimens from as many Isds –These will be found to be 2 or 3 varieties—Each variety is *constant* [emphasis added] in its own Island—. [note (a)] The Thenca of Chatham Isd Albermarle Isd is the same as that of Chatham Isd.— This is a parallel fact to the one mentioned about the Tortoises. These birds are abundant in all parts: are very tame and inquisitive: habits exactly similar to the Thenca.—runs fast, active, lively: sings tolerably well, is very fond of picking meat near houses, builds a simple open nest.—I believe the note or cry is different from that of Chili. (341; see also Keynes 2000:298)

So Darwin thinks he has it: different taxa (he calls them "varieties" here) on some of what ultimately turn out to be the older islands of the archipelago. But note two further points: first, the Galápagos Thencas are very much like their close relatives on the mainland, though Darwin does mention one difference: he thinks the "note or cry" here in the Galápagos is different from what he heard on the mainland. In other words, there is a slight but real difference between the birds on the mainland and those on the Galápagos.

Second, like the Falkland foxes before them (that is, in Darwin's experience), these different "varieties," he says, are "constant" within their own islands. In other words, the characteristics that set the "varieties" from different islands apart from one another do not vary within the populations on the respective islands. Darwin does not say what these features are, but when John Gould had the opportunity to study the specimens back in London on the *Beagle*'s return, he agreed with Darwin's assessment, recognizing four distinct, separate species of Galápagos mockingbirds. The species-level taxonomy of the Galápagos mockingbirds has not changed much in the intervening years.

This pattern is what Darwin later alluded to as a "halo": a group of obviously closely related birds (mockingbirds, Family Mimidae) are endemic to the Americas, and most diverse in South America. Though some species overlap in their geographic distributions, dominant species tend to replace one another in different regions, a pattern that Darwin saw carried through to the Galápagos vis-à-vis the mainland. But within the Galápagos mockingbirds, there was further noticeable inter-island diversification.

And what of the "parallel fact" to the tortoises? In his notes a few pages before the mockingbirds, Darwin (1832–1836b) writes extensively about the Galápagos tortoises: "It is said that slight variations in the form of the shell are *constant* [emphasis added] according to the Island which they inhabit—also the average largest size appears equally to vary according to the locality.—Mr Lawson states he can on seeing a Tortoise pronounce with certainty from which island it has been brought" (328; see also Keynes 2000:291). "Parallel fact" indeed—though Darwin himself did not make the observation originally.

So Darwin had two examples of geographic replacement, the mockingbirds being the more complete, as they were obviously closely allied to species he well knew from the mainland. Both involved closely allied taxa on different islands, each one being "constant" on its own island: a pattern of variation among islands, not within islands.

As is by now notorious, when Darwin (1836b) wrote the summary Ornithological Notes somewhat less than a year later, as the *Beagle* was headed for home (but perhaps while it was diverted back to Brazil for some additional measurements Captain FitzRoy felt he needed), he put it all together in what has become the consensus agreement that he had accepted the fact of transmutation while still on the *Beagle*:

> In each Isld. each kind is *exclusively* found: habits of all are indistinguishable. When I recollect, the fact of the form of the body, shape of scales & general size, the Spaniards can at once pronounce from which Island any Tortoise may have been brought. When I see these Islands in sight of each other, & possessed of but a scanty stock of animals, tenanted by these birds, but slightly differing in structure & filling the same place in Nature, I must suspect they are only varieties. The only fact of a similar

> kind of which I am aware, is the constant/asserted difference—between the wolf-like Fox of East and West Falkland Islds.—*If there is the slightest foundation for these remarks the zoology of Archipelagoes—will be well worth examining; for such facts [would] undermine the stability of Species* [emphasis added]. (73; see also Barlow 1963:262; Kohn et al. 2005:645)

Darwin inserted the word "would" in the same ink, probably right after finishing the sentence. He still had the captain to worry about. But "would" was not there when first he wrote this passage.

And, of course, I agree with most recent historians that this sentence does confirm Darwin's all but formally declared adoption of transmutation.

But as we have seen, this was not his first such declaration while on the *Beagle*. In "February 1835," Darwin characterized natural patterns of replacement of extinct species by closely related species—as he had in fact already done in the fall of 1832 while still in Bahia Blanca.

But there was nothing about geographic replacement of closely allied still-living species in "February 1835," even though Darwin's notes reveal that he had been thinking about such patterns from mid-1833 on. This passage on the Thenca in the Ornithological Notes redresses the balance, doing for geographic replacement what he had already done for temporal replacement patterns. As we shall see shortly, he was to put those two patterns—geographic and temporal patterns of species replacement—together in one powerful sentence in the Red Notebook sometime early in 1837, after the *Beagle* returned home.

But first there is the important matter of the "Galápagos finches," later often called "Darwin's finches." Historians have rightly pointed out that it was a myth that Darwin saw the finches on the Galápagos and immediately leapt to the conclusion that life has evolved. And it is the truth that Darwin did not keep notes or accurately labeled specimens on the ground finches that he found on all the Galápagos Islands he visited.

Here is what Darwin (1832–1836b) wrote in his notes while at the Galápagos: "Far the preponderant number of individuals belongs to the Finches & Gross-beaks—There appears to be much difficulty in ascertaining the Species" (340; see also Keynes 2000:297).

Darwin (1836b) is a bit more explicit in Ornithological Notes: "Amongst the species of this family [the finches] their reigns (to me) an inexplicable confusion . . . moreover a gradation in the form of the bill, appears to me to exist.—There is no possibility of distinguishing the species by their habits, as they are all similar, & they feed together (also with doves) in large irregular flocks" (72; see also Barlow 1963:261).

Darwin was confused by the very sort of transformation in characters that would have delighted Lamarck. He saw that this confusing mélange of finch morphology, making it impossible for him to sort out discrete species, was present over and over on all the islands. He literally was throwing up his hands in dismay. As my colleague David Kohn once remarked, Darwin did not keep notes or accurately labeled specimens of these finches from the separate islands, as he saw little point in doing so. To Darwin, the locality of these finches was the Galápagos archipelago in general. He was looking for discrete species, and expecting to see differences among closely related species on the different islands. The confusing mélange of finches on each island, from the perspective of what Darwin expected to see, was something of a nightmare.

Darwin's "failure" to make much of the Galápagos finches is beautifully clear evidence that there were two forms of transmutation current in the decades preceding Darwin's experiences on the *Beagle*. One was Jean-Baptiste Lamarck's vision of gradual integration linking species both temporally and geographically. The other was Giambattista Brocchi's image of the births and deaths of species and their successive replacement through time.

Darwin was focused nearly completely on Brocchi-style patterns, rather than the imagery suggested by Lamarck. That is why the Galápagos finches confused him, and he did not bother to make note of the islands from which his and the other crew members' collections came. And adaptation in general did not loom large in his transmutational thinking at the moment.

At the very outset of our 2008 trip to follow Darwin around South America, I had the privilege of speaking at a meeting of evolutionary biologists in Pucón, Chile. The invitation had come from Rodrigo

Medel of the University of Chile in Santiago. Pucón is overshadowed by the magnificent and highly dangerous Volcan Villarica, one of the two most active volcanoes in South America.

At the meeting, I found that evolutionary biologists familiar with the South American biota often express surprise that Darwin somehow overlooked some obvious and remarkable examples of adaptation in different species, such as the giant bee of Chile. Indeed, they are hard to miss, and I am sure Darwin must have seen them too. But they didn't figure into his earliest transmutational thinking. This is yet more, albeit anecdotal, evidence that Darwin was not spending too much time thinking about adaptation, or even about intergrading anatomical series, at that point in the development of his evolutionary thinking. By the time he made it to the Galápagos, the Bahia Blanca snake had been studied three years earlier.

The earliest aha! about the finches in Darwin's mind was indeed about adaptation through natural selection. In the second edition of the *Journal of Researches* (1845:379), Darwin published a famous illustration of the profiles of the heads of four species of Galápagos finches (again, in the post-*Beagle* taxonomy of Gould). And here he added another hint about his transmutational views. After a paragraph discussing the finches, Darwin writes: "Seeing this gradation and diversity of structure in one small, intimately related, group of birds, one might really fancy that from an original paucity of birds in this archipelago, one species had been taken and modified for different ends" (380).

But when Darwin fetched up in the Galápagos, he had little use for concepts of variation as he strove to pinpoint patterns and causal explanations of the births and deaths of species seen as discrete and stable entities.

And so we leave Darwin on the *Beagle*. He went on to the South Seas, New Zealand, and Australia. He met with John Herschel in Cape Town, and saw how the scrubby vegetation of southern Africa could indeed support large numbers of a diverse array of large mammals—important to his thinking about the extinct megafauna of South America. He wrote up Animal and Ornithological Notes. And he opened up the Red Notebook, recording latitudes and longitudes, and a few other relevant (that is, to this narrative) thoughts, such as the already mentioned conclusion that the rocks of Punta Alta and Monte Hermoso are indeed contemporaneous, and that seeing that the scrubby lands of southern Africa can support rhinos makes finding mastodon fossils in Patagonia no longer surprising.

But it was in the second half of the Red Notebook, written sometime in the first half of 1837 (the *Beagle* had arrived home the previous autumn, nearly five years after setting sail), that next we see Darwin discussing transmutation.

THE RED NOTEBOOK

I have already mentioned my conclusion that the famous, explicitly transmutational passages, written around March 1837 in the second (post-*Beagle*) half of the Red Notebook, are essentially a rewrite of "February 1835." There are some differences, to be sure. For example, Charles Darwin uses "camels" (*Macrauchenia*, the camel-like South American litoptern he excavated at Port St. Julian, along with the living guanacos) as an example of replacement of an extinct species by a presumed descendant. This example usurped the original example of the living Patagonian cavy and its putative extinct smaller relative from Mt. Hermoso after Richard Owen proclaimed the fossils to be remains of an extinct species of tucutucu.

But conceptually, Darwin is the same species-replacement transmutationist he was in 1832—and certainly as he expressed his views in "February 1835." He remains utterly faithful to the two cardinal points of Giambattista Brocchi: (1) that species have natural births and deaths just as do their component organisms; and (2) species extinction, again, like that of organisms, is a function for the most part of internally built-in longevities.

What *is* somewhat new is the explicit equation of geographic with temporal patterns of species replacement, deeply implicit but not directly made in Zoological Notes and Geological Notes. Another interesting point is that Darwin reveals himself as a saltationist: species that replace one another are "inosculate," and do not meld gradually into one another. Thus when one species diverges from another, ancestral one, it must be sudden—*per saltum*. This is especially arresting as Darwin, with ample justification, has come down to us as the father of gradualism: the idea that species change slowly and gradually through time. Though it was in fact Jean-Baptiste Lamarck who was the father of gradualistic notions of transmutation, it was indeed Darwin who left what proved to be an indelible impression of gradual change as the signature pattern of the evolutionary process. Darwin and Lamarck differed, of course, on the causal process underlying patterns of gradual evolutionary change.

That the transmutational passages of the Red Notebook are essentially a summary of what Darwin had been thinking on the *Beagle* is, I think, borne out by the fact that they occur together in a fairly coherent stream of notebook pages (with one coda thought added a bit later). The passages in the Red Notebook (1836–1837) are not as explicitly essay-ish as is "February 1835," but they do reveal a coherent thematic development:

> [127] Speculate on neutral ground of 2. Ostriches; bigger one encroaches on smaller.— change not progressif<e>; produced at one blow. If one species altered: Mem: my idea of Volc: islands. Elevated, then peculiar plants created, if for such mere points; then any mountain, one is falsely less surprised at new creation for large.—Australia's = if for volc. Isld then for any spot of land.= Yet new creation affected by Halo of neighboring continent: ≠ as if any [128] creation taking place over certain area must have peculiar character: . . . [129] Should urge that extinct Llama owed its death not to change of circumstances; reversed argument. Knowing it to be a desert.— Tempted to believe animals created for a definite time:—not extinguished by change of circumstances: [130] The same kind of relation that common ostrich bears to (Petisse. & diff. kinds of Fourmillier); extinct Guanaco to recent; in former case position, in latter time. (or changes consequent on lapse) being the relation.— As in first cases distinct species inosculate, so must we believe ancient ones: (therefore) not *gradual* change or degeneration.

> From circumstances: if one species does change into another it must be per saltum—or species may perish. = This representation of species important, each in its own limited & represented.—Chiloe creeper; Furnarius . . . Calandria: inosculation alone shows not gradation. . . . [133] There is no more wonder in extinction of species than of individual. [153] When we see Avestruz two species, certainly different. Not insensible change.— Yet one is urged to look at common parent? Why should two of the most closely allied species occur in same country? (see also Herbert 1987:61–70)

Even the mockingbirds are there (Calandria, assuming Darwin meant to write Callandra), along with the "halo" referred to on an earlier page neatly delineating Darwin's view of the hierarchically nested pattern of geographic replacement of non-intergrading, closely allied species in different regions on the mainland, extending out to archipelagos at sea, with further subdivision of the pattern with those different mockingbird (and fox) taxa on different islands.

To my mind, the Red Notebook is a summary of Darwin's Brocchian, taxic transmutation brought to life with his experiences with both the extinct and living biota as recorded in his contemporaneous *Beagle* notes, and somewhat more cryptically captured in "February 1835." It is profoundly important as the first explicit admission (save the mockingbird passage in Ornithological Notes) of Darwin's acceptance of the basic fact of evolution, with the geographic replacement of closely related species added to the older notions of temporal replacement and the origin of the modern fauna. Given what the Red Notebook is, it is unimaginable to maintain that Darwin was not a transmutationist by "February 1835," and, I would say, by fall of 1832 at Bahia Blanca.

Indeed, when Darwin makes the analogy between geographic and geological species replacement patterns explicit, we have the beginnings of Darwin's geographic speciation theory. Darwin's analogy—that replacement of species in space and the replacement of species in time mirror one another—stands alongside Brocchi's analogy as cornerstone concepts in the early history of evolutionary biology.

THE END OF AN ERA

The passages in the Red Notebook, brief as they are, were to be Darwin's very last recorded transmutational thoughts written with the thorough-going, and nearly sole and entire, perspective of a Brocchian-style transmutationist. The Transmutation Notebooks (Notebooks B–E), first begun in mid-1837, from the opening pages of Notebook B record a sudden adoption of interest in adaptation and gradational change, even before he formulated natural selection itself. Vestiges of his original taxic perspective persist, and to some degree offer a tension, especially in Notebook B, between the theme of continuity of character states (as in Lamarck) versus seeing species as discrete, and their origins via sudden, brief episodes of saltation. Geographic isolation is mentioned often, and what can only be described as a much-condensed vision of geographic (allopatric) speciation is developed, once again, in the earliest pages of Notebook B, but now with an overlay of adaptive differentiation directly associated, much as we have seen in the passages cited earlier, one each from the first and second editions of *Journal of Researches*.

That first of edition of the *Journal of Researches* (1839), which Darwin was writing while also making notes in Transmutation Notebooks, is a marvelous amalgam of Diary and his scientific notes. It cannot be read for an exact chronology (the reader is led to believe, for example, that Darwin visited Bahia Blanca only once—in 1833, holding back the thrilling discoveries of 1832, presumably for the purposes of a coherent storyline, and saving them for his second visit). There is no mention of "February 1835," of course: for Darwin, though privately a fiercely committed transmutationist openly writing about an astonishing array of evolutionary phenomena, is in hiding mode with the public—a stance he maintained until jolted out of his conservative reluctance to expose his radical views in public by that fateful letter from Alfred Russel Wallace that arrived in mid-1858.

And though Darwin mentions the rheas and the Galápagos "productions" from time to time in his notes and early evolutionary essays of the 1840s, the *Journal of Researches* is really the last hurrah for many of the key players in the development of Darwin's ideas. Darwin made sweeping generalizations on the nature of transmutation based on relatively few examples. All involved taxa endemic to the Americas, and many to the plains of Patagonia. The fossil cavy from Monte Hermoso, held to be an extinct species somewhat smaller than the extant Patagonian cavy with which, Darwin proclaimed, it was congeneric, was Darwin's first—and while on the *Beagle*, only—example of replacement of an extinct species by a modern, congeneric species. As briefly mentioned, when Richard Owen proclaimed the Monte Hermoso fossil rodent to be a tucutucu, not a cavy, Darwin used the fossil *Macrauchenia*, erroneously identified as a distinct camel, as the forerunner for the still-living guanaco, in the Red Notebook. But in the *Journal of Researches*, he restores the example of what he now calls tucutucus, fossil and recent, one replacing the other. It became a prime example of the law of succession: "existing animals have close relation in form with extinct species." It was Darwin's way of discussing the replacement of extinct by modern species of endemic taxa neutrally, without having to mention evolution itself. He summarizes George Waterhouse's (1837) list of edentates and includes "living and extinct Edentata" as an additional, broader example of the law of succession. The law of succession is a fig leaf for what he has long since confided, if only to himself, about what is really going on.

And likewise with the rheas replacing one another in space, along with other examples of birds, and the halo effect, so clearly seen in the Galápagos mockingbirds and also first in the Falkland foxes. Just a few examples, but with powerful implications in Darwin's mind. All of these players are paraded out and almost lovingly discussed in the *Journal of Researches*, together with the poisonous snake from Bahia Blanca, the pesky Galápagos finches (by now put in order by John Gould, who could indeed see separate species, making Darwin regret not having more carefully collected and labeled them). But this really is a "last hurrah" for nearly all these species, as Darwin, once he opens the Transmutation Notebooks, starts out on what is to prove to be a lifetime's assiduous pursuit of relevant examples from the literature—examples that are soon to take prominence over Darwin's own field experiences, and even the pronouncements of identity by the likes of Richard Owen and John Gould, of the material Darwin collected while on the *Beagle*.

And finally, Darwin discusses extinction. He gives an excellent account of his reasons for rejecting climate change as an underlying cause of extinction. He ends up concluding that species die natural deaths, much as individuals do, citing the fruit tree analogy, and once again saying "I do not wish to draw any close analogy" (a nod to the language of "February 1835") as evidence of internally regulated deaths, in this instance of many buds that Darwin saw must be (in modern terms) genetically identical. And so we have both components of Giambattista Brocchi's analogy, as Darwin concludes his passage on extinction with: "as with the individual, so with the species, the hour of life has run its course, and is spent."

But the real icing on the cake is a stunning declaration that Darwin made to Leonard Jenyns in a letter written on November 25, 1844. In that year, Darwin had begun to tell several of his closer friends of his transmutational ideas; one of them apparently had been Jenyns, though no correspondence of the first communication survives. But in what seems like a follow-up letter of further explication from Darwin to Jenyns, Darwin (1844b) clarifies what it was that had convinced him of transmutation in the first place:

> With respect to my far-distant work on species, I must have expressed myself with singular inaccuracy, if I led you to suppose that I meant to say that my conclusions were inevitable. They have become so, after years of weighing puzzles, to myself *alone*; but in my wildest day-dream, I never expect more than to be able to show that there are two sides to the question of the immutability of species, i.e. whether species are *directly* created, or by intermediate laws, (as with the life & death of individuals). I did not approach the subject on the side of the difficulty in determining what are species & what are varieties, but (though, why I shd give you such a *history* of my doings, it wd be hard to say) from such facts, as the relationship between the living & extinct mammifers in S. America, & between those living on the continent & on adjoining islands, such as the Galapagos— It occurred to me, that a collection of all such analogous facts would throw light either for or against the view of related species, being co-descendants from a common stock.

Nothing more need be said. Here, Darwin himself tells us that he was looking for a natural causal explanation for the "creation" of species "as with the lives and deaths of individuals," rather than trying to gauge the degree of difference in characters between species or "varieties" (subdivisions of species). And he says it came from the relation of the fossil to the living South American mammals, as well as geographic patterns, including among those species living on islands and others on the mainland.

We know now, from his recorded notes while in South America, that he had begun this process long before he arrived home and openly acknowledged (at least in his private notes) his acceptance of transmutation. He had had an education steeped in the radical thinking of pro-transmutational Edinburgh. And in any case, his notes starkly reveal he had begun the process of thinking of the transmutational implications of his South American scientific adventures all the way back in 1832 while in Bahia Blanca for the first time.

Darwin did what he said so many times that he did. And he began the process while on the *Beagle*.

3

ENTER ADAPTATION AND THE CONFLICT BETWEEN ISOLATION AND GRADUAL ADAPTIVE CHANGE, 1836–1859

Up to the mid-1830s, in the minds of those actively seeking a non-miraculous explanation for the origin of the species of the modern fauna, lay a fundamentally taxic perspective: species come and go, as Robert Grant put it so eloquently in 1828, while Giambattista Brocchi's analogy between the births and deaths of individuals, and the births and deaths of species, lay at the heart of early evolutionary theory. And no one had explored the empirical and theoretical aspects of this taxic perspective more than Charles Darwin had.

But in that very same passage, Grant had also said that "each species has its peculiar form, structure, properties, and habits, adapted to its situation, which serve to distinguish it from every other species; and each individual has its destined purpose in the economy of nature."

Species, according to Grant, are adapted, such that each of its component individual organisms plays a particular, specifiable role in the economy of nature. Adaptation, in other words, was already on the table, along with the births and deaths of species, arguably as a coequal component of any complete theory explaining the origin of the modern fauna.

As his comments on the "more perfect organ" of the rattlesnake over the simpler tail of his poisonous snake (fer-de-lance) at Bahia Blanca in 1832 reveal, Darwin must have been acutely aware of adaptation all along. For Darwin discussed not only the comparative morphology, but also the behavioral use of these "organs." Although we have long been accustomed to think of Darwin's evolutionary theory as fundamentally a theory of adaptation through natural selection, there are two interconnected reasons why the often exquisite morphological and behavioral properties of organisms, matching them up with their economic roles in their local ecosystems, could not themselves have played a direct role in the arguments leading up to the general acceptance of the very fact of evolution.

The first reason was William Paley's book *Natural Theology; or Evidences of the Existence and Attributes of the Deity, Collected from the Appearances of Nature* (1802). Later in life Darwin ([1876]) 1950) said that the "logic" of this book, which was widely read in its time, "gave me as much delight as did Euclid"—both of which he mastered so completely as to graduate Cambridge with a decently high degree. He also said that he was "charmed and convinced by the long line of argumentation" (27), though Darwin prefaces this with his comment that he did not bother to examine "Paley's premises," which he says he took for granted.

And the book indeed is a well-written exposition of Paley's thesis: that the often intricate designs of organisms bespeak the handiwork of a Creator. Paley's famous analogy, of course, involved watches: just as the existence of a watch, with its intricate internal workings, implies the existence of a watchmaker, so too do the exquisite adaptations of organisms bespeak a Creator. The watchmaker metaphor runs throughout Paley's initial chapters outlining his argument, which encompasses even apparent examples of incomplete or inferior design. In a phenomenal piece of twisted logic, Paley says God used "contrivances" such as eyes, rather than just granting some perfect capacity of vision, so that we could appreciate His creative intelligence at work.

Paley's book mostly consists of detailed accounts of what we would now call adaptations. Paley himself freely used the word "adaptation" in conjunction with the design and construction of watches, *and* of human eyes. In a very real sense, Paley beat the future transmutationists to the punch by saying, in effect, "look at this organism, with these features, evidence of the handiwork of the Creator." Paley basically prevented naturalists from appealing to the same data and saying "consider the human eye, so beautifully adapted for vision—obviously the result of a natural process of transmutation."

And that is the second, real reason why adaptation could not be used as the de facto evidence for the simple truth of evolution: for not only did Paley co-opt the argument (at least in heavily religious Great Britain), but it is simply impossible to look at a static bit of morphology or behavior, no matter how intricate, no matter how well it seems to serve the economic or reproductive needs of a single organism, and say anything, on first principles, about how that condition arose. Clever as Paley's analogy with watchmaking may have been, his argument, his "logic," was simply a rhetorical appeal. There is no way to test such an assertion, any more than there is a way to look at the eyes of a human or an eagle and proclaim that, with no additional evidence, they are the product of a natural process of evolution. What it all boils down to is the question: Is biological design the product of the direct actions of the Creator, or do we ascribe them to the actions of natural causes?

Only when morphologies and behaviors can be compared to a series of "closely allied" species do such data support the notion of inner connectivity, of genealogical continuity between species in the history of life, with concomitant anatomical and behavioral change. Unlike the Galápagos finch example, in which divergent beak sizes and shapes define a classic "adaptive radiation," what is needed is a "morphocline," an apparent interspecies gradation from one anatomical configuration, through one or more intermediate states, to the final state of the series. Apparent gradations of morphology within a series of "closely allied" species would at least seem to agree more with a natural process of descent than with the postulate that God created species rigidly fixed—and, as the philosopher/creationist William Whewell (1837) put it, "a transition from one to another does not exist" (626). As encountered briefly in chapter 2, some of the entries in the second volume of Edward Griffith's *The Animal Kingdom* (1827) are among the comparatively rare occurrences of a call for a natural explanation for the apparent morphological intergradations among a series of taxa.

As we have seen, Darwin in fact did moot such a comparison, at least once, on the *Beagle*, when he wrote that the tail of his fer-de-lance "marks the passage" between the simple tail of an adder and the "more perfect organ" of the rattlesnake. And I do think this episode bolsters the argument that he was thinking along transmutational lines at Bahia Blanca, triggered by the fossil discoveries. Perhaps if Darwin (or anyone else) had encountered more apparently linear examples like the tip-of-the-tail morphology of his fer-de-lance compared with its presumptively more primitive and advanced states in other members of the pit viper clan, adaptive change might have emerged earlier and more clearly as a repeated pattern begging for secondary causal explanation.

Thus the simple description of the properties of an organism cannot, in and of itself, constitute "proof" of the handiwork of God any more than, in and of itself, it can "prove" that life has evolved through secondary, natural causes. For adaptation to grow into a full-fledged, and indeed vital, component of evolutionary theory, something else was needed: independent evidence that life has evolved. And that "something" was the growing conviction that natural processes did indeed account for the progressive "younging" of the fossil faunas, as species appeared, became extinct, and were quickly replaced by other, slightly different species that, anatomically, progressively approached the species of the modern fauna.

Thus the conviction that there is indeed a discoverable, natural causal process of transmutation, I submit, came through patterns of species replacement in time and in space, primarily in the mind of Charles Robert Darwin. Jean-Baptiste Lamarck's transmutational arguments caught the attention of many savants who believed that natural processes account for natural phenomena. But Lamarck's thesis of progressive intergradation lacked empirical credibility. In contrast, Brocchi's analogy seemed to agree much better with the actual spatiotemporal patterns observed in nature. This eventually led to Darwin's conviction that an explanation of the origin of modern species and of the entire diversity of life through natural causes was at hand.

Only then did adaptation become a problem to be tackled. For if species have evolved through natural processes, how then *do* we explain all those exquisite (and not-so-exquisite) adaptations? And how in general does anatomical diversification occur in conjunction with the births of species?

A few months after he essentially rewrote the gist of "February 1835" in Red Notebook in early 1837, explicitly revealing (if only to himself) that he was a convinced transmutationist, Darwin opened the first of the Transmutation Notebooks: Notebook B. There he toyed with ideas of speciation (births of species), anatomical diversification, and adaptation in a fascinating, and interconnected, mélange of thoughts.

ADAPTATION AND NATURAL SELECTION IN SPECIATION AND GRADUAL PHYLETIC CHANGE: DARWIN'S TRANSMUTATION NOTEBOOKS, 1837–1839

Toward the end of Notebook B, Darwin (1837–1839) writes: "With belief of transmutation & geographical grouping we are led to endeavour to discover *causes* of change.—the manner of adaptation (wish of parents??) instinct & structure becomes full of speculation & line of observation" (227).

Just so: once transmutation is accepted, then the next job is to tackle adaptation—and to specify a causal mechanism for it. And indeed, from the very first pages, Darwin is seeking such an explanation, returning to his grandfather Erasmus's *Zoonomia* (1794, 1796) at least for thematic inspiration.

Sensibly, Darwin turns in Notebook B to the fundamentals of what is known about reproduction ("generation"), recognizing two kinds: "coeval" (what we would now call "clonal," or "asexual") and "ordinary" (in our terms, "sexual") (1). Darwin says that the young of species seem to vary "according to circumstance," such that "generation here seems a means to vary, or adaptation" (3). Thus adaptation is linked to, and perhaps in some way caused by, generation, with its twin components of heredity and heritable variation.

But throughout his notes, Darwin worries that this statement is just a beginning, hardly a completely satisfactory theory of adaptive change. For though the young of species might be said to demonstrate variation "according to circumstance," by and large offspring resemble their parents. Indeed, at one point Darwin stops to define the term "species," saying: "one that remains at large with constant characters, together with other beings of very near structure" (213).

Together with his two empirical criteria of recognizing species in nature (absolute knowledge that one species has descended from another, which Darwin says is rare, and, especially, "when placed together they will breed" [122; see also 212]), Darwin's definition is essentially the same as the "biological species concept" developed by Theodosius Dobzhansky and Ernst Mayr some hundred years later. And throughout Notebook B, Darwin reminds himself that species are fairly constant, though they may vary geographically, especially in "clinal" situations such as distributions up the slopes of mountains.

So generation per se is simply not sufficient to account for adaptation to changed, or newly encountered, environmental conditions ("circumstances"). As far as other factors that may underlie the process of adaptation are concerned, Darwin at one point calls "Lamarck's 'willing' doctrine absurd" (216), after earlier commenting that "changes not result of will of animal, but law of adaptation as much as acid and alkali" (21); that is, adaptation cannot be a conscious wish or desire, but arises instead from a natural law-like process. Yet Darwin himself waffles on this, writing, as in the quote that opened this section ("wish of parents??") and also a few pages earlier: "Can the wishing of the Parent produce any character on offspring? Does the mind produce any change in offspring? If so adaptations of species *by generation* explained?" (219).

The upshot of Darwin's speculations on a general theory of the causal factors of adaptation emanating from the processes of generation is hardly definitive or satisfactory to him. He settles for a prescription on how to proceed:

> My theory would give zest to recent & Fossil Comparative Anatomy, & it would lead to study of instincts, heredetary. & mind heredetary, whole metaphysics.—it would lead to closest examination of hybridity <<to what circumstances favour crossing & what prevents it—>> & generation, causes of change <<in order>> to know what we have come from & to what we tend.—this & <<direct>> examination of direct passages of ‹species› structure in species, might lead to laws of change, which would then be main object of study, to guide our <past> speculations. (228)

Darwin in fact did discover natural selection, his much-longed-for general law explaining adaptation, less than a year later. He did it not by the inductive pathway of adding more data or examples as outlined in this passage, but by recognizing the general applicability of the Malthusian cap on population sizes—which, when added to the twin principles of generation (organisms resemble their parents; variation is heritable), yields the beautifully simple law of natural selection. Indeed, there are passages in Notebook B in which Darwin explores reproductive rates and the concept of possible limits of growth in numbers, at least for species. With his modest mathematical bent, Darwin seems to have possessed a mind already prepared for the lessons of Thomas Malthus.

Geographic Speciation

Yet though he lacks a coherent and satisfactory explanation for adaptation, Darwin does have a causal theory for the origin of new species. Once again, it is essentially the same as the theory of allopatric (or geographic) speciation outlined by Dobzhansky and Mayr some hundred years later.

In conjunction with his argument that species are "individuals," biologist Michael Ghiselin (1987) has argued that allopatric (or simply geographic) speciation is indeed in itself a law: when parts of a species are separated, and especially if one segment is in novel environmental circumstances, it will diverge anatomically from the parental species and, given time, become reproductively isolated. As we shall see in chapter 4, ever since Dobzhansky and Mayr, such anatomical divergence is generally considered to be adaptive and effected by natural selection. Lacking a fully satisfactory theory of how adaptation occurs, and thus perforce considering the processes underlying adaptation to be a "black box" at the time he was writing Notebook B, nonetheless Darwin's version of geographic speciation, with the dual recognition of the importance of isolation, and the as-yet not fully explained "black box" process of adaptation to new circumstances, is equally law-like.

Easiest to see, of course, is the effect of pronounced, unarguable isolation, especially on islands separated by a goodly distance from the mainland, and in islands in archipelagos, separated one from another:

"According to this view animals, on separate islands, ought to become different if kept long enough— <<apart, with slightly different circumstances.—>> Now Galapagos Tortoises, Mocking birds; Falkland Fox—Chiloe, fox,—Inglish & Irish Hare" (Darwin 1837–1839:7).

Darwin is using his direct observations in South America, observations that led to the recognition of patterns of replacement of allied species, to frame this nearly full statement of geographic speciation. Only the "Inglish and Irish Hare" pairing forms an additional example (this one gleaned from the literature), a beginning of Darwin's nascent tendency to choose the data and insights of others over his own field observations.

And Darwin found it also easy to imagine how this might happen as a sort of "Gedanken experiment": "Ægyptian cats & dogs ibis same as formerly but separate a pair & place them on fresh isld. it is very doubtful whether they would remain constant" (6). Here, once again, is the sacred Ibis, acknowledged as not having changed over the relatively brief interval between the Old Kingdom and modern times, but now speculatively linked to the process of geographic isolation and adaptation to the conditions of a "fresh island."

But there is that problem of different "inosculant" species, in the case of the rheas with abutting geographic ranges, yet with no obvious isolating barrier, past or present, in view: How to explain them? Darwin is not sure, but makes several comments, such as: "As we thus believe species vary, changing climate we ought to find representative species; this we do in South America closely approaching—but as they inosculate, we must suppose the change is effected at once,—something like a variety produced" (8).

Thus Darwin still supports saltational change, however it is generated, to explain the origin of closely related, geographically adjacent species with no obvious barriers to have produced one species from the other. Of the rheas, he repeats his comment from the Red Notebook: "I look at two ostriches as strong argument of possibility of such change,—as we see them in space, so might they in time As I have before said *isolate* species especially with some change probably vary quicker" (16–17).

So at least in this instance, Darwin sees replacement patterns in space as equivalent to those in time; and in this instance he also seems to be suggesting that the two rheas diverged through isolation in the geological past. If so, the argument is a bit different from the pure saltationist leap envisioned in the Red Notebook and the passage a few pages earlier in Notebook B.

Darwin has many more things to say about geographic speciation, including failures of complete speciation to occur. But his speculation on the geographic implication of the point origin of species through isolation bears closer examination: "If species made by isolation; then their distribution (after physical changes) would be in rays—from certain spots" (155). (David Kohn renders this word as "sports" in his transcription of Notebook B; I read it as "spots," which would seem to make more sense.)

This realization provides the entry to explain all biogeographic patterns, including the all-important theme of endemism. But it is more. This and related passages are the quintessence of the "taxic" perspective: to Darwin, in Notebook B, species are very much real entities, with discrete and sometimes rapid births, followed by often less eventful histories. For example, he writes: "A species as soon as once formed by separation or change in part of country repugnance to intermarriage settles it" (24). And: "With this tendency to vary by generation, why are. species are [*sic*] constant over whole country" (5).

So isolation leads to "repugnance of intermarriage" (in more modern terms, an early phase of actual "reproductive isolation" à la Dobzhansky and Mayr). And Darwin answers his query in the second quoted sentence, based as it is on the long geographic ranges of many mammalian species he saw in southern South America, by appealing to interbreeding that would counteract tendencies to variation and diversification.

Thus species tend to be stable over space and through time—a fact that persistently worries Darwin. On the apparent long-term stability of some species through geological time, he writes: "Those species which have long remained are those ?Lyell? (*sic*), which have wide range and therefore cross & keep similar. But this is difficulty; This immutability of some species" (170).

Interbreeding may keep a species stable, but he is still worried about this "immutability of species." He remains so until he resolves the issue (discussed later) in a stunning passage toward the end of Notebook E.

But then there is this interesting comment, with Darwin's rejoinder added later (in boldface): "Fox tells me that it is generally said. = How came first species to go on.— **There never were any constant species**" (177e).

There is no telling how much later Darwin wrote this stark pronouncement that the constancy of species is in essence nothing more than a myth, and therefore a "Scheinprobleme"—and thus nothing to worry about.

One final point to emphasize how closely Darwin's ideas on speciation through geographic isolation and adaptive divergence approaches the core of the ideas developed, first by the German zoologist Moritz Wagner, and then much later on by Dobzhansky and Mayr: "If *species* generate <<other *species*>>, their race is not utterly cut off:—like golden pippen. if produced by seed go on.—otherwise all die" (72).

In other words, when a descendant species splits off an ancestral species, the ancestor (the parental species) is by no means destined to die. Indeed, it usually persists, yielding a pattern of two species where once there was but one.

And what about extinction—the eventual deaths of species? Darwin speculates that species whose adaptations are no longer viable in a changing world might go extinct. And he repeats this thought in the Red Notebook that, in good Brocchian fashion, the deaths of species is no stranger than the deaths of individuals, and "As all the species of some genera have died; have they all one determinate life dependent on genus" (29e)—meaning that species may have ingrained lifespans (again, originally Giambattista Brocchi's idea), and furthermore, all species within a genus may share a typical lifespan.

Yet, at one point, Darwin expresses reservation on this explanation of species extinction: "Weakest part of theory death of species without apparent physical cause" (135).

Thus two concepts that seem, at least at first glance, to be comfortably interspersed with one another are first the point origins of stable species through adaptive change in geographic isolation, and second Darwin's general theory of adaptive change modifying entire species through time. Yet tensions between these two rather different visions of the evolutionary process are soon to appear in this and subsequent Transmutation Notebooks.

DARWIN'S TWO SEPARATE MODELS OF ADAPTIVE CHANGE

That species are real, constant in their characters (a component of Darwin's very definition of "species"), have point origins, replace one another in space and time, and eventually succumb to extinction, are fundamental themes latent in Darwin's Geological and Zoological Notes from his days on the *Beagle*. They are quintessentially taxic statements. Here, in Notebook B, Darwin adds adaptation (caused by a partially explained black box process involving generation) and geographic isolation as components of a law-like causal explanation of the origins of species, and begins to question Giambattista Brocchi's idea of the innate longevities of species that he had embraced wholeheartedly in "February 1835" and in the Red Notebook. These are major steps ahead—but nonetheless still fundamentally taxic in their very nature. Species are real, with explicable births, histories, and deaths. And adaptation in isolation explains their births, and the differences (which he says may be sometimes anatomically slight) between closely related species. That's the taxic perspective that Darwin, wittingly or not, got from Brocchi and the Edinburgh commentators of the 1820s, now developed to a highly sophisticated degree.

Brocchi remains in still other passages, most strikingly when Darwin (1837–1839) writes, "They die; without they change; like Golden Pippens, it is a *generation of species* like generation *of individuals*" (63), Brocchi's analogy once again (as well as another analogy with "Golden Pippens," a variety of apples produced clonally by grafting). And also this strong statement, from Brocchi via those Edinburgh intermediaries: "Absolute knowledge that species die &. others replace them" (104).

Yet along with the consideration of adaptation in general, rather than the specific use of the concept to explain divergence in speciation through isolation, comes an overt embrace of one of Jean-Baptiste Lamarck's most striking and controversial views. This is something very new for Darwin—and the Lamarckian theme of evolutionary pattern we encounter in Notebook B remains a cornerstone of Darwin's evolutionary writings for the rest of his lifetime. Indeed, Darwin's adoption of Lamarckian spatiotemporal patterns of continuity and gradual intergradation seem very familiar to the modern eye, schooled as we have been in Darwin's thinking that has come down to us from the sixth edition of *Origin of Species*.

The Lamarckian theme concerns patterns of smooth morphological intergradation both within, and even between, species. Consider this statement, early on in Notebook B, right after Darwin's first passage on geographic speciation through isolation: "Species according to Lamarck disappear as collection made perfect.—truer even than in Lamarck's time. Gray's remark, best known species, (as some common land shells) Most difficult to separate. Every character continues to vanish, bones instinct &c &qc &c" (9).

Darwin is saying that Lamarck's claim that the better the sampling of species in nature, the more they appear to blend into one another, appears to be truer in the late 1830s than it was when Lamarck made these claims, especially in 1809, coincidentally the year Darwin was born.

And this is remarkable as it flies in the face of Darwin's comments on the constancy of species, their point origins through isolation, and the pattern of their replacement of one another both in time and in space. It is tempting to suggest that Darwin is drawn to Lamarck's position on the smooth intergradation of species to the point at which discrete species no longer exist, simply because it makes a great rhetorical case for the fundamental claim that transmutation is a real phenomenon. But there is more to it than that.

In another passage, Darwin does say that "my theory very distinct from Lamarck's" (214), alluding to branching patterns and perhaps to the more general pattern of speciation in isolation.

Yet this theme of blending, of seeing what Darwin in the following passage calls "insensible" gradations, actually gains a real foothold in these notes, without explicitly citing Lamarck as its inspiration:

> [85] In some of the lower orders a perfect gradation can be found from forms marking good genera—by steps so insensible, that each is not more change than we know *varieties* can produce.—Therefore all genera MAY have had intermediate steps.—But it is other question, whether there [86] have existed *all* those intermediate steps especially

in those classes where species not numerous. (NB in those classes with few species greatest jumps strongest marked genera? Reptiles?) For instance there never may have been grade between pig & tapir, yet from some [87] common progenitor,—Now if the intermediate ranks had produced infinite species probably the series would have been more perfect, because in each there is possibility of such organization.

Darwin claims that some groups, on the level of differences between mere "varieties," do indeed show such slight differences between genera that they form insensibly graded series. And as for the groups with apparent great gaps between members, Darwin speculates that the lack of intermediates comes from the fact that they diverged from a common ancestor; or perhaps the "intermediate ranks" had failed to produce "infinite species," for had they done so, "probably the series would have been more perfect." In other words, extenuating circumstances account for the lack of predicted, expected, insensibly graded series—for Darwin has already decided that such are the normal patterns of evolutionary history. And later on Darwin will simply attribute most apparent gaps between existing species to extinction of intermediates.

There are other such passages on gradations, but a single additional one will suffice, all the more important as it introduces what will soon become a familiar mantra in Darwin's evolutionary writings henceforth: "Now a gradual change can only be traced geologically (& then monuments imperfect) or horizontally & then cross breeding prevents perfect change" (239).

So here are two more reasons why we might expect not to observe the predicted patterns of gradual evolutionary change: one is the disruptive effects of crossbreeding, and the other the imperfection of the fossil record (that is, "monuments imperfect"). And interbreeding, as we have seen, in Darwin's eyes is bound to counteract a gradational pattern of differentiation within a species over a large geographic area. The charge especially that the fossil record is too imperfect, too full of gaps, will become a persistent theme, ending up as a full chapter in Darwin's culminating statement of evolution, *On the Origin of Species by Means of Natural Selection*.

It is striking that, in contrast to Darwin's derivation of a complete theory of geographic speciation via isolation and local adaptation from actual observational data—most of it his own—there is a dearth of data and examples in his invocation of a sort of general process of adaptation that will gradually modify species though time or in space.

The closest thing to real examples of such intergradational patterns in modern species in Notebook B are within-species clinal variation up the slopes of mountains or approaching desert conditions—and even here, no actual examples are cited:

> [235] It may be argued against theory of changes that if so in approaching desert country or ascending mountain you ought to have a gradation of species, now this notoriously is [236] not the case, you have stunted species, but not such as would make species (except perhaps in some plants & then a chain of steps is found in same mountain).
>
> How is this explained by law of small differences producing more fertile offspring.— Ist. All variation of animal is either effect or adaptation, [therefore] animal best fitted to that country when change has taken place, Nature . . .

The remainder of this section has been lost: though most of the pages that Darwin excised (for later use) from Notebook B have been located, unfortunately pages 237 and 238 have not been found. Darwin is saying here that a true (inter)gradation of species is not observed—just "stunting"—and goes on to pose explanations for the pattern. The pattern does not make new species, except perhaps in some cases of plants. There are no specific examples cited.

And again, in another passage, Darwin acknowledges that gradual clinal variation up mountain slopes is insufficient to produce new species:

> [209e] The reason why there is not perfect *gradation* of change in species, as physical changes are *gradual*, is this if after isolation (seed blown into desert) or separation by mountain chains &c the species have not been *much* altered they will cross (perhaps more fertility & so make that sudden step, species or not. [210e] A plant submits to more individual change, (as some animals do more than others,

> & cut off limbs & new ones are formed) but yet propagates varieties according to same law with animals??
>
> Why are species not formed, during ascent of mountain or approach of desert?—because the crossing of species less altered prevents the complete adaptation which would ensue.

Thus Darwin achieves two things with these passages on within-species clinal variation. Though he does not say so, it seems a fair conclusion that isolation would still be required to disrupt the smooth intergradation and produce separate species. On the other hand, such (not specifically cited) examples do tend to establish the reality of patterns of gradual change, of smooth intergradations—albeit only within species.

And suffice it to say, Darwin produces no examples whatsoever of progressive gradual change from the fossil record.

Mountain slopes and semi-desert ecotones aside, the other general passages on Lamarck-like gradual intergradations (cited earlier) seem very much like theoretical expectations derived not from data, but as an expected pattern of change given his not-yet-complete theory of the adaptive process in general.

That explicit rumblings of gradual adaptive change actually preceded the development of Darwin's law of natural selection I personally find somewhat surprising. It is a testimony to Darwin's conviction that there is such a general process of adaptive modification, as yet unfound, that will prove to be applicable beyond the specific, somewhat narrow arena of change in isolated populations. And he must have felt a certain confidence that he would soon have his satisfactorily complete theory of adaptation: the law of natural selection.

Thus in Notebook B, two rather different views of evolutionary patterns are held simultaneously, with little indication that Darwin is troubled by the juxtaposition of such contradictory viewpoints. Were the emphasis on continuity restricted to within-species patterns of variation (as is indeed discussed in the two passages involving mountain slopes and deserts), it would be, as already acknowledged, possible to claim that Darwin was looking at within-species pattern, which would not conflict with the among-species pattern of the point origin of new species. But Darwin is manifestly looking for patterns of graded series connecting species, thus genera and so on. And he is already hard at work explaining why examples of such insensibly graded series are relatively hard to find. Darwin's embrace of Lamarckian patterns of continuity and blending (osculation with a vengeance!) is not only a departure from his by-now accustomed taxic way of looking at evolutionary pattern and process, but it also marks the beginning of a theme that will soon come to dominate his thinking. Darwin does not explicitly link such patterns of progressive, insensible change—either geologically or geographically—with the general concept of adaptation. But that is soon to come.

On the other hand, and admittedly with the benefit of hindsight, Darwin's conclusion that, given the acceptance of the basic fact of transmutation, adaptation must now be confronted and explained; that generation must have something to do with the process of adaptation—though heredity, and heritable variation, are in themselves not quite sufficient to do the job; that closely related species are typically geographically disjunct, often separated by barriers, implying that isolation is a key factor in the emergence of new species from pre-existing species; that perhaps above all, the process of adaptation, whatever its details may be, is at work when a portion of a species becomes isolated and diversifies to adjust to "new" circumstances; and finally, that newly minted species in isolation become stabilized while the "parental" species also typically survive. This all constitutes a coherent theory of the origin of species that, in my opinion, is an absolutely brilliant, masterful, and totally original addition to Darwin's thinking.

I say this because, although it took a hundred years to again emerge, this simple, concise vision of the origin of new species has been the core of the canonical theoretical model since the mid-1930s and early 1940s. It is the way evolutionary biology sees the origin of most species in the biota today, therefore throughout geological history. Yet this vision was to become progressively muted in Darwin's thinking through the 1840s and 1850s.

NATURAL SELECTION

Darwin continues to remark on the panoply of issues raised in Notebook B, issues that fall into contradictory images of geographic speciation versus a general spatiotemporal process of gradual adaptive evolutionary change in the succeeding Notebooks C, D, and E. All are fraught with eye-catching observations, but little is truly new in the same dramatic way that much of the content of Notebook B is dramatically novel.

Notebook C consists mostly of Darwin's notes on the observations of other naturalists that bear on his problem of developing a coherent theory of transmutation. Otherwise, interbreeding (under the general rubric of "hibridity") occupies much of his thoughts here.

But there are some dramatic moments nonetheless; for example, in what seems very like an anguished fit of near-despair, Darwin (1837–1839) writes: "Extreme difficulty of TRACING [emphasis in original] change of species to species <<although we see it affected>> tempts one to bring one back to distinct creations" (64).

It is ambiguous whether Darwin means by "distinct creations" literally to abandon the search for secondary causal explanations for transmutation or the less extreme position of instead reverting to a point model of the origin of species in isolation. But he expresses frustration at not finding direct evidence supporting the gradual change of one species into another.

A few pages earlier, Darwin said that isolation frustrates nature's mechanisms (such as crossbreeding) to stabilize the inherent tendency of species to vary and change: "One is tempted to exclaim that nature conscious of the principle of incessant change in her offspring, has invented all kinds of plan to insure stability; but isolate your species her plan is frustrated or rather a new principle is brought to bear" (53).

Here species are frustratingly stable, and that stability is broken by isolation—a remark not only presaging the rise of geographic speciation in the mid-twentieth century, but also of later work centering on the notion of "punctuated equilibria." Isolation still has its virtues in Darwin's nascent theory. A few pages later, Darwin claims that adaptive change, even in isolation, must be slow: "Local varieties formed with extreme slowness, even where isolation, from general circumstances effecting the area equably" (59).

Two further points of interest from the perspective of this narrative should be singled out from Notebook C. The first is that Darwin speculates that the sacred Ibis and the other Egyptian animals collected as mummies and depicted on tomb walls were actually domesticated, thus shielded from change through external conditions: "The Ægyptian animals domesticated <<??>>, & therefore Most especially under care of Man. & external circumstances not variable" (153).

And he takes a whack at William Paley and his example of the complexity of the human eye being a sure sign of the workings of the Creator: "We never may be able to trace the steps by which the organization of the eye, passed from simpler stage to more perfect. preserving its relations.— the wonderful power of adaptation given to organization.— This really perhaps greatest difficulty to whole theory" (175).

It may never prove possible to trace all the intermediate stages of the evolution of the vertebrate eye, but Darwin leaves no doubt he is convinced that it happens through secondary, natural causes—that is, the "wonderful power of adaptation."

Yet of course the truly big development in Darwin's post–Notebook B work is his addition of the third Malthusian component, which, when added to the facts of heredity and heritable variation, yields natural selection.

Darwin does not name "natural selection" in either Notebook D or E. In Notebook D, in one striking passage, he paints a vivid, in places almost poetic, picture of what is implied by the Malthusian realization that there are inherent, natural checks on unlimited population growth:

> 28th [September 1838] I do not doubt, every one till he thinks deeply has assumed that increase of animals exactly proportional to the number that can live.—We ought to be far from wondering of changes in numbers of species, from changes in nature of locality. Even the energetic language of Malthus DeCandoelle does not convey the warring of the species as inference from Malthus—increase of brutes must be prevented solely by positive checks. excepting that famine may stop desire. In Nature production does not increase,

whilst no checks prevail, but the positive check of famine and consequently death.

Population in increase at geometrical ratio in FAR SHORTER [emphasis in original] time than 25 years—yet until the one sentence of Malthus no one clearly perceived the great check amongst men—Even a few years plenty, makes population in Men increase, & an ordinary crop. causes a dearth then in Spring, like food used for other purposes as wheat for making brandy. Take Europe on an average, every species must have same number killed, year after year, by hawks. cold, &c—, even one species of hawk decreasing in number must effect instantaneously all the rest. One may say there is a force like a hundred thousand wedges trying to force into every kind of adapted structure into the gaps in the oeconomy of Nature, or rather forming gaps by thrusting out weaker ones. The final cause of all this wedgings, must be to sort out proper structure & adapt it to change—to do that, for form, which Malthus shows, is the final effect (by means however of volition) of this populousness, on the energy of Man. (134)

As Darwin points out in this passage, the prevailing, unexamined assumption by naturalists had been that organisms tend to reproduce more or less in numbers that would replace themselves. That's why Malthus's point about the inherent geometric increase implicit in sexual reproduction in general, and the incontrovertible fact that populations do *not* exhibit steady geometric increase (and that therefore there must be some natural external checks on that inherent rate of increase) is so important. So simple, yet so important.

Darwin begins Notebook E (insofar as is known, since the first two pages, excised, have not been found) with a passage in which he paraphrases a passage from Thomas Malthus (1826) that, in part, reads:

It accords with the most liberal! Spirit of philosophy to believe that no stone can fall, or plant arise, without the immediate agency of the deity. But we know from *experience*! that these operations of what we call nature, have been conducted *almost*! invariably according to fixed laws: And since the world began, the causes of population & depopulation have probably been as constant as any of the laws of nature with which we are acquainted. (529)

On which Darwin remarks: "This applies to one species—I would apply it not only to population & depopulation, but extermination & production of new forms—their number and correlations" (3).

Though Malthus was indeed writing of but a single species (our own, *Homo sapiens*), the wording of this passage suggests a more general principle affecting species "since the world began." In any case, Darwin is saying that he will be applying it to all species, and with it account for their origins and deaths, as well as their population numbers.

Later in Notebook E (begun in October 1838), Darwin put the whole thing together, adding the Malthusian insight as point three, after the fact of heredity and the fact that heritable variation exists in species. He did this in an absolutely marvelous, succinct, three-sentence syllogism:

> Three principles will account for all
> (1) grandchildren like grandfathers
> (2) tendency to small changes (especially with physical change)
> (3) Great fertility in proportion to support of parents. (58)

Or (1) heredity, (2) heritable variation, and (3) the Malthusian cap on population growth, implying that not all organisms born each generation will survive and themselves reproduce.

Given these parameters, natural selection is an ineluctable law. Darwin had the contents of his little black box at last, though his conviction that there was such a process, as we have seen, was so strong that he could already treat that little black box as if it were a law, so convinced was he that there was a natural causal explanation for adaptation.

And indeed, much as allopatric speciation, encompassing a theory of adaptation as a component, is legitimately considered a law, so too is natural selection: an ironclad law governing the origin, stability, and eventual fates of the adaptive properties of organisms.

Thus what Darwin later (in "Pencil Sketch" [1842]) named "natural selection" constitutes a general theory of adaptation. Whether it was to be used mostly in a temporal context as a theory of slow gradual progressive change producing patterns of transformation of entire species into descendants, or as the explanation for adaptation in isolated populations encountering novel environmental conditions—or, of course, for both—remained an open question. Indeed, as we shall see, Darwin had more things to say on this in other passages in Notebook E.

The remarkably succinct three-point passage laying out the gist of natural selection occurs on page 58, opposite where Darwin makes the already cited exclamation: "Herschel calls the appearance of new species the mystery of mysteries & has grand passage upon problem! Hurrah.—'intermediate causes'" (59).

In Notebook E, Darwin was certain he had specified a natural law of adaptation, in concert with the passages he quotes from both Malthus and John Herschel, on the need for understanding natural phenomena, not as the directions of the "deity," but as the outcome of natural secondary causes that can be gathered into natural laws.

THE DIE IS CAST

Notebook E has much on within-species (inter-organism) variation: the idea of natural selection demands a closer look at such variation than Darwin had so far included in his thinking. Also, Darwin begins to see the rhetorical and intellectual utility of the analogy between artificial selection in domestic breeding and natural selection in the wild.

And perhaps because he has natural selection firmly in his grasp, in Notebook E Darwin tends more and more to emphasize that patterns of slow, continuous gradual change *ought to be* an expected outcome of his theory and vision of the adaptive evolutionary process. He continues to see that the central parts of far-flung species will tend to remain constant, and that selection in isolated populations is an important source of new species. Yet nonetheless, Darwin says, we really ought to be seeing examples of slow, steady, progressive adaptive change in the fossil record. On this, Darwin (1837–1839) first remarks once again on the predicted slowness of such gradual change:

> It cannot be objected to my theory, that the amount of change within historical times has been small—because change in forms is solely adaptation of whole of one race to some change in circumstances; now we know how slowly and insensibly such changes are in progress. It cannot

> be objected to my theory, that the amount of change within historical times has been small—because change in forms is solely adaptation of whole of one race to some change in circumstances; now we know how slowly and insensibly such changes are in progress.
>
> Those who have studied history of the world most closely, & know the amounts of change now in progress, will be the last to object to this theory on the score of small change.—on the contrary islands separated with some animals, &c.—if the change could be shown to be more rapid, I should say there was some link in our chain of geological reasoning, extremely faulty.
>
> The difficulty of multiplying effects & to conceive the result with that clearness of conviction, absolutely necessary as the foundation stone of further inductive reasoning is immense.
>
> It is curious that geology. by giving proper ideas of these subjects. should be *absolutely* necessary to arrive at right conclusion about species. (4–5)

He says we *know* how slowly such changes are now occurring (witness the stability of the sacred Ibis!), and so it would be expected that the changes seen in fossils in the geological record likewise should be slow, steady, gradual, and insensible—by what we would now call a logical extrapolation of observed rates in the present. Something would be wrong if we observed faster rates (Darwin's aside on isolated islands may record an exception—but probably not—as earlier he said that even with isolation, change is necessarily slow and gradual).

Indeed, Darwin's third sentence ("The difficulty of multiplying effects") seems to be a recipe for making further predictions about what ought to be observed given a "clearness of conviction" from previous experiences, indications, and results. That is a recipe for imagining what patterns of change ought to look like under the moving force of natural selection.

And Darwin finds it "curious" that the data of the fossil record should be "*absolutely* necessary to arrive at right conclusion about species." Curious indeed—the more so that Darwin, according to my interpretations of his work on the *Beagle*, actually began his direct encounters with data bearing on transmutation with his experiences with fossils at Bahia Blanca in 1832.

But now Darwin is clearly wishing that species would show gradual change through time. Ever honest, Darwin is openly disappointed about what the fossil record seems to be saying about species. Species are disconcertingly stable, even through what seem to be considerable periods of time: "My very theory requires each form to have lasted for its time: but we ought in same bed if very thick to find some change in upper & lower layers.—good objection *to my theory*: a modern bed at present might be very thick & yet have same fossils. does not Lonsdale know some case of change in vertical series" (6).

Even allowing for the ingrained constancy of species, particularly in the accustomed habitat at the core of its geographic range, Darwin says that under his theory (that is, progressive adaptation tracking environmental changes), there really ought to be some evidence of change. And he is hoping that geologist William Lonsdale will provide some examples of what he is looking for.

A few pages later, he returns to the problem: "Species not being observed to change is very great difficulty in thick strata, can only be explained, by such strata being merely leaf, if one river did pour sediment in one spot, for many epochs—such changes would be observed" (17). Darwin is reverting to the image of strata as leaves in a torn-up book: many pages are missing, and in this instance Darwin, rather forlornly, is hoping that even thick strata are but single leaves, and therefore, despite their thickness, formed over relatively brief intervals of time.

But I find the following final passage on the problem of gradual change and the fossil record to be the most compelling of all—indeed, I personally find it gripping:

> If separation in horizontal direction is far more important in making species, than time (as cause of change) which can hardly be believed, then, uniformity in geological formation intelligible. (135)

And there it is! Darwin says that time, meaning natural selection tracking the inevitable change in environment, *must be* more important than geographic isolation plus adaptive divergence in the evolution of new species! He acknowledges the stability of species through rock formations ("uniformity in geological formation"), and says it is "intelligible" if "separation in horizontal direction" (allopatric speciation) is actually the more important of the two evolutionary options. And he finally sees he has to make a choice between these two essentially conflicting images of the evolutionary process.

And it is at this point that I must inject an anachronistic, and highly personal, comment. One of my earliest papers trying to connect evolutionary theory with the fossil record went by the rather ungainly title "The Allopatric Model and Phylogeny in Paleozoic Invertebrates" (1971a). The paper stressed that species in the fossil record are for the greatest part empirically seen to be very stable over long periods of time, usually from the first recorded instance right up through the last. I suggested that new species probably arise, not through the selection-effected gradual changes long supposed (ever since Darwin!) to be the evolutionary source of new species, but through geographic speciation. This I learned from the work of Theodosius Dobzhansky, Ernst Mayr, and by the 1960s and 1970s, many other biologists: geographically isolated populations can become adaptively modified via natural selection, and thence reproductively isolated. I thought that paleontologists simply were not taking advantage of this newly established process of allopatric speciation (as it turns out, re-established!). I saw that the process must take place in periods of time that are not so brief as to be saltational, but in any event far shorter than the total lifespans of the presumed ancestor and its presumed descendant. This is the core of the idea that, in the following year, Stephen Jay Gould and I republished, embellished, and named "punctuated equilibria."

Much more on this in chapter 5. But consider: Darwin is weighing the two options here and chooses the option that gradual change through time *must* be more important than geographic speciation in producing new species over geological time. He bases this assessment on a priori principles derived from his newly minted, and very true, theory of natural selection, rather than on actual data. In Notebooks B through E, he even has an explanation for the observed stability of species (which we called "stasis"): species indeed are constant in the central parts of their ranges, and such constancy can persist. Moreover, way back on the *Beagle*, Darwin had broken with Lyell over the causes of extinction. Darwin was convinced that the same austere, dry Patagonian scrubland of the modern day was in place back in the day when now-extinct ground sloths and glyptodonts were wandering around. So much for the inevitable environmental change over time accounting either for extinctions, or for that matter, the origin of new species. Recall Darwin's fossil cavy, replaced by its successor, the modern Patagonian cavy.

Such is the power of projection from a theory of process. True, at some unknown point, Darwin did return to that passage and remind himself of the importance of geographic speciation, writing: "No. but the wandering & separation of a few, probably would be most efficient in producing new species; also being reduced in numbers, but not so much these, because circumstances" (135).

Which mitigates the full force of this passage as originally written. But when Darwin wrote these lines, he clearly believed them. And as time went on, he was to emphasize gradual evolution through geological time far more than geographic isolation as the main mode of production of new species. That's why Moritz Wagner objected in the 1870s, and Dobzhansky and Mayr had to reinvent the wheel and mint anew the theory of geographic speciation in the 1930s and 1940s.

Darwin was downplaying, if not wholly discarding, the model that saw point origins of species in geographic isolation as the locus of the adaptive process in evolution in favor of a general model of gradual, adaptive transformation of entire species through long periods of geological time. In so doing, he was dropping Giambattista Brocchi in favor of Jean-Baptiste Lamarck. He was leaving the taxic perspective that brought him to transmutation in the first place, in favor of the transformational perspective of slow, steady, progressive adaptive change within entire species. It was the perspective that the biological world has long since associated with his name—with "Darwinism."

I never saw these words of Darwin until I read David Kohn's (1987) transcription of Notebook E sometime around 2004, although Gavin de Beer had published his earlier transcriptions of the Transmutation Notebooks roughly a decade before I published my paper. I had no idea that I was faced with the identical choice that Darwin had posed for himself, and that we had chosen differently. I was twenty-eight the year my paper was published; I am seventy as I write these words now. I still think I am right.

DARWIN'S EVOLUTIONARY TEXTS

"Pencil Sketch," 1842

Darwin evidently wrote no more about his transmutational ideas until May and June 1842 in his first essay on transmutation since "February 1835." "First Pencil Sketch of Species Theory" (or simply "Pencil Sketch")—part essay, part notes—is Darwin's first attempt to spell out his evolutionary theory in a coherent, discursive form. That it basically succeeded for him is made abundantly clear since the structure, order of arguments, and topics covered are all forerunners of his later, much longer "Essay" (1844a), and ultimately the *Origin* itself, first published in 1859.

"Pencil Sketch" was "Written at Maer & Shrewsbury during May & June 1842." Maer was Emma Wedgwood Darwin's girlhood home, while nearby Shrewsbury was Darwin's ancestral digs. The two had been married in 1839 and had been living on Upper Gower Street in London for the past several years. But "Pencil Sketch" was written in a quieter, more familiar location. When the Darwins actually moved from Upper Gower Street, they may still have been technically within the political confines of London. But they were way out in the countryside, in Downe Village. Trains had arrived, and travel was far easier. They moved to Down House in 1842.

Darwin finished the Transmutation Notebooks in London in 1839. He wrote "Pencil Sketch" in Maer and Shrewsbury, quiet and familiar places. Everything else that he did was done at Down House, in the independent quiet of the countryside.

I have never seen Gower Street, nor, for that matter, have I ever visited the countryside homes in Shropshire. But I have been to Down House. And to this day, it is very quiet (save when school kids parade through on a day's tour). And there is a lot to be learned from a visit or two.

I've been to Down House several times, starting with the need to learn (and, all-importantly, secure loans for) our American Museum exhibition *Darwin: Discovering the Tree of Life*. The house has been preserved and restored so that a sense of the Darwins' living there reaches all but the least imaginative of visitors.

Take Darwin's study, for example. There is his wheeled chair, for easy access to the microscope at the window in pre-electricity days; his clutter of notes and things to do on his fireplace mantle; and his gerrymandered washroom for times of gastric distress.

And the rack of small shelves on the back wall on the right side (as glimpsed roped off on the far side of the room). I learned from Randal Keynes, great-great-grandson of Charles Darwin, and (along with David Kohn) a man absolutely instrumental in our gathering such a wonderful collection together for the American Museum exhibition, what that little piece of shelving was all about. Randal told me that each shelf had contained an individual chapter of *On the Origin of Species*. And he also said that it was clearly a hangover from Darwin's days on the *Beagle*, where his quarters had been so cramped that he needed desperate measures simply to stow away the mélange of specimens, notes, and the like, that he was compiling virtually daily. Our reconstruction of Darwin's study in the exhibition was so faithful to the original that when the Down House curator arrived to help us finalize the installation, she had a dizzying moment when she felt she did not know where she was. Probably jet lag, but that's how close we got it.

But what I really loved was Darwin's "Sandwalk," his thinking path that he frequented every day that he was home (which was most days—he didn't travel much) and the weather was conducive to perambulations. The Sandwalk is still coated with a crushed mixture of Cretaceous chalk rocks and fossils from the formation exposed as the "chalk cliffs of Dover." The neighboring fields at the rear of the Darwins' property are still agricultural; they were apparently the site of where Darwin and a small team recruited from the household staff essayed what has to have been one of the earliest "biodiversity surveys" of plant life.

And yet it is the peace of the place that matters most. Sure, you can fantasize that Darwin's own thoughts still lurk there—much as I have facetiously maintained that American jazz trumpeter Clifford Brown's solos come, intact, in every old Blessing Super Artist trumpet from the early 1950s. I have owned several Blessing Super Artist trumpets over the years. And I have walked the Sandwalk in solitude several times over the years. Neither experience makes you a Clifford Brown or a Charles Darwin.

But as for the cliffs at Bahia Blanca, just being where Darwin had been is inspiring. Darwin did pace his Sandwalk and ponder things. Sometimes it was issues revolving around grocery lists, or perhaps legal affairs he had to attend to as a local magistrate. Other times it was the sad things of life, such as illnesses and even the deaths of some of his children. He must have retreated to the Sandwalk as well after receiving the letter and manuscript from Alfred Russel Wallace in 1858. And who knows? He may even have thought great thoughts as he went around the familiar path with his dog.

This was the setting in which Darwin wrote the rest of his life's work, including the manuscripts in which he developed his mature thoughts on the processes of evolution. The sequence—from 1842 through publication of the *Origin*—is of course profoundly important. But for the purposes of this narrative, which focuses on the development and history of the importance of species in evolutionary biology, these works, while still retaining vestiges of the "taxic perspective," are in fact Darwin's manifesto of his "general theory of adaptive change"—his version of slow, intergradational transformation of entire species, and the picture that in fact he left to succeeding generations of evolutionary theorists. With perhaps the exception of the principle of divergence developed in the mid-1850s, the main interest of these narratives, then, is the development of Darwin's transformational perspective, driven by natural selection, and the collateral eclipse of the importance of geographic speciation in isolation, to the point at which the subject had to essentially be reinvented (not even resurrected) by Dobzhansky and Mayr in the 1930s and 1940s.

Nor are there any real surprises in "Pencil Sketch" (1842), as the contents of the Transmutation Notebooks (1837–1839) are essentially rearranged into what Darwin saw as a logical order for their presentation. Darwin begins with three short "sections" on variation and artificial selection, then natural selection (herein named for the first time), and finally variation on instincts "and other mental attributes." He does so, rather than by starting with the evidence for evolution itself: grand patterns in nature, such as species replacement/distributions in space and time, plus others now added to his discourse, such as homology ("unity of type"), embryological development, and the Linnaean scheme of classification that links taxa according to propinquity of descent. This was to remain his approach—the basic structure of his essays and finally his famous book on evolution—right on through all the editions of the *Origin*, expanded as each iteration was simply to add more data drawn from the literature, and arguments to counter criticisms either anticipated or actually experienced.

The first section emphasizes continuity—among series of homologous organs, individual organisms, races, and eventually species and beyond. For example, Darwin (1842) ends the section on natural selection with the following rhetorical appeal:

> The gradations by which each individual organ has arrived at its present state, and each individual animal with its aggregate of organs has arrived, probably never could be known, and all present great difficulties. I merely wish to show that the proposition is not so monstrous as it first appears, and if good reason could be advanced for believing the species have descended from common parents, the difficulty of imagining intermediate forms of structure not sufficient to make one at once reject the theory. (54)

This call to entertain total intergradation despite the absence of a complete series of intermediate forms is very like Darwin's even better remembered explanation for missing series of intergrading forms as one ascends sequences of fossil-bearing rocks. As he says near the beginning of part II (in sections iv and v of "On the Evidence from Geology"): "Our theory requires a very gradual introduction of new forms, and extermination of the old. . . . The extermination of the old may sometimes be rapid, but never the introduction. In the groups descended from common parent, our theory requires a perfect gradation not differing more than breeds of cattle" (60).

And though some good examples of such intergradation are known, in general they are rare; but: "If geology presents us with mere pages in chapters, towards end of a history, formed of tearing out bundles of leaves, and each page illustrating merely a small portion of the organisms of that time, the facts accord perfectly with my theory" (63).

Already familiar from the notebooks, Darwin's theory in 1842 is dominated by his general model of adaptive change in time and in space.

The vagaries of fossilization and the haphazard preservation of time in geological sequences, both undeniable, are to Darwin sufficient to explain the general absence of actual examples of insensibly graded series.

Likewise the absence of the Lamarck-predicted complete series of intergrading forms within and among species in the modern biota. Though not as well developed in 1842 as in later discussions, Darwin seems to hint in the following section ("Extermination") that competition among individuals, varieties, and species can lead to the extermination of the old in favor of the new. So in the modern data, it is a dynamic process of extinction, rather than mere missing data, that leads to the failure to find the patterns "expected" from his theory.

These themes of predicted gradual transitions, while seldom if ever observed, dominate Darwin's continued promulgation and defense of his general theory of adaptation through natural selection. They remain his main explanation of the evolution of life throughout all his evolutionary expositions—up to and including, of course, the *Origin of Species*.

But what, then, of his ideas on the point origin of species through geographic isolation, in which the engine of adaptation works to modify isolated parts of an ancestral species, matching it to the novel or changed environmental conditions in which it finds itself, and with the bulk of the ancestral species remaining unchanged? How do these profoundly taxic ideas fare in 1842, in Darwin's first attempt at a general exposition of his evolutionary ideas?

Isolation is indeed there, although reduced to a decidedly cameo role and swamped by the opening salvo of gradational adaptive evolutionary change of entire species. Consider the first use of the word "isolation" in this essay: "change of external conditions, and isolation either by chance landing of a form on an island, or subsidence dividing a continent, or great chain of mountains, and the number of individuals not being numerous will best favour variation and selection" (68).

Isolation in the obvious examples of islands, continental foundering, or mountain tops, is an ideal trigger for variation and adaptive change. But even here, Darwin mitigates this as the passage continues: "No doubt change could be effected in same country without any barrier by long continued selection on one species" (68). To my mind, this reveals the gist of the reason why Darwin opted to go the route of vastly preferring a general model of slow entire-species transformation under natural selection through eons of geological time: because there are vastly more species on large swathes of continental terrain than there are on all the islands and mountaintops in the world. And once again, he did not have knowledge of the kinds of habitat-altering environmental (especially climatic), hence ecological, changes that are now known to be the very heartbeat of the continental environmental experience through time.

And if isolation does occur, creating two species from an ancestor on a continent, it shows us why species on the same continent would have "closest affinities," which is of course the rule.

Finally, in a note on the back of page 34 of the original manuscript, Darwin once again reminds himself that it is "as well to recall advantages of" isolation, just as in his reminder note appended to the passage in Notebook E in which he announced his decision to back the general theory of gradual species adaptive change over his geographic isolation model, so much as to say "be careful! Isolation is indeed important!"

And that's about it for isolation in "Pencil Sketch."

"Essay," 1844

Darwin's much longer "Essay" (1844a) is, in essence, a more smoothly composed version of "Pencil Sketch." Ironically, it forms a nearly perfect bridge between its predecessor and its eventual linear descendant, the *Origin* itself. Would that Darwin could confidently point to such continuously graded series in the modern and extinct biotas of the world as he produced in his series of progressively larger written expositions of his theory! Darwin had no intention of publishing this essay immediately, but he did have a "Fair Copy" made by an amanuensis, suitable to be sent to a printer. He left instructions in his will that the "Essay" be published in the event of his death.

Isolation of course appears in this essay, but by now almost solely with reference to islands.

From the perspective of this narrative, one section of the "Essay" stands out. It not only (once again) summarizes Darwin's unshakable commitment to a gradualist perspective but also raises an interesting doubt and elicits an intriguing comment from Darwin's son Francis, who first published this essay in 1909:

> I need hardly observe that the slow and gradual appearance of new forms follows from our theory, for to form a new species, an old one must not only be plastic in its organization, becoming so probably from changes in the conditions of its existence, but a place in the natural economy of the district must [be made,] come to exist, for the selection of some new modification of its structure, better fitted to the surrounding conditions than are the other individuals of the same or other species. (Darwin 1844a)

Francis Darwin was the first to discover the footnote to this passage. In the "Fair Copy," Darwin had inserted a note that reads: "Better begin with this. If species, really, after catastrophes, created in showers over world, my theory false."

What prompted Darwin to suddenly start worrying about patterns that had more or less been dropped (or at least deeply muted) from geological discussion for at least twenty years—patterns of multiple species extinction followed by episodes of multiple species "creation," most closely associated with Georges Cuvier's writings in 1812? Even his teacher Robert Jameson was disinclined to accept Cuvier's "multiple turnover," catastrophist vision of the history of life on earth. Something must have happened to jog Darwin's thinking and to trigger his worries over this possibility.

Just speculation, perhaps, but I think it must have been Darwin's old nemesis Alcide d'Orbigny who planted this particular worrisome seed of doubt. A highly skilled naturalist, d'Orbigny's explorations of South America had preceded Darwin's by some six years. Darwin ([1832] 2008, [1835] 2008) mentioned d'Orbigny twice in letters home to John Henslow—worrying in essence that the older Frenchman, the more seasoned naturalist, would steal much of Darwin's thunder, whether collecting specimens or deciphering South American geology.

And d'Orbigny indeed amassed prodigious collections of marine and terrestrial species, including the "avestruz petise," the more southerly of the two South American "ostriches." English ornithologist John Gould had named this species *Rhea darwinii*, but it was later found to have been described earlier by d'Orbigny, ironically removing Darwin's name from a species that had been so pivotal in his early evolutionary thinking.

On his return home, d'Orbigny started publishing his results, which eventually came to a total of eleven profusely and beautifully illustrated volumes. I have often wondered if Darwin's four-volume barnacle monograph was perhaps inspired in some small measure, at least, by the prodigious output of his rival.

But also, not long after returning home in 1833, d'Orbigny had returned to his main scientific interests and was hard at work on a monumental study of French rocks and fossils—work published in eight volumes between 1840 and 1860. His parallel works on molluscan fossils and stratigraphic zonation of the world, and another outlining his theoretical conclusions, were equally monumental in scope.

I first heard of d'Orbigny as a marine micropaleontologist who had made important contributions, not only to paleontology, but more intriguingly to the theoretical aspects of the relation of fossils to the sedimentary rocks in which they are preserved. Somewhat ironically, I did not encounter d'Orbigny's name in any of my American books on stratigraphy and paleontology, nor even from the otherwise compelling and demanding lectures of Marshall Kay, our professor of stratigraphy at Columbia. Instead, it was Malcolm McKenna, who taught the graduate course on fossil mammals, who first brought d'Orbigny's name to my attention. Malcolm had a deep and abiding interest in the history of the paleontological side of geology, and he took the pains to introduce the theoretical constructs and conclusions of these early European paleontologists, most of whom (like d'Orbigny himself) specialized, not on fossil mammals, but instead on Jurassic ammonoids. Thanks to McKenna, I was introduced to a world of analysis and theory that helped me change my approach to thinking about evolution while contemplating the fossil record.

Oddly, in retrospect, most if not all of these mid-nineteenth-century European biostratigraphers were either opposed to evolution, or simply agnostic. D'Orbigny's mentor had been none other than Georges Cuvier. After his return to France, d'Orbigny's work on the geological distributions of fossils revealed a series of what he called "stages" (*étages*): divisions of rocks defined by discrete assemblages of fossils, most seeming to become extinct at about the same time, and then replaced by newly appearing species that defined the succeeding stage. D'Orbigny thought his stages were worldwide (*globe entier*) in extent. D'Orbigny originally published these thoughts in 1842, a full thirty years after Cuvier had published his catastrophist views. In many ways, his "stages" were an intellectual inheritance from Cuvier. And although he was wrong in his claim that his *étages* in fact encircle the entire globe ("world over" in Darwin's worried phrase), they remain recognized as empirically valid time stratigraphic units: units of time as demarcated by rock sequences, most especially defined and recognized by their fossil content. D'Orbigny saw the pattern of many component species within a stage becoming extinct roughly synchronously, followed by a radiation (or immigration in from elsewhere) of many different, often "new," species to start up the succeeding stage.

In point of fact, d'Orbigny's stages are the foundational empirical basis of modern-day extinction/evolutionary theory based on the near-simultaneous extinctions and subsequent evolution of species in the fossil record. Such patterns, as we shall later see in this narrative, go by terms such as "turnover pulse" or "coordinated stasis."

Darwin had a copy of d'Orbigny's *Paléontologie française* (1842). Darwin did not annotate his copy (once again disappointing us!), and there is no hard and fast evidence that he had even possessed his copy of d'Orbigny's book prior to writing "Essay." Yet given his footnote on the "Fair Copy" of "Essay," it is hard to believe that he hadn't in fact read through d'Orbigny's discussion of his *étages*.

Four letters from d'Orbigny to Darwin survive, two from 1845, and two from 1846, after Darwin had scribbled his anxiety that creation of species "world over" threatened to falsify his theory. D'Orbigny had been identifying many of Darwin's South American marine fossils, concluding among other things that Darwin was correct that all the fossil shells Darwin collected at Bahia Blanca indeed do belong to species still living along the coast. In another letter, d'Orbigny remarks on the apparent sudden death from abrupt changes that took the lives of a group of slipper shells, the death of organisms, not necessarily of a species, though the passage is somewhat unclear.

In any case, insofar as I am aware, Darwin never "began with this." To my knowledge, he never again really mentions the possibility of worldwide extinctions and subsequent bursts of "creation." One of the most active areas of taxic-centered evolutionary theory today concerns the reconciliation of the empirically recognized patterns of regional and global mass extinctions and subsequent bursts of evolution with mainstream evolutionary theory. Darwin's "theory" is "falsified" only if that theory is the extreme insistence that most evolution entails the slow, steady transformation of species and the gradual splitting of lineages through time.

On the other hand, Darwin's concept of the point origin of species in isolation neatly fits these empirical data from the history of life. Though he saw such possibilities earlier in the Transmutation Notebooks, Darwin had by 1844 long since ceased to see that adaptive change of populations in isolation held the key for explaining the commonly observed fact that species, as the overwhelming rule, do not exhibit much change throughout their often long durations in time, implied by their persistence in thick bodies of rocks. For if most adaptive change can be accounted for in brief episodes of adaptive change in isolated populations, the implied corollary—as Darwin himself had earlier seen—is that, throughout the vast bulk of a species' history, little or no adaptive change would need be expected.

One further note on Cuvierian-style faunal extinctions and subsequent proliferations in relation to evolutionary theory in the early nineteenth century is warranted here. This is the celebrated case of the Scottish writer Patrick Matthew, whose book *Naval Timber and Arboriculture* (1831) includes a rather stunning eight-page passage on selection-driven evolution, near the end of the appendix that concludes his book. There is, by historical consensus, no evidence to suggest that either Darwin or Wallace had seen the publication until shortly after the publication of Darwin's *On the Origin of Species* in 1859.

The essay is astounding in several respects, including Matthew's (1831) simple characterization of the adaptive response of species to changing environmental conditions as "the self-regulating adaptive disposition of organized life" (348)—a phrase he uses to introduce his concept of "selection by the law of nature" (387). And it is indeed remarkable that Matthew blended heredity, heritable variation ("sports"), and a firm grasp of the Malthusian principle of geometric increase of population size to explain adaptation (he uses the term, as well as "fitness") and indeed the history of life. His grasp of Thomas Malthus (though he does not acknowledge Malthus by name) is so vividly strong that he presciently also saw the rapidly increasing threat posed by out-of-control human population growth to the very survival of most of the rest of the world's species—a very modern-sounding, twenty-first-century aperçu.

Historians have long known about what even Darwin and Wallace later acknowledged was a clear case of "anticipation" of their evolutionary views in Matthew's appendix. And though some have suggested that Darwin, in particular, had actually co-opted Matthew's ideas, it seems to me highly improbable. Recall that Darwin knew that adaptation must be explained, and that there is in fact a natural causal explanation for adaptation that must involve "generation" (including heredity) as well as heritable variation, as early as Notebook B in late 1837. It took him another year or so to find Malthus and the geometric growth of populations that supplied the third, until-then missing component of a complete theory of selection articulated in Notebooks D and E. Had Darwin read Matthew, he would have seen the whole thing—all three components of natural selection—in one concise statement.

Less well appreciated (until the recent work of geologist Michael Rampino [2011]) is that Matthew's theory of adaptive evolution through selection was intended to explain patterns of mass extinction and reproliferation—catastrophism—rather than what Darwin came to insist was the gradual modification of species through time.

Neither Cuvier nor, later, d'Orbigny left any direct hints that they may ultimately have favored a natural causal explanation for what they saw as the dominant pattern of extinctions and subsequent proliferations in the geological history of life. In contrast, Matthew embraced the Cuvierian, catastrophic vision of the history of life, even though Cuvier's main translator and commentator in Great Britain, Robert Jameson, at least by 1827, had ceased to support the view that the history of life was in fact a succession of extinctions and proliferations.

Matthew, in the most general, non-specific terms, accepted that past geologic eras supported forms of life more or less characteristic of each division ("epoch")—species that were destroyed in large numbers in global catastrophes, in Matthew's (1831) words "destroying nearly all living things" (382–383). Modern species, Matthew says, have had a certain conformity (constancy) for the past forty centuries, and geologists find a similar conformity of the very different species characteristic of the succeeding epoch. And while Matthew does not couch his discussion explicitly in terms of the advent of species of the modern fauna, he does say that lower forms give rise to higher.

Episodes of global destruction would leave "an unoccupied field . . . for new diverging ramifications of life" (Matthew 1831:383) derived from the relatively few surviving species. The "diverging ramifications" would proceed at a more leisurely pace, and through his "selection by the law of nature" (387), finally settle down to produce the "regular specific character" (that is, the "species conformity") present throughout the remainder of time represented by the "fossil deposit." Until, that is, the next episode of destruction comes along.

Matthew speaks in generalities, with no penchant either for specific, empirical examples, or for worrying about how the processes might work to produce such patterns in any detail. And again, acknowledging that Matthew's thought had arguably no impact whatsoever on the subsequent history of evolutionary thinking, it is nonetheless remarkable (and to me somewhat gratifying) to find that someone back in the early nineteenth century tried to explain Cuvierian-style patterns of extinctions and subsequent proliferations in natural causal terms. Indeed, Matthew's choice of terms for causal processes bore a striking resemblance to Darwin's (and later, Wallace's) theory of natural selection and even the principle of divergence (discussed later), which is indeed the vision of evolutionary mechanisms that we in the twentieth and twenty-first centuries have inherited. Matthew, as far as I am aware, was the first thinker to try to achieve what so many of us have been working on in the modern era.

Francis Darwin (1909), who prepared the transcription of "Essay" (1844a) from the "Fair Copy," also observed in that same footnote in which Darwin expressed his alarm at the thought of global catastrophes and the creation of species "in showers world over," that "in the above passage [the footnoted passage quoted earlier] the author is obviously close to his theory of divergence" (145). He might well have been right in thinking that. The principle of divergence is a component of the *Origin*, of course, but appears first in somewhat more complete form in the ill-fated "Big Species Book."

Darwin's Principle of Divergence, 1856–1858

Darwin's "Big Species Book," which he intended to name *Natural Selection*, was written from 1856 to 1858. The first two chapters were eventually published as *Variation of Animals and Plants Under Domestication* (1868), while the remainder served as the manuscript from which the *Origin* was composed as an "Abstract." As is well known, the *Origin* was written in haste, triggered by the arrival of Wallace's letter and manuscript at Darwin's home in mid-1858. The books differ in minor details, including the fact that there are even more examples, drawn in the main from Darwin's extensive reading, in the unpublished predecessor than in the *Origin*.

The organization of both manuscripts mirrors the order of topics discussed in the "Pencil Sketch" (1842) and "Essay" (1844a)—with variation, selection, and hybridization, in a domestic as well as natural, context, dominating the first ten chapters of the "Big Species Book," and the eleventh dedicated to geographical distribution. In the *Origin*, two chapters are added on the geological record: the first on how deficiencies in the preservation of time, thus ancient life, explain why the record tends *not*, in detail, to support Darwin's vision of how the evolutionary process works. The succeeding geological chapter deals mainly with how the fossil record in broad outline does indeed support Darwin's theory. Following two chapters on geographical distribution in the *Origin*, Darwin adds a discussion on classification, homology, and embryology (as predicted observations if the basic idea of evolution is true—still one of the very best ways to demonstrate the scientific nature of the very concept of biological evolution), and sums up his argument in a concluding chapter.

While Darwin's themes and arguments are abundantly familiar at this juncture in terms of both the sequence of his notebooks and early essays, and the narrative arc of this text, there is one important novelty introduced in the "Big Species Book": the principle of divergence.

The principle of divergence is nothing short of Darwin's theoretical argument on how varieties, species, and even genera and families can arise in large areas (whole sectors of continents, for example, such as southern South America) without the imposition of barriers (though such are mentioned in a few places in the text). In the chapter on natural selection, it is presented as a general discussion (Darwin 1859:111–126; Stauffer 1975:227–250), followed by several entirely hypothetical examples illustrated by the famous lone diagram (Darwin 1859:between 116–117; the predecessor in the "Big Species Book" [Stauffer 1975:236–237] differs only in details).

Darwin's model of divergence hinges on the supposition that all populations, be they varieties, "permanent varieties," species, indeed even genera and families, are limited in the sense that the numbers of their included organisms are at or near their maximum at all times. He sees it as a continual struggle for populations to increase their size, and that the only way such can be achieved is through the adaptive acquisition of the ability to invade new habitats and utilize novel resources. He sees it as a slow but inexorable process, driven by the inherent need of populations to increase their numbers, to continually explore and potentially exploit novel ecological resources, thus lifting the Malthusian cap. Selection, then, favors divergence—and indeed it is the extremes that will lead to success. Citing the zoologist Henri Milne-Edwards, Darwin invokes the metaphor of division of labor.

The principle of divergence, I have been tempted to think, could represent one last attempt on Darwin's part to integrate the older, basically long since forgotten taxic aspects of his theoretical work on evolution, with what had emerged already as an emphasis on continuity and gradually changing species, slowly diverging, insensibly intergrading through time.

And there is indeed a "taxic" feel to the principle of divergence, as what is left after the process has been, in effect, ultimately distinct and separate species (and even genera and families). Moreover—and this is truly a "taxic" proposition—the competition between "permanent varieties," or between closely related species, does indeed smack of much later discussions of "species selection" and "species sorting." Indeed, the very subtitle of *On the Origin of Species by Means of Natural Selection, or the Preservation of Favoured Races in the Struggle for Life*, has a decidedly "group selectionist" ring. That subtitle alone is evidence enough of the importance that Darwin attached to the principle of divergence.

Yet the principle of divergence is hardly a taxically imbued manifesto. Instead, it is the very embodiment of slow, steady, and gradual selection-driven adaptive change underlying the origin of new varieties, and eventually species, and the like. In the simplest case, the process is linear, and though, as Darwin says, perhaps passing through a sequence of short-lived steps, on the whole it is linear and progressive.

The more complicated (and interesting) examples involve the literal multiplication of species, but by no means in the way of point origins in isolated populations. The gradual contiguous geographic divergence within a species leads to variation, selection, and eventually to the establishment of "permanent" varieties, which Darwin views as "incipient species." Varieties eventually become species as selection continually accentuates and favors the more extreme adaptive novelties. The residual varieties—especially those that mark the transition from the primal ancestral adaptive condition of a species, and the emerging extreme varieties—find themselves at a disadvantage and eventually succumb to extinction. As the process continues, the gaps (in terms of adaptive morphology) between the once interconnected varieties, species, and the like become more and more pronounced, such that eventually distinct families of endemic taxa occupying a region can be distinguished.

Thus it is extinction of intermediate taxa, rather than anything inherent about the process of diversification, that tends to create the gaps between closely related species. By this point Darwin is firmly committed to an overarching theory of extinction based on intervarietal and interspecific competition, all stemming, at base, from the struggle for existence among organisms within populations. He acknowledges that environmental change can drive species to oblivion. But in a short section on extinction (just before his discussion of the principle of divergence in both books), Darwin concludes that extinction for the most part occurs through competition with the closest neighboring related taxa, be they varieties or species.

In the *Origin*, Darwin returns to extinction in the second chapter on geological phenomena. First he reports that "the old notion of all the inhabitants of the earth having been swept away at successive periods by catastrophes, is very generally given up" (1859:317), which was true enough—though Darwin almost certainly knew of d'Orbigny's Cuvier-inspired *étages*. So much for "beginning with this!" And then, proclaiming that the "the whole subject of the extinction of species has been involved in the most gratuitous mystery," he immediately goes on to say that "some authors have even supposed that as the individual has a definite length of life, so have species a definite duration"—in itself a gratuitous swipe, without of course any mention of his earlier acceptance, contra Charles Lyell, of Giambattista Brocchi's views. Darwin was correct to discount the possibility of inherent, built-in longevities of species, there being no empirical evidence, then as now, to support such a claim. But Darwin's principle of divergence and "speciation" through the competitively engendered extinction of intermediates has also failed to receive much empirical general support as well.

So in this second discussion in the geology section, what does Darwin say causes extinction? Darwin simply repeats his argument from the earlier section that selection drives change and is inherently competitive at all levels. Newly emerged, adaptively derived entities, be they varieties or species, simply drive their less well-adapted brethren to oblivion. The main engine of extinction is competition between closely allied "races," again, be they varieties within species or species themselves.

And what of isolation? Though Darwin casually alludes to "barriers" in places throughout his development of the principle of divergence, isolation itself is discussed for the most part in a brief passage under factors enhancing or inhibiting natural selection ("Big Species Book"), or, in the *Origin*, merely "circumstances favourable to natural selection." Although Darwin still acknowledges the obvious effects that isolation can have, given the availability of novel or changed conditions, he is at pains in the *Origin* to continually remind his readers that what happens on islands cannot be nearly as generally important as the processes of evolution on larger areas. Famously, Darwin (1859) concludes: "Although I do not doubt that isolation is of considerable importance in the production of new species, on the whole I am inclined to believe that largeness of area is of more importance, more especially in the production of species, which will prove capable of enduring for a long period, and of spreading widely" (105). This passage immediately precedes his first brief discussion of extinction, followed by his principle of divergence.

According to historian Frank Sulloway (1979), Darwin's downplaying of the importance of isolation in evolution and the origin of species caught the attention, and profound disagreement, of German zoologist Moritz Wagner. Wagner evidently felt that most new species do indeed arise in isolation through more or less rapid phases of adaptive change. If they survive, and are then able to expand their range (hence numbers), the rate of evolution will typically slow down, and species will stabilize: once again, a model close to what (as we shall soon see) was proposed many decades later by Theodosius Dobzhansky and Ernst Mayr ("allopatric speciation"), and then again by Niles Eldredge and Stephen Jay Gould ("punctuated equilibria," with the dimension of geologic time being brought back into the discussion).

Darwin did modify his discussion in light of Wagner's objections. In the sixth edition of the *Origin* (1872)—important, as it was the version of the *Origin* that actually came down as the central canonical text from Darwin's nineteenth-century hand to twentieth-century evolutionary biologists—Darwin writes: "Moritz Wagner has lately published an interesting essay on this subject, and has shown that the service rendered by isolation in preventing crosses between newly-formed varieties is probably greater even than I supposed. But from reasons already assigned I can by no means agree with this naturalist, that migration and isolation are necessary elements for the formation of new species" (81–82).

SUMMING UP

Charles Darwin displayed both tenacity and creative brilliance developing his evolutionary theory throughout his long career (figure 3.1). He read Jean-Baptiste Lamarck's transformational views as a student at Edinburgh and could not but have been equally apprised of Giambattista

FIGURE 3.1 The older Charles Darwin. (By permission, Library of Congress, Prints and Photographs Division)

Brocchi's analogy between the births and deaths of individuals and of species. Robert Jameson and Robert Grant were among his teachers and mentors, and he must have been aware that the explanation of the advent of the species of the modern fauna in terms of natural causes had by the mid-1820s, at least in Edinburgh, become the focusing question of what was to later, and largely through Darwin's own offices, become "evolutionary biology."

Species, it was agreed, seemed to become progressively more and more like modern species as one samples fossils higher up, in progressively younger and younger sedimentary rocks. Thus the question: What is the natural causal explanation for the progressive replacement of species through time? This question was adumbrated by others ([Jameson] 1826, 1827c) and succinctly posed by John Herschel (1830), whose book was dedicated to the proposition that natural phenomena have natural causes—and a book that Darwin had read.

As a young man on the *Beagle*, Darwin found evidence of species replacement in time in his fossils, especially in the relation he thought he saw between the extinct cavy fossil from the beds of Monte Hermoso and the living Patagonian cavy. He seems likely to have been the first one to make his own analogy: just as species replace one another in time (as closely "allied forms"), so too do they in space. The two rhea species of southern South America are perhaps the premier example of this in Darwin's own field experience. And he also saw that the principle of geographic replacement could be extended to offshore islands, especially with archipelagos where the species offshore were allied with, but often distinct from, the mainland species. And he saw that sometimes there seemed to be distinct species (or at least varieties) on separate, nearby islands within the archipelago (Darwin's "halo" of the Red Notebook [1836–1837]). The several distinct forms of the mockingbirds on different islands of the Galápagos were his best example, though he got the idea from the distinct forms of Falkland foxes at least a year before he got to the Galápagos.

Darwin's shipboard essay "February 1835" finds him analyzing the births and deaths of species and adopting Brocchi's analogy with individuals, including the probable cause of extinction being ingrained (somehow) internally within species (there being no evident environmental change he could see in the sediments compared with present-day Patagonian xeric conditions). He assumed that species diversity remains more or less constant through time (attributing this to the will of the Creator, but getting the idea in reality from Charles Lyell)—thus meaning that new species within a lineage must appear at just about the time that its predecessor (that is, ancestor) became extinct.

Darwin returned to these thoughts, changing his examples somewhat and making his pointed analogy between species replacement patterns in time and in space, in the latter part of the Red Notebook, written in early to mid-1837. Though the Red Notebook makes it completely clear that Darwin was a convinced transmutationist, his thoughts on temporal species replacement as early as the fall of 1832, while still in the field at Bahia Blanca, demonstrate that, at the very least, he was thinking very seriously about transmutation even then. He was adding his own empirical observations, and eventually his own thoughts, to whatever he heard and read on the subject at Edinburgh and Cambridge in the mid-1820s and early 1830s.

After explicitly revealing in writing (if only to himself) that he was a convinced transmutationist, Darwin had to confront the issue of adaptation: for if species have origins through natural causes, it immediately follows that their characteristics must also change through natural causes. The four Transmutation Notebooks (1837–1839) reveal his explorations of what such a theory of adaptive change might consist. These four riveting notebooks reveal scientific creativity in action.

In Notebook B, Darwin saw (as did his grandfather before him) that reproduction ("generation") must be part of the answer, as organisms resemble their parents ("grandfathers"). He also understood that there is heritable variation within, as well as among, families. Yet there was a third ingredient missing that would complete the analogy with artificial selection carried out in the barnyard. Still lacking that, Darwin simply asserted that there was a law-like mechanism, details to be specified later, that underlies the modifications of species.

He proposed two models in which that law would be found to be in action: (1) in isolated populations, most readily seen on islands; and (2) more generally, as a slow, steady transformation of entire species to match what he saw as the inevitable environmental change that occurs through geological time.

By Notebooks D and E, starting roughly a year later (mid-1838), he had read Thomas Malthus, and realized that resources and other environmental factors restrict the population sizes of varieties and species to numbers far below what the geometrical increase of sexual reproduction would otherwise produce. He had the governing factor that would control which variations would survive and themselves reproduce, and thus the analogy with selective breeding was complete.

In Notebook E, and despite a later notation warning himself not to downplay the importance of isolation too much, Darwin explicitly saw his two models of adaptive change in evolution as more or less an either/or choice. That he chose to develop his general adaptive model (eventually formulated as the principle of divergence in the 1850s) over adaptation in isolation was clear by the time he wrote "Pencil Sketch" (1842) and "Essay" (1844a). And as we have just seen, the message he left in his mature work (his revision of the *Origin* in the 1870s) was a ringing declaration of the inevitable, if slow, workings of natural selection to modify entire species through geological time. Isolation was not required for the evolution of new species.

In Darwin's final, preferred model, then, species are constantly being modified in time, as well as diversifying in space. New species are simply the product of selective modification of the adaptive properties of their component individuals. And critically, species become separate ("isolated") one from another only through the extinction of less well-adapted intermediate forms. Extinction, moreover, stems as a rule from the competition between different varieties, or species (on what has become Darwin's sliding scale): the adaptive superiority of the more advanced, divergent form eventually eliminates the more primitive, ancestral connecting form. This model is quintessentially transformational, invoking gradual transitions very much in the style of Jean-Baptiste Lamarck.

In Darwin's other model, new species arise as relatively small populations in isolation, and divergence yields enough adaptive change so that they may be told apart. In this model, species are indeed real; they have discrete births and, eventually, deaths. And natural selection produces adaptive change mostly in relatively shorter periods in isolation compared with the longer history typical of well-adapted species. This model, of course, is quintessentially taxic, and is very much in the Brocchian tradition.

The question becomes: Why did Darwin see the two models of the action of natural selection as inherently contradictory? Why could he not allow isolation its due, but also assume that species will continue to evolve gradually and perhaps diverge again as time goes on? After all, this seems to have been the position of Theodosius Dobzhansky, Ernst Mayr, and other theorists starting in the 1930s, after they had resurrected the importance of geographic isolation in the origin of species.

In his alternative model—the one Darwin left on the cutting room floor—a new species comes into existence as a direct derivative of an ancestral species. As Darwin saw, if adaptive change is mostly confined to situations in which relatively small isolated populations are presented with new, or changed, environmental conditions, the way was paved for understanding why most species seem to remain unchanged through "thick formations," meaning considerable periods of geological time. After all, such geographic speciation events must be relatively rare. And if most adaptive change occurs in this context, the implication is that nothing much by way of adaptive transformation is occurring throughout the long periods of time successful species tend to survive. What seems to have worried Darwin is the possibility that the relatively rare process of speciation in isolation would prove to be the context for most adaptive change throughout evolutionary history.

In the end, Darwin simply could not believe that natural selection producing directional adaptive change would not be constantly ongoing, modifying a species continually through geological time. He could not see how isolating barriers could have played an all-encompassing role in the origin of so many species over vast areas, such as the five armadillo species of southern South America. And apparently, he preferred to think of natural selection in terms of its constantly creative powers to mold and shape adaptive change in response to inevitable environmental change, rather than seeing it also as a powerful force for stability. It was simply too likely that environments will continue to change through geological time (despite his early in-the-field conclusion that, in fact, the terrestrial environment had not changed in Patagonia from the days of the now-extinct *Megatherium* to the present).

Thus to keep natural selection as an all-powerful engine of change, Darwin conceived of the principle of divergence, which eliminates the need to see adaptive change relegated to those relatively rare moments of speciation in isolation. Reproductive isolation among closely related, geographically adjacent species is a fallout of the inevitable extinction-through-competition of the less fit intermediates, rather than a condition that arises through adaptive change in isolation from the very beginnings of a new species.

Darwin's principle of divergence, albeit clever, was solely a protracted gedanken experiment, in which his imagination let natural selection go through geological time, producing variant scenarios of change and lineage divergence through time. It is striking that there is nothing in the way of putative, real-world examples in Darwin's development of the principle of divergence (although it must be said that the two species of rheas meeting along the Rio Negro could be explained either way—crucial deciding additional evidence lacking).

The dichotomy that Darwin saw was very similar to my own take on things when I argued for the distinction between the "taxic" and "transformational" views in the late 1970s, discussed in the Introduction. These truly are qualitatively distinct visions of the evolutionary process. In the model that came down to us from Darwin, species are mere "way stations" in the adaptive continuum of life, made discrete only by the extinction of the connecting varieties and species. They are in no sense "real," let alone stable, entities. In his alternative model, species have births, histories, and deaths, as do individuals. Thus species have point origins, histories of stability, and eventual deaths—very much as do individuals.

After Darwin's *Origin* (1859), the taxic view was to lie mostly dormant in mainstream evolutionary biology until the mid-1930s. As we shall see, the resolution of this apparent dichotomy in models of the evolutionary process and conceptualizations of the very nature of species has been achieved decisively in empirical evolutionary science, starting in the mid-twentieth century, and continuing on into the modern era. The taxic perspective has returned in force.

II

REBELLION AND REINVENTION

THE TAXIC PERSPECTIVE, 1935–

4

SPECIES AND SPECIATION RECONSIDERED, 1935–

The manifest tendency of life toward formation of discrete arrays is not deducible from any a priori considerations. . . . A causal analysis of this phenomenon is one of the major tasks of the biological sciences. Strange as it may seem, the predominance of the evolutionary trend of thought in some branches of biology during the last half century diverted the attention of the investigators away from this task rather than toward it, for an evolutionist is frequently more interested in the "bridging the gaps" between groups of organisms than in the nature of the gaps themselves.

—THEODOSIUS DOBZHANSKY, "A CRITIQUE OF THE SPECIES CONCEPT IN BIOLOGY"

By the time Charles Darwin published the sixth edition of the *Origin of Species* in 1872, he had all but abandoned the search for an explanation for the origin of discrete species. Gaps between species became, instead, a missing data problem. The possibility that most adaptive change in evolution might somehow be causally tied up with the emergence of new species in isolation no longer troubled him. Instead, Darwin sought continuities, rather than discontinuities. When he worried about the lack of evidence of continuous progressive change in the fossil record, he placed the blame on the vagaries of the deposition of sediments and the haphazard preservation of the dead remains of ancient plants and animals. He saw the discontinuities between modern species to be the inevitable result of the retroactive extinction of intermediate forms through competition—a raw form of purely eliminative group selection—rather than the direct result of the evolutionary process itself. Thus Darwin's theory, as he left it, and with all its force and "legs" even down to the present day, was essentially immune to empirically based criticism.

True, Darwin did emend his final text, acknowledging the fervid defense of the importance of isolation in evolution emanating from Moritz Wagner. And there were occasional biologists (not so much paleontologists) who, in that fifty-year interval that Theodosius Dobzhansky alludes to in the quotation at the head of this chapter, did maintain the importance of isolation and the understanding of how discrete species in fact come into existence through natural causes. According to Dobzhansky (1937), G. J. Romanes, for example, wrote in the late nineteenth century that "without isolation, or the prevention of interbreeding, organic evolution is in no case possible" (228).

And there were a few others. For example, David Starr Jordan (1905), an ichthyologist who became president of Stanford University, argued for the importance of isolation in evolution. So did biologists J. T. Gulick (1905) and H. E. Crampton (1916, 1932), whose works on patterns of anatomical diversification of land snails on various Pacific islands Dobzhansky also cited. Most important, at least to Dobzhansky, were the writings of the early geneticist William Bateson, who saw species as having properties distinct from "varieties"—a sentiment that Dobzhansky (1935) found sufficiently compelling to quote at the head of his own seminal paper.

But in general, Dobzhansky seems to have been right in his insistence that the importance of discontinuities and the role of isolation in evolution had been a greatly neglected, if not wholly forgotten, theme in evolutionary biology between 1872 and 1935, an interval corresponding fairly closely with the "last half century" Dobzhansky alludes to in his remarks quoted in the epigraph at the beginning of this chapter.

Nor were things any different in paleontology. Darwin's ultimate evolutionary writings did provoke a brief flurry of interest in documenting progressive series of fossils interpreted as consistent with Darwin's prediction of slow, steady, gradual change of species-within-lineages up a carefully collected section of rocks. Thomas Huxley's work on American horse evolution remains the best known, but also prominent in the day were three British studies of evolutionary sequences in invertebrates, providing, if just for the nonce, the sort of data that Darwin, in his notebooks, openly wished "Mr. Lonsdale" would have produced in his work on Silurian fossils—that is, empirical examples of (putative) gradual evolutionary change. The papers included the work of A. W. Rowe (1899) on changes in the heart urchin *Micraster* up the cliff faces of the Cretaceous chalk at Dover; R. G. Carruthers (1910) on Carboniferous rugose coral evolution; and, later, the work of A. E. Trueman (1922) on the gradual evolution of the Jurassic curved oyster *Gryphaea*.

But by and large, paleontologists had little to say about evolution after the publication of the sixth and last edition of the *Origin* (1872). Paleontological data continued, for the most part, to seem out of synch with Darwin's projected patterns of what evolution would and should look like, were the fossil record at all complete. Instead, paleontologists seemed to accept Darwin's write-off of the fidelity of the fossil record as a lamentable, even shameful, fact of life. As a result, the stability of fossil species as an overwhelming generalization was ignored, written off as an illusion—even though Darwin himself, as a young and thoughtful man wrangling with the facts of the matter, was well aware of the inconvenient truth of the stability of most species encountered in the fossil record. As the late Stephen Jay Gould once put it, after Darwin, stasis had become the "trade secret" of paleontology.

But by far the most important series of events in the half-century that Dobzhansky alludes to is the rediscovery of Gregor Mendel's laws, prompting the explosive birth of genetics at the turn of the century. These were, manifestly, extremely exciting times, with the new, fundamental discoveries (such as the existence of discrete genes organized linearly on chromosomes) coming along fast and furiously in the decade or so leading up to World War I.

Naturally, such stunning events that open up entire new vistas of research instantly divert attention. Biology had already developed as a serious science, pursued and taught at universities the world over. The entrance of electricity into the lab at once made the pursuit of biology more legitimately "scientific" in the minds of many of its practitioners.

It also freed the subject from the pejorative identification with prosaic, old-fashioned, and decidedly non-high tech "natural history." If there were quibbles about the veracity of Darwin's natural selection (many of Darwin's critics in the last decades of his life attacked natural selection, rather than the general proposition that life has evolved through natural causes), "Darwinism" was seen by many avant garde biologists at the beginning of the twentieth century as basically synonymous with old-fashioned natural history. It could consequently be ignored, if not openly derided. Overnight, biology had instead become the age of an excitingly revived "functional," as opposed to an "historical," science.

But it was not long before some of these early geneticists started wondering how their new data and perspectives on heredity actually did fit in with the undeniable reality of biological evolution. Columbia University geneticist Thomas Hunt Morgan, at least for a time, thought that the new data of genetics were in and of themselves sufficient to explain evolution. The Dutch geneticist Hugo Devries took a particularly egregious and extreme stance, attributing the origin of phenotypic differences between closely related species to single mutations. And others, such as paleontologist Henry Fairfield Osborn, also jumped into the fray. The resulting Tower of Babel chaos of conflicting evolutionary accounts in the name of the newly discovered genetics is of little lasting interest.

Out of that fray—or, more accurately, as a sane, rational reaction to it—came two things: first, the reconciliation of Mendelian genetics with Darwin's theory of evolution via natural selection, accomplished most notably by mathematically inclined geneticists Ronald A. Fisher (1930), J. B. S. Haldane (1932), and Sewall Wright (1931, 1932) starting in the 1920s, with Fisher leading the way; and second, the reemergence of the "taxic perspective"—the acceptance of the existence of species as independent entities, with births, histories, and deaths, with point origins largely through geographic isolation. It is the latter development that occupies the attention of the remainder of this narrative.

The taxic approach to evolution was first developed by those who sought a natural causal explanation of the advent of modern species through the successive replacement of species within lineages. Paleontology was the point of departure of these early evolutionists: Jean-Baptiste Lamarck, though not per se taxic, Giambattista Brocchi, Robert Jameson,

FIGURE 4.1 Theodosius Dobzhansky. (Courtesy of University Archives, Columbia University in the City of New York)

and Darwin. In stark contrast, the twentieth-century reemergence of the taxic perspective arose from the new genetics. It came from a man who had begun his career in Russia as a systematist, a natural historian of ladybird beetles. Apparently, Dobzhansky became aware of the new genetics through working with the evolutionarily minded geneticist Yuri Filipchenko. This young man, Theodosius Dobzhansky (figure 4.1), immigrated to the United States in 1927 to join the genetics lab founded by Thomas Hunt Morgan at Columbia University.

It was Dobzhansky who singlehandedly applied the new genetics to the real world. In so doing, he instantly confronted the discontinuities between species and related matters that Darwin had seen, and ultimately avoided, that half-century previously. What Dobzhansky started continued to grow, and all the old, forgotten conclusions of the early evolutionists, including the younger Darwin, had to be rediscovered or at least re-acknowledged. These included the natural and causally direct formation of distinct species in isolation and its relationship to adaptive change via natural selection (that is, geographic speciation); stasis (stability of species in the fossil record); the hierarchical organization of functional biological entities (begun with Brocchi's analogy on the births and deaths of individuals and species); and the importance of extinction events, not only in evolutionary history, but also in the actual evolutionary process itself (as a pattern, at least, harking all the way back to Georges Cuvier and Alcide d'Orbigny).

As one might expect, the renewed emphasis on species has gone far beyond the first glimmerings of the taxic perspective as originally developed by our predecessors in the early decades of the nineteenth century. The explosion of work was enormous, beginning with Dobzhansky's brief, yet penetratingly cogent, examination of the nature of species from the viewpoint of the new genetics, as well as their manifestation in the natural world.

In what follows, I can only hope to touch on the main themes of the development and present status of the taxic perspective in evolutionary biology. What is at stake here is nothing less than resolving the issue of just when, where, and how selection-engendered adaptive change occurs in the evolutionary process. And the crux of the matter lies in the issue of the ontological reality of species. Theodosius Dobzhansky began the long, arduous process of restoring what had been left behind in the nineteenth century with his seminal paper in 1935, written while he was at Caltech.

"SYNGAMEONS" AND THE ORIGIN OF THE BIOLOGICAL SPECIES CONCEPT

Dobzhansky's (1935) brief paper reintroduced in one blow the concept of the reality of species whose existence needs to be understood in evolutionary terms. He entitled the manuscript "A Critique of the Species Concept in Biology." Like many similarly titled papers that were to come along later, Dobzhansky says at the outset that he develops his thoughts from a "methodological point of view." Biologists in the twenty-first century are still obsessed with the practical problems of species recognition and "testability" of their determinations—without paying, as a great general rule, any explicit attention to what species are, or even if they do exist in any real, ontological sense.

But here Dobzhansky was different, for despite his insistence that he is adopting a "methodological point of view," in the very next sentence he reveals that the methodology stems from a deeper question of ontology. The only "category" of biological entity that has undeniable ontological reality (he says "significance") is that of the individual organism. He then says that it is, at the very least, useful to recognize species just to reduce the chaos of dealing with the "stupendous number" of highly diverse organisms; thus "the category of species has a clear pragmatic value" (345).

The question then becomes can we go farther and ascertain "whether the species is a purely artificial device employed for making the bewildering diversity of living beings intelligible, or corresponds to something tangible in the outside world." In other words, is the species "a part of the 'order of nature,' or [merely] a part of the order-loving mind?"

Dobzhansky concludes that species are natural, discrete entities. He says they have that very ontological "significance" he poses at the outset, which comes from his viewpoint as a geneticist, as well as his earlier experience dealing with organisms and species in the natural world. He tackles the problem empirically by discussing patterns of continuous and discontinuous variation. If biological systems were continuously variable, classification would, perforce, be entirely arbitrary if nonetheless still pragmatically valuable. And, he says, there are indeed examples of continuous variation in nature such as among the bacteria that bacteriologists still find useful to classify into distinct "species."

Tellingly, Dobzhansky assures his readers that the same is true of fossils, although here his statement lacks any specific examples: "The species of the fossils are readily separable only as long as the known representatives of a given group from the older strata are scarce. New findings fill up the gaps between the species, until a continuous series of forms finally emerges" (346)—there being no paleontologists in contemporary evolutionary discourse to say him nay. And given the existence of a few papers claiming in fact to have documented such gradual progressively changing lineages, Dobzhansky was in fact able to repeat the litany so firmly insisted upon by Charles Darwin. In paleontology, then, the "gap filling" Dobzhansky otherwise laments—even derides—in the opening quotation of this chapter, is perfectly okay when dealing with sequences of fossils.

The significance of Dobzhansky's rote repetition of the myth of the reality of gradual evolutionary change within fossil lineages is not that it is not true: after all, Dobzhansky had no practical experience with fossils. Instead, it was that Dobzhansky was *not* saying that the speciation process was actually the locus of most, or even substantial, amounts of adaptive change in evolutionary history. He was content to let the basic story lie: that some, perhaps even most, adaptive change occurs phyletically within lineages through geologic time.

Yet Dobzhansky goes on to say that, despite the putative examples from the fossil record, and among certain groups of recent organisms (such as bacteria), *dis*continuity, rather than continuity, is the norm. And as we have already seen (in the earlier quotation), Dobzhansky sees this as a major problem ignored essentially since the publication of the last edition of Darwin's *On the Origin of Species*.

Here is where Dobzhansky's perspective as a geneticist enters in. For though all of the categories (we would now say "taxa") ranked higher than species are, if anything, more discrete and discontinuous than species, nonetheless *there is something unique about species*.

In short, "Every discrete group of individuals represents a definite constellation of genes." And:

> If the different groups interbreed freely with each other, a new equilibrium is established in which the different genic constellations become fused into a single one. It necessarily follows that no discontinuous variation can exist in a perfectly panmictic population. Mutatis mutandis, the existence of two or more discrete groups of individuals is a proof that free inter-breeding between them is prevented by some factor or factors. It is a matter of observation that closely similar individuals habitually interbreed, and those less similar do not. Hence, a stage must exist in the process of evolutionary divergence, at which an originally panmictic population becomes split into two or more populations that interbreed with each other no longer. (348)

Dobzhansky acknowledges that Darwin was deeply interested in matters of interbreeding within and among groups. But he also says that it is the new genetics that provides the key to understanding what species are. Citing William Bateson, and especially the Dutch geneticist J. P. Lotsy, Dobzhansky says that it is especially Lotsy's concept of a "syngameon, 'an habitually interbreeding group' of individuals," that holds the key to understanding what sexually reproducing species really are. Lotsy's definition (and term "syngameon") provides "a definition of species which is especially attractive because of its simplicity" (349).

But the definition is not quite good enough, according to Dobzhansky, as it includes only those individuals who can be shown to actually be interbreeding, rather than those others who, by reason of distance or ecological factors, are not actually, but nonetheless are *potentially* capable, of interbreeding. He concludes, "The emphasis should be placed however not on the absence of actual interbreeding between the different form complexes, but rather on the presence of physiological mechanisms making interbreeding difficult or impossible" (349).

This leads Dobzhansky directly to his discussion of "isolating mechanisms." Isolating mechanisms to Dobzhansky are the physiological causes underlying hybrid sterility, or the factors that prevent the formation of hybrids in the first place (with geography, the most important factor, not included within the list of physiological isolating mechanisms). Before considering his critically important analysis of how such mechanisms actually evolve (including the hints of teleology latent in the very term), it is important, I think, to follow out what Dobzhansky saw as the implications of his analysis for the very definition of what species actually are: "Lotsy's definition of what constitutes a species should be modified thus: a species is a group of individuals fully fertile inter se, but barred from interbreeding with other similar groups by its physiological properties (producing either incompatibility of parents, or sterility of the hybrids, or both)" (353).

Soon thereafter, Dobzhansky writes: "Considered dynamically, the species represents that stage of evolutionary divergence, at which the once actually or potentially interbreeding array of forms becomes segregated into two or more separate arrays which are physiologically incapable of interbreeding" (354).

It is a matter of historical importance to point out the close similarity of the contents of these two or three basically identical species concepts of Dobzhansky (derived and inspired initially by Lotsy) to the so-called short definition of species published by Ernst Mayr in *Systematics and the Origin of Species* (1942). It's a familiar definition still memorized by nearly all students in evolutionary biology. Mayr is nearly universally given the credit for the development of both the "biological species concept," as well as the general concept of allopatric speciation. But Dobzhansky was clearly the biologist who actually accomplished both jobs.

Mayr's short definition of species reads, "Species are groups of actually or potentially interbreeding natural populations, which are reproductively isolated from other such groups" (120).

Mayr (1940, 1942) never cites Dobzhansky (1935). His focus instead is strictly on Dobzhansky's book *Genetics and the Origin of Species* (1937). Mayr dismisses Dobzhansky's second definition of species ("the species represents that stage," repeated in 1937), as an excellent description of the speciation process but hardly a satisfactory definition of what a species actually is. To be fair, in the book, Dobzhansky does not include his discussion of Lotsy, from 1935, when he gives his definition of species. Lotsy appears, instead, in Dobzhansky's final chapter: "Species as Natural Units." But it is hard to believe that Mayr never saw Dobzhansky's 1935 paper. Mayr (1940) says that a concept of biological species as a series of genetically interconnected populations was more or less generally accepted (he cites specific examples—criticizing each one, most notably those of A. E. Emerson and N. W. Timofeef-Ressovsky). This offers strong evidence of the influence that Dobzhansky's paper surely had on the biological community virtually from the moment it was published in 1935.

Mayr's objections to all of these definitions, including Dobzhansky's, in the end rest purely on methodological issues: How can we use the definition in question to determine with precision what a particular species is? This is a question of identification of a specific example in nature, rather than an ontological definition of what species in general are.

Insofar as species recognition is concerned, however, Dobzhansky (1935) takes pains to point out that there is a great overlap in the species as defined and recognized by taxonomists (systematists) and what would be construed as a genetic community, actually or potentially interbreeding and reproductively isolated from other such groups. The term "syngameon" never caught on. But that's what species are, to this day, in process theory evolutionary biology (if not in all areas of systematics praxis; see Wiley and Lieberman 2011). They are essentially packages of genetic information neither actually nor even potentially shared with "other such groups."

ISOLATING MECHANISMS

Though Mayr and Dobzhansky differed to some degree in their initial characterizations and "classifications" of what Dobzhansky dubbed "isolating mechanisms," (again, mostly a matter of Mayr sniping at Dobzhansky), both saw that physical isolation was the most obvious cause of ultimately genetic isolation. But permanent genetic isolation can only be asserted with certainty if two species are at least partially sympatric; that is, if they overlap in their geographic ranges and if there is little, if any, evidence of genetic "leakage" through hybridization.

It is the latent teleology of the very term "isolating mechanisms" that is of greatest interest here. Dobzhansky's (1935) first discussion, in the seminal paper that is the focus of this chapter, says, "It is unclear how such mechanisms can be created at all by natural selection, that is what use the organism derives directly from their development. We are almost forced to conjecture that the isolating mechanisms are merely by-products of some other differences between the organisms in question, these latter differences having some adaptive value and consequently being subject to natural selection" (349).

But Dobzhansky could not let the matter lie. By 1940, he was developing a model that saw natural selection acting at areas of overlap and contact between "races" that were to some degree already adaptively differentiated. Selection, he postulated, would then act to prevent the vitiation or even dissolution of these selected-for adaptive complexes. By the time he had thoroughly adopted Sewall Wright's (1931, 1932) imagery of the "adaptive landscape" (especially in the third edition of *Genetics and the Origin of Species* [1951]), Dobzhansky had firmly concluded that selection, the force that shapes adaptations, also acts to conserve those adaptations: selection acts to position and maintain species on their adaptive peaks.

This is what I (Eldredge 2008b) have called "Dobzhansky's dilemma": variation is necessary for evolution to occur; and if selection is acting mostly as a stabilizing force eliminating variation, the risk to survival down the road is heightened. At the same time, though, selection will favor optimality, honing adaptations to fit the "adaptive peak." Dobzhansky saw the tension between the maintenance of ample variation and the countervailing action of natural selection in winnowing all but the most optimal variations.

It was the latter, the tendency for natural selection to mold populations to optimally fit their local environs, that underlay Dobzhansky's thoughts that selection will act to drive partly differentiated populations further apart. Selection will act against hybridization, leading to the emergence of two distinct and genetically isolated species. Much later, South African geneticist Hugh E. H. Paterson called this model of Dobzhansky's "speciation by reinforcement."

I believe, to his credit, Mayr never bought the argument that selection can act *for* isolation. Dobzhansky (1940) himself conceded that the emergence of new species on islands posed a simple, perhaps even fatal, counter-example, in which there is obviously no scope for zones of "reinforcement" to develop, calling into doubt the very existence of selection for reinforcement in the development of "isolating mechanisms."

As far as I am aware, the only conceptual advance made in this entire area since Dobzhansky posited "isolating mechanisms" as integral to the origin of new species was made by Hugh E. H. Paterson in a series of papers in the 1980s and 1990s. Paterson's ideas are worth considering seriously, as they bear not only on speciation, but also on the very nature of what species are—the latter with profound implications for how biological systems are fundamentally organized.

Paterson argues against speciation by "reinforcement" "and especially the idea that speciation and the development and consolidation of "isolating mechanisms" is all about the strengthening and maintenance of already differentiated ecological adaptations. Instead, Paterson (1985) offers a vision of species essentially as syngameons, defining species as groups that share what he calls a "Specific Mate Recognition System" (or SMRS). These are the morphological, behavioral, biochemical, and the like genetically based phenotypic traits that pertain strictly to reproduction. These are the traits that allow prospective mates (whether or not they ever get a chance to meet) to sexually reproduce. This includes everything from mating songs and pelage and plumage patterns among terrestrial vertebrates to the chemical recognition of sperm and eggs among myriad sedentary marine invertebrate taxa, windblown pollination among terrestrial plants, and so on.

Paterson concludes that selection is indeed always at work in the speciation process. Geographically isolated populations do tend to diverge, with selection modifying adaptations to local environmental conditions. But unless the reproductive adaptations diverge as well, simply through selection to maintain continued reproduction in slightly different environments, the ecological adaptations may be lost through hybridization should the two separated populations once again come into contact.

Paterson's geneticist conception of what species are and how they form sees species essentially as packages of genetic information. In the end, "isolating mechanisms" appear not to arise to keep fledgling species apart. Instead, they reflect genetic barriers to reproductive success that arise from separate evolutionary histories in geographically isolated populations. These adaptive evolutionary changes may be ecological, reproductive, or a mixture of both.

But the conclusion seems clear, remaining the current state of thought in the second decade of the twenty-first century. Reproductive isolation arises as a secondary byproduct of divergence in isolation. In that sense, speciation is an accident.

FIGURE 4.2 Ernst Mayr. (Courtesy of Library Services, American Museum of Natural History)

ALLOPATRIC (GEOGRAPHIC) SPECIATION

In his seminal works of the mid- and late 1930s, Theodosius Dobzhansky clearly saw that geographic isolation is the sine qua non, the necessary precondition, for the eventual appearance of new species. Yet Dobzhansky did not devote extended discussion to the actual process of geographic ("allopatric") speciation. That job was left primarily to Ernst Mayr (figure 4.2), who fulfilled it well, especially in the book *Systematics and the Origin of Species* (1942).

Mayr left little doubt that new species almost always arise through a process of differentiation that is triggered by geographic isolation. Speciation has occurred when divergence reaches a point at which hybridization cannot occur, even should the opportunity arise. Though he is aware of chance factors (Sewall Wright's notion of "genetic drift"), Mayr left little doubt in 1942 that most divergence is adaptive in nature: adaptation to local environmental conditions. This, of course, is precisely the view Charles Darwin developed in Notebook B (1837–1838).

But it is nonetheless worthwhile requoting a passage from Wright (1941) that Mayr (1942) quotes with approval:

> If isolation of any portion of a species becomes sufficiently complete, the continuity of the fabric is broken. The two populations may differ little if any at the time of separation but will drift ever farther apart, each carrying its subspecies with it. The accumulation of genic, chromosomal and cytoplasmic differences tends to lead in the course of ages to intersterility or hybrid sterility, making irrevocable the initial merely geographic or ecologic sterility. (158)

This passage embodies the prevailing view on the nature of the speciation process down to the present day. Arguments still flicker over whether or not some form of sympatric speciation (including the "ecologic isolation" of Wright [1941]) may occur. But the vast majority of biologists interested in speciation phenomena agree with Dobzhansky, Mayr, and Wright that geographic isolation is the necessary, if not wholly sufficient, cause of nearly all cases of speciation and the origin of new species from old ones. Mayr's (1942) summary diagram of how the process of allopatric speciation works and what it "looks like" is the simple epitome of the volumes of pages that have appeared describing the ins and outs of geographic speciation (figure 4.3). It's a literature that began in the modern era with Dobzhansky's (1935) paper and carries on actively to the present day.

And thus adaptive divergence—genetically based morphological diversity—is directly linked to the process of geographic speciation. Dobzhansky and Mayr led the way, however unwittingly, in redeveloping one of Darwin's two basic models on the circumstances surrounding the accrual of adaptive change in evolutionary history: the notion of adaptive divergence in geographic isolation. Neither one of them, however, ever went as far as Darwin as to wonder whether this was the actual dominant context of the generation of adaptive change in evolution. Dobzhansky and Mayr explicitly said that both speciation and gradual phyletic evolution were in cooperation on the generation of adaptive evolutionary change. Once new, adaptively differentiated species appear, they will continue to evolve and accrue gradual adaptive changes through time as they enact their now-separate evolutionary histories.

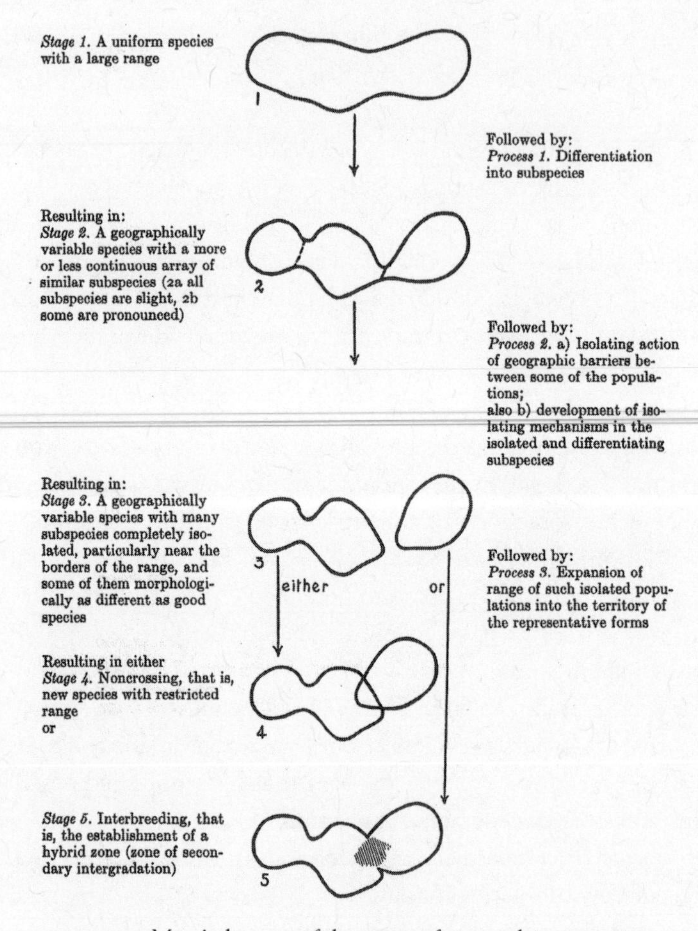

FIGURE 4.3 Mayr's diagram of the stages of geographic speciation. (From *Systematics and the Origin of Species* [1942:160, fig. 16])

NATURE AND REALITY OF SPECIES IN THE 1930S AND 1940S

Both Theodosius Dobzhansky (1935, 1937, 1940) and Ernst Mayr (1940, 1942) thought that species were most definitely "real" entities. Indeed, the final chapter of Dobzhansky's (1937) book was simply entitled "Species as Natural Units." And in 1942, Mayr neatly summarized the received Darwinian heritage—the older Charles Darwin as represented in *On the Origin of Species*—when he wrote at the outset of the chapter "The Species in Evolution":

> Darwin entitled his epoch-making work not *The Principles of Evolution*, or *The Origin and Development of Organisms*, or by some other title which would stress the general problems of evolution. Apparently he considered these titles too speculative and therefore chose the more concrete one, *On the Origin of Species*. To him this was apparently more or less synonymous with these other titles, which is not surprising if we remember that Darwin drew no line between varieties and species. Any pronounced evolutionary change of a group of organisms was, to him, the origin of a new species. He was only mildly interested in the spatial relationships of his incipient species and paid very little attention to the origin of the discontinuities between them. It is thus quite true, as several recent authors have indicated, that Darwin's book was misnamed, because it is a book on evolutionary changes in general and the factors that control them (selection, and so forth), but not a treatise on the origin of species. Obviously, it was impossible to write such a work in 1859, because the whole concept of species was too vague at that time. This has changed in the intervening years, as we have tried to show in the preceding chapter, and we are now in a much better position to examine the role the species plays in evolution and how the origin of discontinuities is correlated with evolutionary changes as a whole. The authors who have devoted themselves to a study of these questions during recent decades have come to the conclusion that the problem of the "origin of species" is one of the cardinal problems of the field of evolution. (147)

Though Mayr's historical reconstruction of Darwin's intellectual grasp of species and the nature of their origin was wildly inaccurate, Darwin does deserve Mayr's critical slam, since his characterization of Darwin's views is accurate regarding the *Origin of Species*. In 1942, Mayr had no way of knowing about Darwin's Transmutation Notebooks of the late 1830s, or that Darwin then would have agreed with Mayr. All anyone in 1942 knew about Darwin was *Origin*, plus the hints found in *Voyage of the Beagle* (1839), and perhaps in some of his letters.

And no one in 1942 had a clue that the fundamental problem of evolutionary biology, as addressed by Jean-Baptiste Lamarck and Giambattista Brocchi and transmitted to Darwin as a student in the 1820s by Robert Jameson and Robert Grant, was *precisely* that: the natural causal explanation of the origin of the species of the modern fauna from extinct predecessors—though *that* pre-Darwinian material had in fact been published.

So species have to be seen as "real" (whether "units," "entities," and the like) in order to write books about the way they originate. Yet there is something more to this conceptual advance than a mere recapturing of what the young Darwin knew, saw, and thought. Darwin did indeed realize that species are reproductively coherent entities, in the end succumbing to the view that competition between closely related "races," subspecies, and the like was the major factor creating the gaps between closely related species. By then he had abandoned his adaptive divergence model in isolation raised in 1837.

But the advent of modern genetics proffers an advanced, far more sophisticated insight into the actual nature of species—what species in fact actually *are*. In the final chapter of *Genetics and the Origin of Species* ("Species as Natural Units") Dobzhansky (1937) returns to J. P. Lotsy and his concept of "syngameons," which Dobzhansky thinks essentially coincide with what he calls "species" throughout his book. Species are collectivities of organisms that can, or at least potentially could, interbreed, *which means that they share a genetic heritage*. Species are lineages (individual species at any one time, with their plexus of past and future interbreeding organisms) that are now—and forever—separated from other entities. If such was implicit in the brilliant early writings of Darwin, so be it. But here, with Dobzhansky and his immediate predecessors (especially Lotsy) whom he cites, we now have a better grasp of what species actually *are*.

We have not been simply rediscovering some bare essential truths about the evolutionary process that Darwin grasped, initially appreciated, and in the end basically abandoned. Regaining the aperçus of Darwin and his predecessors has triggered more than a simple revival of things long forgotten. It has provided a springboard for a much deeper and, I think, accurate description of what biological nature is, how it is organized, and how it evolves than was indeed open to anyone in preceding centuries, as the final two chapters of this narrative make clear.

5

PUNCTUATED EQUILIBRIA

Speciation and Stasis in Paleontology, 1968–

With the return of the taxic view to evolutionary biology through the pioneering effort, this time not of a paleontologist, but that of a naturalist-turned-geneticist, the question then becomes: When and under what circumstances did the taxic perspective make its way back into the work of paleontologists? Theodosius Dobzhansky's (1937) early works emphasizing the genetic nature of species and the importance of isolation in the evolutionary process caught the attention of other geneticists and systematists, creating new avenues of research, and prompting a flurry of attention, even of excitement, in the new (or at least revivified) field of speciation studies. Dobzhansky started a new line of inquiry in evolutionary biological research that very much continues to this day.

What then of paleontology? The simple answer is that, with the exception of a few studies cited in the previous chapter, most paleontologists—especially invertebrate paleontologists ensconced in geology rather than biology departments—looked steadfastly away from evolutionary issues, preferring to focus on systematics and the use of fossils to correlate rocks and determine the nature of ancient environments.

The big exception to this generalization is the work of George Gaylord Simpson (figure 5.1), a vertebrate paleontologist specializing in mammals, who spent the lion's share of his career on the curatorial staff of the American Museum of Natural History. Simpson was well aware of the efforts of Dobzhansky and Ernst Mayr. Dobzhansky was working only

FIGURE 5.1 George Gaylord Simpson. (Courtesy of Library Services, American Museum of Natural History)

forty blocks north at Columbia University, while Mayr was down the hall on the fifth floor in the American Museum's Ornithology Department. Whatever the precise nature of their personal interactions might have been, they surely read one another's work, and Mayr and Simpson must have become quickly aware of the impact of Dobzhansky's work in the latter half of the 1930s. I know for a fact that all three attended the weekly lectures on fossil reptiles and amphibians given by the then-young curator Edwin H. Colbert for one year in the 1940s.

One can read Charles Darwin and his predecessors and hope through their writings, and those of others who have studied them, to understand them, if not really to get to know them. And, of course, one can read Dobzhansky, Mayr, and Simpson. But with the latter three (and many of their contemporaries), it is much better. When I read their words, I hear their voices. I knew Mayr best, meeting occasionally at conferences, and for a time serving on a committee together at the American Museum—Mayr as a trustee, me as a junior member of the curatorial staff. Mayr had long since departed the American Museum for Harvard, and by the time I had arrived there in 1964, so had George Simpson.

I never had the pleasure of having had a real discussion with Simpson. But we had some correspondence on substantive issues. He did not like my essay, written in the 1980s to serve as a new introduction to a reprint edition of *Tempo and Mode in Evolution* (1944). Were I he, I probably would also have insisted on providing my own instead of letting some junior person make a hash of my work. But we did have a polite debate on some issues, including "punctuated equilibria," which Simpson did not much like. Unlike some of my actual teachers, Simpson gave the strong impression that not much had happened, or would likely happen, of a positive nature in the fields of endeavor that he had labored in for so long. It wasn't just punctuated equilibria; Simpson had nothing good to say about the new theory of the earth, plate tectonics.

I had even less direct contact with Theodosius Dobzhansky, widely reputed to have been the most accessible, encouraging, and friendly of the three of these titans. But I did hear him give a lecture one day on the Columbia campus, invited by Marvin Harris, then chair of anthropology. I was entranced with his rich, rumbling voice, and his to my ears deeply inflected Russian-accented English. I'll never forget the way he growled the name *Drosophila melanogaster*.

Putting a living, breathing person to their words means everything. I grew up, as a student in the 1960s, very much wanting to follow their example as explorers of the evolutionary process. I remember gazing into a mirror while shaving one day, and thinking precisely that: I wanted to be like Simpson, Mayr, and Dobzhansky.

Back to species, now as seen by evolutionary-minded paleontologists. Simpson never did think much of the nature and putative significance of the "biological species concept," or indeed of the importance of the speciation process as detailed by Dobzhansky, Mayr, and others. His first major contribution to evolutionary thought was *Tempo and Mode in Evolution* (not *Paleontology and the Origin of Species*, which would have mimicked first Dobzhansky's [1937] and then Mayr's [1942] titles preceding his). Simpson was in Patton's army and alludes to the difficulties of producing a book during wartime. Indeed, there are two somewhat incommensurate sets of nomenclature for the patterns of evolution that Simpson delineates in his text.

In *Tempo and Mode in Evolution*, Simpson devotes little space to the wrangle over a genetically based definition of what a species might be. But through the course of his career, he made it clear that species are in fact genetically internally coherent, adding the further point that they constitute lineages persisting through non-trivial amounts of time. For example, Simpson defines what he called "evolutionary species," in *Principles of Animal Taxonomy* (1961), as "a lineage (an ancestral-descendant sequence of populations) evolving separately from others and with its own evolutionary role and tendencies" (153).

Such lineages are indeed evolving separately from other such lineages, so are of course reproductively isolated from other such groups. Simpson insisted on a temporal dimension to species concepts, while implicitly retaining the sense of their internal genetic cohesion, as well as their genetic separateness from other "closely allied" lineages. In more recent years, both biologist E. O. Wiley (1978) and paleontologist William Miller III (2006) have offered similar concepts of species.

Like geneticist Dobzhansky, though, paleontologist Simpson throughout his career maintained that species lineages continue to evolve gradually through geological time, and that new species arise slowly and inexorably from their predecessors by the transformation of adaptive morphological features via natural selection.

Tempo and Mode in Evolution is a hotbed of ideas—some fresher than others. Connecting Simpson's narrative vision with those of his close colleagues Dobzhansky and Mayr, Simpson (1944) at one point writes in the confines of a single sentence that "the evolutionary significance of the distinction between genetically non-isolated and isolated population units is great and has been sufficiently stressed" (199). This claim is an exaggeration, as "isolation" is never really discussed in his text, and the term does not even appear in the index.

By far the dominant theme in his book, Simpson sees evolution as fundamentally a process of adaptation within and among "adaptive zones" It's a term introduced late in his narrative, seemingly taking the place of Sewall Wrightian "adaptive peaks," which he utilized earlier. Simpson's characterization of "speciation," couched along with the other "modes

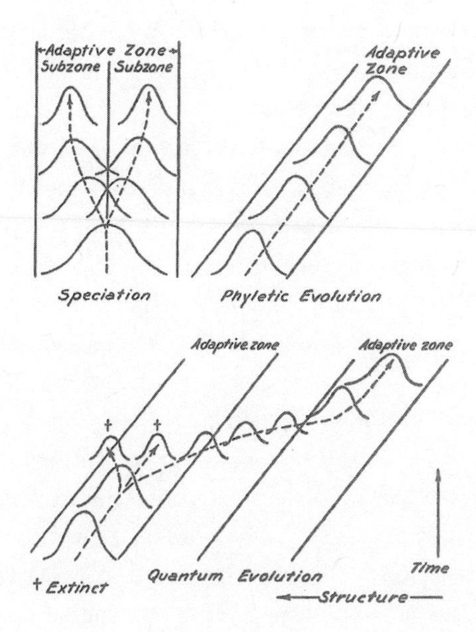

FIGURE 5.2 Simpson's diagram of "characteristic examples" of his three modes of evolution. (From *Tempo and Mode in Evolution* [1944:198, fig. 31])

of evolution" ("phyletic evolution" and "quantum evolution") as evolutionary change within and among "adaptive zones" (figure 5.2), is more about relatively trivial amounts of adaptive change within lineages (as in subdividing an adaptive zone into adaptive subzones). Though such minor amounts of adaptive change might sometime lead to the appearance of new species, they need not.

Instead, new species most commonly arise as arbitrarily defined and recognized subsets of continuously evolving lineages. This is Simpson's "phyletic evolution" within an adaptive zone. Earlier in the manuscript, arguing with Richard Goldschmidt's ideas on "macromutations" and "hopeful monsters," Simpson (1944) writes in a footnote: "To those who have done much work on good phyletic series of fossils it will hardly seem necessary to make such an obvious statement as that good species and genera arise in this gradual way, whether or not they always do" (58n.9).

Here we see Simpson's general assertion of the truth of gradualism as propounded by Darwin and maintained as well by Dobzhansky and Mayr when considering patterns of evolutionary change *after* the origin of new species through a geographic isolation event. Simpson, like Darwin and in contradistinction to his colleagues up the street and down the hallway, never really integrated a view of the importance of geographic isolation as a necessary, if not sufficient, precursor to the origin of taxa at any level: species, to be sure, but even the "higher categories."

Later, in the updated and expanded edition of *Tempo and Mode* (1944), *The Major Features of Evolution* (1953), Simpson is a bit more candid in admitting: "In spite of these examples, it remains true, as every paleontologist knows, that *most* [italics in original] new species, genera and families and that nearly all new categories above the level of families appear in the record suddenly and are not led up to by known, gradual, completely continuous transitional sequences" (360).

Throughout the remainder of his career, Simpson opposed a biological explanation for the sudden appearance of new species, genera, and families. Absence of good supporting evidence was to be attributed primarily to the vagaries of fossilization and the gappiness of the preservation of continuous time in the rock record. He held these views despite a discussion of species and speciation slightly more in harmony with the perspectives of colleagues in *The Major Features of Evolution*.

But Simpson *did* suggest an evolutionary theoretical explanation for the sudden origin of full-blown higher taxa, which in 1944 and 1953 he was still characterizing as higher "categories." That explanation was Simpson's (1944) original characterization of "quantum evolution," his third distinct "mode" of evolution.

Quantum evolution was devised originally to explain the abrupt, evidently "sudden" origins of taxa. Rereading Simpson's (1944) text, I was surprised to find that quantum evolution could be applied to the sudden origins of taxa at any level—*including species*:

> Perhaps the most important outcome of this investigation, but also the most controversial and hypothetical, is the attempted establishment of the existence and characteristics of quantum evolution. A "quantum," in the sense more general than but including the definition of physics, is a prescribed or sufficient quantity. The term is applicable in situations in which subthreshold actions produce reactions of definite

> (not necessarily equal) magnitude (this magnitude being strictly the quantum involved). For the sake of brevity, the term "quantum evolution" is here applied to the relatively rapid shift of a biotic population in disequilibrium to an equilibrium distinctly unlike an ancestral condition. Such a sequence can occur on a relatively small scale in any sort of population, and in any part of the complex evolutionary process. In may be involved in speciation or phyletic evolution, and it has been mentioned that certain patterns within those modes intergrade with quantum evolution.
>
> It is the present thesis that quantum evolution also occurs on a larger scale and in clear distinction from any usual phase of speciation and phyletic evolution. Attention will here be focused on this more distinctive and more important sort of quantum evolution. Like the other modes, it can give rise to taxonomic groups of any size, and the sequences involved can be (subjectively) divided into morphologic units of any desired scope, from subspecies up. It is, however, believed to be the dominant and most essential process in the origin of taxonomic units of relatively high rank, such as families, orders, and classes. It is believed to include circumstances that explain the mystery that hovers over the origins of such major groups. (206)

In other words, the apparent gaps between species and genera in the fossil record for the most part reflect problems inherent in the formation of the fossil record. But the abrupt appearance of taxa such as bats and whales, for the most part without any evidence of connecting intermediates from obvious ancestors, reflects a real phenomenon that should be addressed in evolutionary theory.

I admit to a degree of surprise in reading Simpson's statement that even new species can arise in this manner. Along with the rest of his readership, I have long tended to think of quantum evolution as a theory specially devised to explain the abrupt appearances of full-blown higher taxa, with their hallmark adaptive characteristics (wings of bats and flippers of whales) already fully formed. And that is, in fact, Simpson's intention, to take the abrupt appearance of higher taxa as a real evolutionary pattern, a true "signal," begging for explanation. Still, it is intriguing that Simpson says here that new species can arise from this process that underlies the rapid production of new taxa in general.

Simpson's version of quantum evolution as developed originally in 1944 involved a loss of adaptation of the ancestral "population unit" (the "inadaptive phase" caused by random inadaptive mutations), followed by a "preadaptive phase," which, by chance, allows selection to take over and drive the population into a new adaptive zone. Extinction is always looming, so it happens in a hurry if it happens at all. But when the new "adaptive phase" is reached, stability is achieved and a new taxon has evolved into existence.

Simpson's book was criticized in a review by Sewall Wright (1945), and in Simpson's (1953) revision and update, he toned down his genetically based theory of quantum evolution, preferring instead to characterize it simply as an "extreme and limiting case" of phyletic evolution.

But Simpson stuck to his guns, his original point that "quantum evolution" was designed to explain in theoretical, mechanistic terms a very real evolutionary pattern. This lesson was not lost on those of us in the succeeding generation who came along and further developed the taxic point of view. This we did largely by integrating the insights of Dobzhansky and Mayr on the importance of geographic isolation in the evolutionary process of speciation with Simpson's notion that paleontological patterns can be taken seriously, even literally, and not written off as mere artifacts of poor preservation of time and life in the fossil record. What was new with us was the rediscovery of the existence of stasis—the overwhelming tendency of species not to change in any one concerted directional fashion after they arise.

But before getting to this next phase of the story, there is one further, important point about George Simpson's writing to mention. In 1963, Simpson wrote a paper in which he drew a sharp distinction between the terms "category" and "taxa." Categories, in the Linnaean hierarchy, are nested grouping names: species are within genera, genera within families, and so forth. "Taxa," in contrast, are actual groups of organisms with proper names. *Homo sapiens*, for example, is the name of our species (and incidentally, our genus name is included by tradition as well). Species are levels in a classification; it is actual species—taxa—that evolve. Hence the theme of this book: the "taxic" view in evolutionary biology theory revolves around the notion that species are real and have births (through what we now call "speciation"), histories, and deaths (extinctions) that, from time to time, may give rise to descendant species. Simpson's contribution in this regard is more than semantic; it really cuts to the chase of a core ontological proposition of what actually exists in nature.

Finally, of course there were other paleontologists, including some of my mentors, such as Norman Newell, Bobb Schaeffer, and Edwin H. Colbert at the American Museum, as well as E. C. Olson and H. J. MacGillavry, who made important evolutionary contributions from a paleontological perspective in the 1940s, 1950s, and 1960s. It is especially noteworthy that some of the earliest contributors to the newly minted journal *Evolution* included some of these paleontologists.

But the next big thing that was to come along was punctuated equilibria. And here is where my narrative becomes sharply personal. For I can do no better in tracing the reemergence of the taxic perspective in paleontology than to look back on my own history, working alone and with colleagues, relying on memories of course (and chary of anthropologists' time-honored caution about accepting the personal narratives of their "informants" at face value). I will rely as well on manuscripts (both published and unpublished), just as I have done for my predecessors, and will continue to do for those who have enriched the taxic perspective in the years following the initial publications on punctuated equilibria. Fortunately, there is a rich paper trail documenting the early days of the development of what came to be known as "punctuated equilibria."

THE GENESIS OF PUNCTUATED EQUILIBRIA

In its simplest form, the concept of "punctuated equilibria" is the integration of geographic speciation theory with the empirical observation of stasis—the common, indeed predominant, pattern of stability of species that survive for long periods of time once they appear, usually abruptly, in the fossil record. To be clear, the term is alternatively often rendered as "punctuated equilibrium," a neologism apparently coined by my colleague Stephen Jay Gould—ironically, the very one who dreamed up the name "punctuated equilibria" in the first place. I have always stuck to the original terminology.

But there was nearly a decade of run-up to the publication of our paper in 1972, in which the terms "punctuated equilibria," "stasis," and "phyletic gradualism" first appeared in print. I took the introductory geology sequence in my sophomore year, collecting my first fossils at an outcrop in eastern Pennsylvania in the spring of 1963. Among them was the Middle Devonian trilobite species *Phacops rana*, which was later to become the focus of my doctoral research, and the ur-example of the pattern ultimately named "punctuated equilibria."

My interest in fossils and evolution intensified while in Brazil looking at the Pleistocene fossils in the sandstone "reef" at the Arembepe harbor. When I got back I began to take advanced courses in paleontology in the fall of 1963. I immediately encountered a significant influx of entering graduate students who had enrolled in Columbia's Geology Department to pursue their doctorates in paleontology and/or stratigraphy. Two others came at the same time to the Zoology Department to study vertebrate paleontology. Of these seven or eight new graduate students, one, H. B. Rollins, was to prove especially significant, as he taught me all the practical field and laboratory techniques in paleontology; in addition, Bud's insights into ecology and evolution have been invaluable over the years. Bud had already done significant research, coming with a fresh Master's degree in geology (paleontology) from the University of Wisconsin.

But there was another entering student who was to have an immediate, and profound, impact on me: Stephen Jay Gould.

From the first few days after I met Steve, I immediately saw how unremittingly intellectually engaged he was; how determined and committed he was to learning just about everything; and how ambitious and what a pure workaholic he was. I started thinking that I had never met anyone as smart as he who worked so hard—and that, I think, remains true to this day. We would often commute between the Columbia campus and the American Museum of Natural History (where most paleontology courses were taught and seminars were held), and he was never at a loss for words, usually in the form of a continuous, spontaneous speech about some intriguing, evolution-imbued piece of research that he had just read about. It was that knack of acquiring all sorts of information and his ability to see the implications and weave it into a narrative that led me to recommend him, some years down the road, to replace the outgoing columnist for *Natural History* magazine.

Steve came from Antioch College, where he had encountered boxes of Bermudan fossil land snails of the genus *Poecilozonites*, deciding one day to analyze them for his doctoral dissertation research. The sediments (a mixture of lithified windblown sands and red soils) represented the last 300,000 years of the Bermudan Pleistocene. The fossils turned out to belong to the species lineage *Poecilozonites zonatus*—a species still very much alive in Bermuda today. Because these were pulmonate land snails, they had evolved in isolation on this one island: Bermuda. Steve had a perfect controlled evolutionary experiment (when he published his study in 1969 he aptly dubbed it an "evolutionary microcosm"). He had an endemic lineage with a fossil record leading up to the present day.

Though neither of us knew it at the time, Steve had a wonderful, young Darwin-like study of the origin of a modern species. So it is tempting to point to Steve's doctoral research as an early example (in the modern era) of taxic thinking in paleontological evolutionary theory.

To a degree, that is true (though I ended up recasting his doctoral research in specifically taxic terms when we published the punctuated equilibria paper in 1972). But Steve was first and foremost a morphologist, fascinated by the relationship between development and evolution, or "growth and form," as the British zoologist D'Arcy Thompson had put it. One of Steve's teachers—a true mentor—at Antioch had been the paleontologist John White, who introduced Steve to the allometric growth equation $y = bx^k$. Steve's first scientific publication (in 1965 with White as senior author) was on the biological meaning of b in that equation.

As a symbol of his intellectual curiosity and ambition, Steve took a year off from his doctoral research to write a review paper on allometric growth and evolution for the journal *Biological Reviews*. Many of his fellow students were appalled that he was taking this detour. Steve saw it as an opportunity and even a prestigious honor to be entrusted with the task of writing a review paper on this, his favorite subject matter in evolutionary biology. When I asked him about delaying his dissertation research while writing this review paper, he spoke of the opportunity the invitation afforded and asked rhetorically why anyone should wait until he was sixty to write general, theoretical review papers. Ironically, he was sixty when he died in 2003.

Steve's first book was *Ontogeny and Phylogeny* (1977)—and Steve continued to write, off and on, on development and evolution throughout the remainder of his career. Indeed, one of his best ideas was the suggestion that relatively small-scale mutations in the genetic regulatory apparatus could quickly lead to large-scale changes in adult morphology—one of the central tenets of today's "evo-devo" (developmental evolutionary biology).

And it bears emphasizing that Steve (and, to a degree, me, although it was largely guilt through association) was commonly considered to be anti-adaptationist, and perhaps even a skeptic about the nature, role, and power of natural selection itself. Punctuated equilibria, for one thing, was at least initially considered by many biologists and paleontologists as anti-Darwinian heresy. And while it is true that Steve was always intrigued by newer ideas, including possible alternatives to a strictly selectionist account of the process of adaptation, he was always at heart (as were we all) a neo-Darwinian, very much including natural selection as the very core of understanding the evolutionary process of adaptation. I'll have more to say on this as this narrative unfolds; suffice it to say here that Steve, in his heart of hearts, was always, fundamentally, a morphologist.

In any case, when Steve and the rest of the eager, bright new graduate students enrolled at Columbia in the fall of 1963, I was thrilled to be accepted by them though I was a mere undergraduate (junior), young (I was just twenty) aspiring paleontologist. They let me tag along on a marvelous fossil-collecting trip to the Miocene cliffs of the western side of Chesapeake Bay in Maryland—memorable for revealing another knack that Steve had: of the thousands of fossil shells of the snail *Turritella plebeia* that were everywhere around us, Steve found one (perhaps two) specimens that were showing abnormal growth (they were "uncoiling," meaning that the whorls that normally just wrap around in touch with the previous volution, were beginning to show open space between the whorls). This led Steve to study true, full-blown uncoiling in an unrelated group of snails of the genus *Vermicularia*—his love of morphological growth patterns once again coming to the fore.

"February 1965"

But the most signal event was the founding of a wholly student-run seminar in evolution. I don't have the entire syllabus. But, miraculously, I do have the handwritten essay (on now-crumbling yellowing ruled paper), as well as the 3 × 5 file cards that formed the basis of my own presentation in February 1965. It's by no means as significant as Charles Darwin's first evolutionary essay, "February 1835." Yet I'm glad I saved this, as it is, in its own exceedingly modest way, my own first "pencil sketch" on evolution. The topic was the synthetic theory of evolution with reference to higher categories. (See "Notes" and the bibliography for the DOI digital address for this and the following documents.) For the most part, the contents are utterly conventional, following closely especially George Gaylord Simpson's (1953) and Ernst Mayr's (1963) books. I remember having bought Mayr's book in the winter of 1963/1964 in a Greenwich Village bookstore. But before I rediscovered my bibliography for this seminar, I do not recall having bought or read Mayr's earlier *Systematics and the Origin of Species* (1942). I also recall reading (and struggling with it, a bit overawed as I had earlier been first trying to read Darwin's *Origin*) Simpson's *Major Features of Evolution* (1953), though Simpson's *Tempo and Mode in Evolution* (1944) did not appear on my reading list. Nor was there mention of Dobzhansky, but Sewall Wright was included.

On the whole, my discussion on the issues surrounding the evolution of "higher categories" reads more like a book report than an original essay. Included with the seven-page handwritten manuscript are extensive notes on relevant sections of Simpson's and Mayr's books. But it is interesting to see what had caught my eye, from the standpoint of what was later to come.

I was especially entranced with how Simpson's ideas on quantum evolution seemed to fit the facts of the fossil record. I liked the notion that rapid evolution in small populations, often involving areas ecologically different from the ancestral "adaptive zone," fits the apparently real pattern of abrupt appearance of new taxa rather well. It also explains why, for the most part, evolutionary transitions between higher taxa are not readily encountered in the fossil record (though I also say that "none would seriously deny, I think, that reptiles arose from amphibians through a gradual process rather than a saltation"). But "fast phylogenetic change, small pops., & localities are items which are particularly important for us to take note of, when we come to examine the record"—a postulate that seems to dominate the pages of this rather redundant brief discussion.

I also maintain, in apparent homage to Mayr, that "the fundamental unit of evolution is the species," but (presumably following Simpson) "speciation in its strictest sense, the formation of allopatric species from a parent species (cladogenesis) is not the model, but rather a phylogenetic evolution (or anagenesis) is probably the more common case is important, but not an essential point here" (1965:3). So at least I was aware of allopatric speciation, even though I denied its central role in the evolutionary process. And I saw it (as Mayr had depicted it in 1942) as the origin of two daughter species from an ancestral species through geographic isolation. This is a bit different from the picture that later emerged in punctuated equilibria, in which a new species arises from an ancestral species through geographic isolation—and, in most cases, both parent and daughter species survive, usually unchanged. Here, in "February 1965," there is no glimmer of stasis, which came along later as these thoughts were further developed, based in large measure on empirical research on species in the fossil record.

In stressing Simpson's essentially non-taxic vision of the evolutionary process over Mayr's emphasis on geographic speciation as the essential ingredient in the origin of new species, these early thoughts of mine are hardly "taxic." Yet the idea is there; patterns of abrupt appearance of new "categories" as commonly seen in the fossil record reflect the way the evolutionary process works. The "categories" are actually higher taxa, yet rooted at base in some form or other of the evolution of new species. I preferred it to the simple explanation favored by the later, published Darwin, and Jean-Baptiste Lamarck before him, that the preservation of animal and plant remains—indeed, of time itself—is so incomplete that the patterns are in fact not real at all. These views formed the very core of what was soon to emerge as punctuated equilibria. That the fossil record presents more signal than noise in terms of evolutionary pattern has remained the core postulate of my thinking ever since. Indeed, it is the very basis, the core supposition, of the entire field of "paleobiology."

"April 1968"

In the summer of 1964, I accepted an NSF-sponsored undergraduate research opportunity, working with Roger L. Batten, a gastropod (snail) specialist working as a junior colleague with Norman D. Newell on the curatorial staff of what was then known as the Department of Fossil Invertebrates at the American Museum of Natural History. After an abortive attempt to raise living snails in hopes of determining the chemical conditions of sea water that might affect the development of aspects of the "ornamentation" (spines, ridges, and the like) on the shell exterior, I abandoned the project. I did not have the remotest equivalent of a "green thumb" for the care and feeding of live marine animals.

So Batten instead offered me a completely different project, one more closely fitting my emerging passion to study evolution using large samples of marine fossil invertebrates. The study involved Upper Paleozoic (Pennsylvanian) gastropods from Texas and Oklahoma. Specifically, Roger knew that two species of snail, belonging to two completely different families, had "converged" on one another, so that most paleontologists, unless they were specialists looking closely, wouldn't know the difference between them. He also introduced me to simple univariate and bivariate statistical analysis, an interest that was quickly to expand into various multivariate analytic techniques, run on Columbia's mid-1960s newly installed IBM 7090/7094 system. I kept working at the AMNH on this project for several years (I have kept a desk at the museum continuously for forty-eight years as I write these words), trying to analyze the evolutionary changes within the two separate species-lineages, and to see under what circumstances which species converged on the other. I finally published the results—titled, accurately if unsurprisingly, "Convergence Between Two Pennsylvanian Gastropod Species: A Multivariate Mathematical Approach" (1968b). It was my first published scientific paper.

But what I had forgotten about was my little, never-published essay "Some Aspects of Species-Level Evolution in Paleontology" (1968a). It is a photocopy of a typescript, with no vestige of the original or its handwritten predecessor(s). And it is dated "April 1968." It is the next link in understanding the genesis of punctuated equilibria: it turns out that I had used my work on the Pennsylvanian snails to explore the application of geography and isolation (allopatric speciation) as evolutionary factors to patterns of variation and evolution in space, in addition to time, in the fossil record.

After saying in the introductory paragraphs that paleontologists were increasingly looking at evolutionary patterns at the species level, I dipped a rhetorical toe in the water, essaying an alternative to a strictly phyletic transformational vision of species-level evolutionary patterns in the fossil record:

> What these and other similar studies have in common is a stress on a phyletic-model of species differentiation; paleontologists tend to view the origin of new species as gradual, progressive change through time. This is fine as far as it goes, since a species existing at any one point in time has three possible fates:
>
> 1. Extinction
> 2. Persistence as is
> 3. Change into something a systematist recognizes as sufficiently distinct to warrant calling a new species.
>
> Of these, extinction is the commonest fate, and persistence without change the rarest, approaching zero if a large enough time span is considered. (1968a:1–2)

So here is at least the beginning of my break with a purely gradualistic model of species evolution: the ultimate fate of any species is of course extinction. But stasis—long-term species stability—is basically denied in this passage. Since the early days of full-blown punctuated equilibria, I have been convinced that if a species is to persist at all, it will do so for the most part unchanged (albeit variably oscillating within and among populations, in genetically based morphological terms, around some mean). And I have long since come to reject as based on no data and also theoretically implausible the Lamarck/Darwin claim that entire, far-flung species can, and regularly do, become slightly modified progressively though time: "Change into something a systematist recognizes as sufficiently distinct to warrant calling a new species." There has never been any credible evidence for the latter—and in April 1978, I was merely mouthing tradition.

But then I raise another possibility: the importance of geographical isolation. The passage, I believe, is worth quoting in full as it signals the arrival of Theodosius Dobzhansky and Ernst Mayr's concept allopatric speciation into paleontological evolutionary thinking, another necessary forerunner for punctuated equilibria to emerge. The passage just quoted continues:

> But the enormous amount of literature amassed by neontologists in the last 30 years or so has focused on a different model; if the paleontologist has emphasized the time component of a species' history, the neontologist has perforce emphasized the spatial distribution of species at a single time transect—the present.
>
> To summarize all this literature in a few words, neontologists feel that for a new species to originate and take its place [! compare the rhetoric of 1820s Edinburgh!] beside its ancestor, a population of the parent species must spend some time in geographic isolation, enough time to allow some barrier called an isolating mechanism to develop which would prevent interbreeding with, hence resorption into, the parent species, should the geographic isolation break down. Apart from some recent suggestions that geographic isolation may not be as crucial in all cases as previously thought [alluding to the ever-present and still shaky possibility of sympatric speciation] this principle seems firmly established in evolutionary theory.
>
> It is already clear from my brief characterization of the past work in evolutionary paleontology that this so-called allopatric model has not been extensively tested and applied to fossil organisms. But the development of an evolutionary theory stressing the role of populations of varying organisms coupled with the advent of electronic computers and a sophistication of statistical models have paved the way for paleontologists—particularly those who study marine invertebrates—to

> analyze the shifting patterns of variation in populations of organisms in a spatial as well as a temporal sense. In other words, the techniques and interests of paleontologists are now suited to a detailed examination of the fossil record to test the general applicability of the allopatric model of speciation and conceivably to expand this model by documenting what actually does happen to different races, say, or a species through time. (1968a:2–3)

Stirring words—and the allusion in the final line to looking at what happens to, for example, "races," as well as entire species through time after speciation is a harbinger of actually looking at patterns of post-geographical speciation patterns of change—and, as it turns out—stability.

But here we have the next part of the puzzle. If "February 1965" echoes Simpson's (1944) call to take patterns of abrupt origin of taxa in the fossil record seriously, asking if they reflect underlying evolutionary processes rather than (simply) just bad data occasioned by the vagaries of the preservation of both time and organic remains, "April 1968" explicitly adds the "allopatric model" as a theoretical evolutionary process that needs to be taken seriously by everyone—including paleontologists.

I wish I could report that the rest of the essay, with the data and discussion on Pennsylvanian gastropods, lived up to the promise of the introduction, where geography is added to the mix of causality in patterns to be seen in the data of paleontology. It doesn't, really—though I gave it a good shot. The study on convergence between two species of Pennsylvanian gastropods was not constructed with the idea of explicitly looking at the relative importance of, and interplay between, time and space that would have been required to fill the bill. But my next study was set up precisely to do that—a study of the Devonian trilobite lineage *Phacops rana*—and was also designed for a very different, additional reason: it was my doctoral dissertation.

Phacops rana: *Evolution in Time and in Space, 1966–1969*

The two most difficult courses I took in graduate school (and perhaps in my entire educational career) were "Fossil Fishes" with the great paleoichthyologist, later colleague and friend, Bobb Schaeffer; and the two semesters of "North American Stratigraphy" taught by Marshall Kay. Until plate tectonics had come along in the 1960s, Kay was the acknowledged author of the reigning (if unnamed) theory of the development of the earth's crust. After plate tectonics began to appear in several different quarters, largely developed by geophysicists, it was Kay who organized the "Gander Conference" in Newfoundland, and famously asked: Now that we have a theory of a dynamically mobile earth, how do we modify the older views (that is, his own) to integrate with all this fabulous new information and theory? Kay's and Schaeffer's courses were encyclopedic in their demands for mastery of course content, but they both transcended mere rote memorization of "facts." They in fact demanded a grasp of the integrating, including theoretical, work that underlay the generation and, most importantly, the interpretation and meaning, of that otherwise stupendous mountain of facts that each demanded mastery of.

Kay expected his students to know the geography of North America like the backs of our hands. He expected us to produce, on his exams, a vertical section of rock sequences anywhere in North America: from the flat-lying "layer-cake" stratigraphy of Ames, Iowa, for example, to the vastly more complex geological terranes of Winnemucca, Nevada, or the central volcanic belt of Newfoundland. Anywhere. He also required us to be able to produce, from memory, a cross-section across a geographic region showing thickening and thinning of rock sequences. He especially loved the complex structure of mountain ranges, usually where cross-sections would reveal the inner carbonate-dominated continent side (the "miogeosynclines") versus the outer, usually volcanic eugeosynclinal belt. Kay had already made the leap to analogize with places like Japan, with its volcanoes and deep oceanic trenches comprising the outer belt, and the shallower, sedimentary Sea of Japan between Japan and the Asian continent being the inner, miogeosynclinal belt. It is still mind-boggling to me that Darwin saw precisely this structure when he crossed back and forth over the Andes. In any case, as students we had to know our geology, but it was always in a completely geographic context.

I designed my doctoral research deliberately to document patterns of geographic and stratigraphic (that is, temporal) variation. I chose an arthropod: the Middle Devonian trilobite species lineage *Phacops rana*. In the Appalachian basin, *Phacops rana* was known to have persisted throughout the entire Hamilton Group, a now-lithified body of sediments that was thought to have been deposited over an 8- to 10-million-year interval (now thought to be closer to 6 or 7 million years). The time span centered on around 380 million years ago. Geographically, stratigraphers had long since documented the existence of thinner sequences of Middle Devonian rocks east of the Appalachian Mountains—temporal equivalents of portions of the Hamilton sequence bearing the same basic fossil taxa (over two hundred species) as are found in the Hamilton Group of New York State. The Middle Devonian rocks in question run along the Appalachians and the basin immediately to the west from central New York down through Virginia, and extend westward through southern Ontario, Ohio, and Michigan, and all the way to Iowa.

Arthropods are complex organisms in general, so it should theoretically be a simple matter to document evolutionary change, given their disparate body parts and intricate sculpturing of fine anatomic detail. And trilobites readily preserve. Like arthropods generally, trilobites grew by molting, so it is possible in most circumstances to be able to study their ontogenies (their patterns of growth as they continue to molt and to grow larger). My friend, fellow graduate student Bud Rollins, grew up in the vicinity of Hamilton, New York, graduating from Colgate University there. He was already well conversant with the region's Middle Devonian rocks and fossils. It was Bud who showed me just how common and easily collected *Phacops rana* is, so statistically significant sample sizes looked, in prospect at least, to pose no further problem.

So *Phacops rana* it was to be. I spent the better part of the summers of 1967 and 1968 cruising up and down the Appalachians, and visiting all the known Middle Devonian outcrops of Ontario, Ohio, and Michigan, accompanied variously by my wife, Michelle, and my younger brother, Rick—both of whom were somewhat more adept at spotting good specimens than I. For the most part, we obtained sample sizes of more than twenty usable specimens from each locality/geological horizon. Most of them were isolated heads (cephala), tails (pygidia), and portions of the eleven free-segmented middle region (thorax). Complete specimens were more difficult to find, but find them we did.

Back at the lab at the American Museum of Natural History, I learned how to free the specimens from the covering matrix, using primarily dental instruments. When I needed help on particularly intractable, but intriguingly promising, specimens, I enlisted the aid of Frank Lombardi—our ace fossil preparator.

But where I needed help the most was actually making sense of these attractive, complex fossil animals. At first I was almost completely lost. I learned that specimens look different depending on whether or not the shell material was intact or not: internal molds—the more or less accurate impression of the inside of the body—revealed little of the fine structure of the ornamental bumps and tubercles that covered the animal's external surface. On the other hand, they did clearly show the sites of major muscle attachment ("apodemes") and sometimes, as I finally grew to understand, details of the arrangement of muscles inside the head: muscles that supported the stomach, providing a window into the feeding behavior of these little beasts, as well as clues to the deeper evolutionary relationships among the major subgroups of phacopid trilobites. Indeed, investigation of the musculature provided an entire separate chapter in my thesis, and a later separate publication.

But what about evolutionary patterns in time and space—after all, the initial and still central raison d'être of the entire protracted piece of work I had jumped into. I was increasingly alarmed, that first summer, when we left New York State for southern Ontario, Ohio, and Michigan. The Paleozoic limestones and limey shales of the North American eastern mid-continent contain some of the most exquisitely preserved fossils to be found anywhere, and we had no problem collecting large, gorgeous specimens of *Phacops*. But the farther we went, and the more we collected, the more alarmed I became, because everything looked exactly alike. All the phacopid trilobites, regardless if they were preserved in limestones or sandy siltstones (bespeaking different environments); as intact specimens or internal molds; if they were from the bottom of the pile of sediments (hence older), or some 4 or 5 million years younger; and if they were from Michigan instead of central New York: they *all looked the same to me*.

And that was terrifying—as this was my Ph.D. dissertation, and I was expected to turn up with some positive results. I was expecting through time to find some gradual change as my samples of *Phacops rana* got younger and younger: after all, that's what everyone was expected to find if they looked at the fossil record hard enough. That's, after all, what Darwin said we should find. And, insofar as I then knew, nearly everyone, including Dobzhansky, Mayr, and perhaps especially paleontologist George Gaylord Simpson, agreed with Darwin that slow, steady, gradual change was indeed the underlying pattern to be revealed during the bulk of the history of any given species. (True, there were saltationist naysayers—like Richard Goldschmidt and Otto Schindewolf—but they were widely derided in the American mainstream of evolutionary thinking.) As my first little paper exploring these topics for that seminar with the graduate students made clear, I thought that was exactly right: that given the passage of millions of years of time, and given the ready availability of excellent specimens of a complexly anatomically configured animal such as *Phacops*, that passage of time should have shown obvious morphological differences between start and finish (those 6 million years), regardless what patterns of variation geographically and in different environments might also show.

I seemed headed to failure—the nadir coming in a launderette in Alpena, Michigan, when I pulled a perfect, rolled up specimen of a *Phacops* out of my jeans pocket, and, with a sinking heart, concluded that it was identical with everything I had already seen.

There were, as it turns out, two problems here. One indeed was my inexperience working on trilobites: I had no mentor who was schooled in them, and my previous work had all been on fossil snails, where I had had plenty of guidance. It turned out that, back in the lab, with the aid of a good quality low magnification stereoscopic microscope, I did start to see differences, including the embarrassing fact that the launderette specimen from the Alpena limestone was actually a member of a distinct species: *Phacops iowensis*. But nearly the only way you could tell these two species apart was from differences on the details of the anatomy of their eyes, for which you needed good light and a decent microscope.

The hundreds, often thousands, of lenses in the eyes of most insects, crustaceans, and other arthropods are tiny. They only show up under high magnification, and are best seen, actually, when using a scanning electron microscope. The same is true for most trilobites. But not phacopid trilobites, where the lenses are many times larger, and actually covered with their own "cornea." There are at most only a few hundred lenses in the eyes of the largest-eyed species of the extended group of phacopid trilobites. And the two species of *Phacops* I was collecting actually seldom have more than a hundred lenses per eye.

It turns out that Euan N. K. Clarkson (1966; later to be our host in Edinburgh), a young paleontologist who also worked on *Phacops*-like trilobites, had himself focused on the eyes of his somewhat older (Silurian) acastid trilobites for his own Ph.D. work—duly published, just in time to help me out, in the journal *Palaeontology*. Clarkson had developed a simple way to quantify the number and organization of the lenses on his specimens: first of all, he saw the lenses simply arranged as vertical columns (he called them "dorso-ventral files"), rather than as a series of swirling, oblique rows distributed over the slightly bulging visual surface of the eye.

The latter arrangement is also true, and neatly described by the famous Fibonacci mathematical pattern. But, it turns out, viewing the arrangement of lenses as a Fibonacci series is less easy to use and make sense of. Think of the hexagonal compartments of a honeycomb: you can look at them as straight up and down parallel rows of compartments, or you can see them in the more hypnotic manifestation of curved rows. It turns out that Clarkson's focus on the simple parallel up-and-down columns—those "dorso-ventral (d-v) files"—held the key to making sense of the evolution of the eye, at least in my *Phacops* samples.

Clarkson would simply count, from front to back, the number of d-v files. And he would count the total number of lenses in each d-v file, which sum up to the entire complement of lenses in each eye. He would present the results for each specimen. As an example, relevant in a generic sort of way to my *Phacops rana* specimens, he would write 233 434 343 434 343 32 (17, 55): 17 columns of lenses, with a total of 55 lenses on the eye.

Nothing better to do, and totally stumped with my intransigently seeming invariant samples, I simply aped Clarkson's approach. I had meanwhile, apart from just ogling my specimens, devised a series of measurements of the various body parts (especially the head) and had begun cranking them through the high-speed computers at Columbia, using "factor analysis," as well as a multivariate version of standard bivariate statistics. No patterns, whether of gradual phyletic change through time, or even anything interesting from a geographic point of view, emerged. Sure, some "trends" (for example in eye length in relation to total head length) were there, but the hell of it was that they seemed, in the short-term long run, ultimately to reverse themselves. In terms of what I expected to see, there was nada, zilch to allow me to see some concerted and lasting evolutionary change.

Except (yes!) in the eyes. One day I found myself staring at the pencil-written 14 × 11-inch data sheets, and finally my own eye dropped onto the (17, 55) style data counts. I had one, two, or three such sheets for each place we had collected samples with enough heads to do the work. I saw that, with very few exceptions, each one had that first number (for example, "17") for the d-v files, exactly the same for each place. (And usually, if the number was discrepant for any one specimen, rechecking showed the other eye to have the "correct" number for that time and place).

That was cool, because I knew some of the samples had a regular number of 18 d-v files—while a few others had 15, and, a few, 13. Hmmm.

I literally ran to get out some of Marshall Kay's old blank maps of North America outlining the forty-eight contiguous states of the Union. I added the d-v file numbers, locality by locality, in the right geographic spots, and kept separate maps for each of the five subdivisions of geological time that were easily recognized throughout the geographic extent of the Middle Devonian Hamilton Group. (Well, checking back in on the maps, at some point, I stopped filling in the entire picture, as it was already abundantly clear what the pattern was that was staring me in the face.)

Turns out that "17" was the number during the first 2 to 3 million years in central New York, pretty much all the way down the Appalachians. But rocks of the same age in Ontario, Ohio, and Michigan—gorgeous, limey rocks with gorgeous, grinning fossil *Phacops*—all had 18 d-v files. There was a geographic disconnect between the *Phacops rana* in central New York Appalachians and the Appalachians to the south, and the *Phacops* living in the shallow lime-rich seas on the continental interior.

It was as if they were replacing one another in space.

I found no other reliable anatomical markers that allowed me to say that these two taxa, these two species defined on the basis of the dorso-ventral file counts, were distinct. But the eye counts suggested that—along with Kay's maps.

Nor were these d-v files counts simple one-off "characters." I had discovered that the very smallest specimen in any sample had fewer d-v files (and, unsurprisingly, total numbers of lenses). As growth continued, more lenses were added. And the number of d-v files also increased, up to the final, stabilized number typical for the population I was sampling—and as it turns out, for entire sub-lineages within the *Phacops rana* lineage.

By now it had become clear that there were a number of constant subsets of the *Phacops rana* lineage that could easily be recognized as separate species within the lineage (though I chickened out and called them "subspecies"). The ontogeny of the eyes of those roughly 380-million-year-old trilobites was clearly a complex, genetically controlled process. Given that these ancient Devonian trilobites left us no genes to consider, that was the best you can say.

But from a geographic angle, there were two especially intriguing populations on my data sheets. Both were from New York State. The most compelling was from the earliest Hamilton-aged fossils I collected there, in a little cow pasture "borrow pit" near Morrisville, New York. Trilobites are very rare there, and over the years we only amassed a few. Some had 18 d-v files—while the others had only 17. One of the "18" had only a single lonely lens at the top of what would have been the first d-v file.

All the rest of the *Phacops* we collected in New York and southward down the Appalachians had, consistently, 17 d-v files.

So I was looking at three patterns:

1. The 18 d-v file in the Midwest was replaced in the East by the 17 d-v file. Given that the closely similar, slightly older specimens of *Phacops* from Germany (and, I was learning, from North Africa) all had 18 d-v files, and given that many other species (including non-trilobitan mollusks and brachiopods) in the Hamilton Group had obvious

affinities with slightly older Old World (that is, European and North African) species, it seemed obvious that the original Hamilton fauna not only migrated in from the Old World Province, but that, as well, in my specific study case, 18 d-v files was the original, ur- "primitive" condition for *Phacops rana*.

That further implied that the ancestral form persisted in the Midwest, while the derived, d-v file form evolved in the East.

2. The evolution of the 17 d-v file form from the ancestral 18 d-v file form seemed to have happened at the geographic margin of the ancestral species range.
3. The 17 d-v file form, empirically, persisted for some 5 to 6 million years in the Appalachian basin.

Interesting stuff, made sensible only by looking at allopatric speciation theory and by confronting the big gorilla in the room: stasis. In my thesis, I advocated thinking about allopatric speciation as a way to understand the origin of this derivation of new species from old (there was a second event, in which the d-v files were reduced from 17 to 15, near the end of this stratigraphic interval).

I also pointed out the stability of the evolutionary derivation of the new eye configuration for millions of years. But I did not make much of this stability (more apparent in my handwritten thesis outline than in the thesis itself). After all, it was a doctoral dissertation—consecrated to the proof of being capable of designing and successfully prosecuting respectable scientific research, rather than to challenging orthodoxy. As it was, one of the faculty members on my dissertation committee objected to the picture I was painting of evolution in Paleozoic invertebrates. I managed to hold my ground without caving in.

Along with the efficacy of citing geographic speciation as an underlying generator of evolutionary change in the history of life, I also was transfixed by that stability in the complex characters that distinguished the species within the lineage, in space, as well as in time. When the 18 d-v file form became extinct when the seas withdrew from the continental interior midway through the Hamilton interval, they were eventually "replaced" by the surviving derivative, the 17 d-v file form, when the seas once again flooded the North American continental interior. The 17 d-v file form persisted, unchanged, for at least 4 to 5 million years.

But I had little to say about that stability—that non-evolution, that lack of gradualism—in my thesis. My goal was the degree. I soon changed all of that.

"The Allopatric Model and Phylogeny in Paleozoic Invertebrates," 1971

My dissertation was completed in 1969, and my degree duly granted. My doctoral dissertation defense actually went off without a hitch, teaching me the important lesson that, if you know your material inside and out, you can usually prevail over others, even if they number among your mentors and teachers, and even if they have a bone or two to pick with you. No one knows the material better than the candidate who is being poked, challenged, and grilled.

My top priority, along with assuming my new role as assistant curator in the Department of Fossil Invertebrates at the American Museum, then became the publication of my *Phacops rana* patterns of stasis, and evolution, in time and space. I circulated a draft in the spring of 1970 (Eldredge 1970a), and submitted it to the journal *Evolution* in June. It was in due course accepted, and the slightly revised version was published in March 1971 (Eldredge 1971a).

Most of the commentary I sought on the initial draft was from friends and teachers, some of whom were already familiar with my dissertation. Among them was Stephen Jay Gould, who had already assumed his role at Harvard. Surprisingly—and disappointingly—none of the comments I got back (with one exception, a lightly edited version of the manuscript), along with the comments of the peer reviewers (some of which I remember as being highly critical) seem to have survived, in sharp contrast with the thick folder of correspondence and reviews I have on hand for the punctuated equilibria paper I began writing with Steve Gould in the fall of 1970. I do remember that the editor of *Evolution*, Ralph Gordon Johnson, was sympathetic to my efforts. Ralph was an invertebrate paleontologist, and from that perspective an unusual (and, for me, lucky) choice to be the editor of a journal that had long since become the favorite place for evolution-minded geneticists to publish their work.

In retrospect, it does seem a bit strange that I would have opted to send my paper, intended primarily for the eyes of paleontologists, to *Evolution*. After all, no neontologist would object to my invocation of geographic speciation, which was by then, and still remains, the preferred "model" of thinking about speciation in general.

But I did know, in my heart of hearts, that the paper would prove to be controversial, among evolutionary biologists as well as paleontologists. I was worried, of course, about the anti-Darwinian-seeming postulate of entrenched, pervasive morphological stability—what Steve and I were soon to be calling stasis. Of course, paleontologists and evolutionary biologists alike were indeed (as expected and even feared) duly riled up over my harping on morphological stability as a common, indeed, *the* common pattern, dominating the known histories of most of the species that have ever lived on earth. That controversy is most evident in the papers that survive surrounding the publication, a year later, of our paper "Punctuated Equilibria: An Alternative to Phyletic Gradualism" (Eldredge and Gould 1972a)—as well as, of course, in the torrent of literature that quickly amassed following the publication of that second paper. Compared with all the ink that was spilled over "Punctuated Equilibria," my more modest (and less bombastic) initial foray in *Evolution* attracted very little notice. It's just that, of those who saw it before and after publication, many were not pleased.

My argument in this eleven-page paper is pretty succinct. I start with a near platitude, saying that time "mitigates, to a degree," the "disadvantages inherent in the fossilization process." But then I jump right in, claiming that, while paleontologists regularly emphasize the importance of time, "the concept of gradualism" (citing Simpson 1970) "has permeated paleontological thought to the extent that all phylogenetic change is generally conceived to occur by small increments over vast periods of time. This dominantly *phyletic* model of transformation . . . has underlain most paleontological discussions of the origin of new taxa, including species" (1971a:156 [italics in original]). I then add that the model is also the basis for most discussions of subsequent divergence after a splitting event.

I then say that, at the species level, "the only such level in the taxonomic hierarchy where a taxon can actually be said to exist in nature" (meaning, of course, that species are real entities),

> the gradualistic view is at odds with accepted views of speciation derived from studies of the modern biota. And though undoubtedly phyletic transformation has no doubt led to the appearance of descendent taxa reasonably construed as new, descendant species [a view I have long since completely abandoned], nevertheless a model that allows for one or multiple speciation events during the same period of time would be the more satisfactory for the explanation of the diversity of life since the Cambrian. On probabilistic grounds alone, we must conclude that the overwhelming majority of metazoan species that have appeared on earth's surface arose through some process of splitting. (156).

This, unbeknownst to me, is very close to one of Darwin's two visions on how a law of adaptation would act, as developed in Notebook B (1837).

I then discuss the preference among most biologists for the allopatric (geographic) model of speciation, in which a period of geographic isolation between conspecific populations is generally deemed necessary (though not in all cases wholly sufficient) for reproductive isolation to be established, and for new species to evolve. And I then "suggest that the allopatric model (geographic speciation) be substituted in the minds of paleontologists for phyletic transformism as the dominant mechanism of the origin of new species in the fossil record, and that the allopatric model, rather than gradual morphological divergence, is the more correct view of the processes underlying cases of splitting already documented by numerous workers" (1971a:156–157).

After a brief discussion of environments of the Paleozoic shallow-water epicontinental ("cratonal") seas, I then present a four-alternative description and analysis of the dominant evolutionary theories that have been routinely applied to explain the succession of *non-intergrading*, yet closely similar, and arguably ancestral/descendant species in the fossil record (figure 5.3). The arguably closely related species, in other words, replace one another as one ascends the column of rocks. The first option—a saltationist one—simply postulates a rapid transformation of one species into another in a singular evolutionary event. As I already remarked, saltationism was frowned upon (scorned, actually) by Simpson and his contemporaries, and despite some later rhetoric trying to link what was soon to become known as "punctuated equilibria"

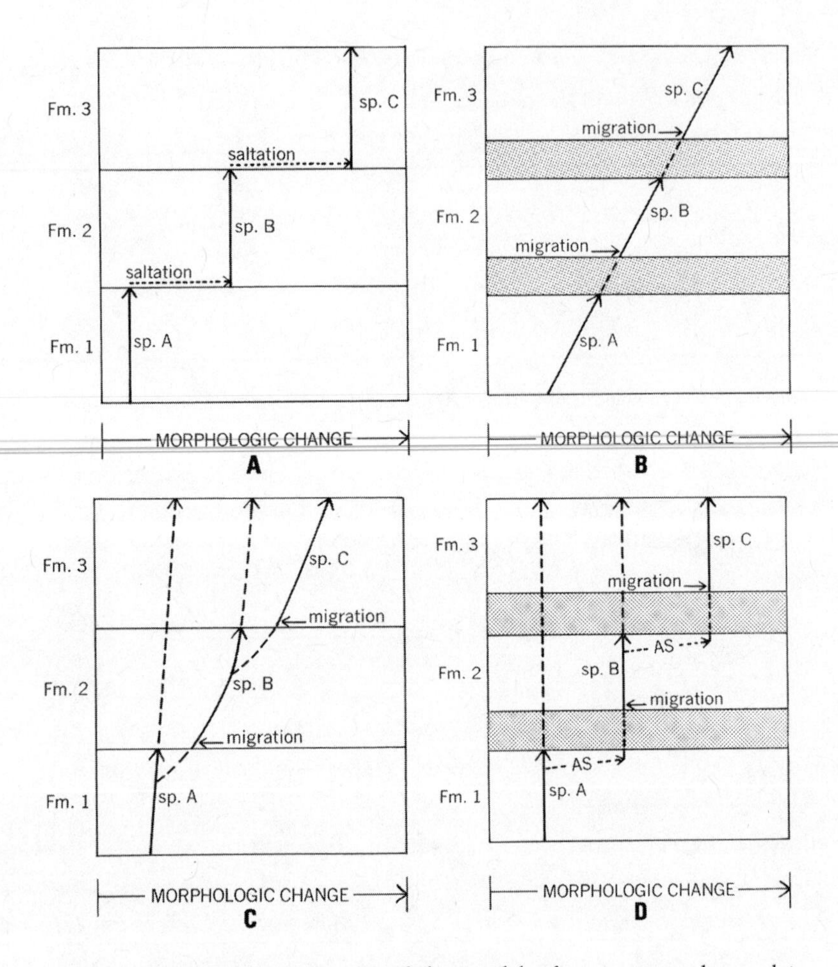

FIGURE 5.3 Niles Eldredge's diagram of "four models of speciation used to explain the occurrence of three species (A–C) in three successive bodies of strata (1–3) in one local area." (By permission, from "The Allopatric Model and Phylogeny in Paleozoic Invertebrates" [1971a:159, fig. 2])

with saltationism, this was clearly—and explicitly—rejected in my 1971 text.

It was clear that saltations (literally "jumps") between species could not be at work, if for no other reason than that the vertical replacements invariably occurred with the reappearance of seaways in major transgressions; these new seaways, with their biota containing a mixture of surviving taxa, and new species, reappeared after a long-term temporal hiatus. Because seas wax and wane, at times thoroughly flooding the continental interior, and at other times withdrawing to the continental margins, the vertical sequence of rocks in the continental interior (places like Ohio, for example) is likely to be full of temporal gaps; and these gaps correspond exactly to the apparent "jumps" in a sequence of replacement species.

So that alone suffices to rule out saltation as an evolutionary process explaining patterns of temporal (vertical) replacement of closely similar/closely related, yet non-intergrading species. But at least the saltational model acknowledged—and took as real pattern—the non-intergradation among temporal replacement of species of a lineage.

So if not saltation, what else was there? Well, if the missing time were acknowledged, either of two gradualistic possibilities—the then-preferred explanations—could be nominated as the explanatory model of choice. These would be the linear phyletic gradualism model, or the tad more modern-sounding splitting followed by continued adaptive change à la, for example, Theodosius Dobzhansky and Ernst Mayr. Species would be imagined simply to keep on changing gradually through that missing time period—obviously living somewhere else while the seaway, hence their marine habitat, simply was not in that one place (again, such as Ohio) for that longish bit of time.

But the problem of these two models was simply that no such gradual change could be documented through even the "thick formations" representing considerable amounts of geological time, as lamented in the Notebooks by the young Charles Darwin. "Mr. Lonsdale" never did show up with the goods.

Such examples, at least for Paleozoic invertebrates, simply do not exist. So what's left? Into the breach I ride with my *Phacops* study, presented as an empirical example of the fourth remaining explanatory model: allopatric speciation, in which a descendant arises somewhere on the periphery and then, when favorable environmental conditions resume, comes in as an ecological, rather than an evolutionary replacement event to produce the pattern of vertical succession of non-intergrading species in the rocks of the continental interior.

Recall that my *Phacops* story included two samples from the continental margin that appeared to catch a speciation event (the earlier one, for example, involving a reduction of d-v files from 18 to 17). The 17 file, putatively descendant species, remained in the marginal basin in the east, while its putative ancestor—the 18 file form—survived in the midcontinent for several millions of years. In other words, the 17 file form (species; though I hedged back then and called them "subspecies," later causing some hoots of derision from George Gaylord Simpson, who really hated punctuated equilibria) replaced the 18 file species in the east. But the two lived side by side allopatrically for at least 2 million years. Geographic speciation led first to geographic replacement; the 17 file species survived in the east (the seas were still there), but the 18 file form succumbed to extinction (at least it was never seen again) when the seas rolled back from the continental interior. Only when the seas returned to the midcontinent and the original 18 file form had disappeared could the derivative 17 file species invade the interior, which it did and lived presumably happily on for another couple of million years. I summarized this example in a diagram, contrasting events in the marginal seas, in which two episodes of speciation apparently occurred, with the events in the continental interior—where after extinction of the existing species, the more recently evolved species migrated in from the marginal basin when suitable habitat was restored to the continental interior (figure 5.4). Net result: through time, successions of discrete species replacing one another in the continental interior.

Such is the story I told, and at least no evolutionary biologist would find such a story unfamiliar. It was, indeed, the same type of interpretation that had been reinstated ever since Dobzhansky (1935, 1937) initiated it and Mayr (1942) followed up, in a one-two punch sort of way.

Ah, but what about this gorilla in the room, this seemingly anti-Darwinian (and anti-Dobzhansky, Mayr, and Simpson) insistence that there is little or no change to be seen in the (presumptively adaptive) morphological features that allow one to distinguish different taxa, different species, with their different spatiotemporal distributions? What about this "morphological stability" (as I called it in 1971—this "stasis," as Steve Gould and I more euphoniously dubbed it a year later)?

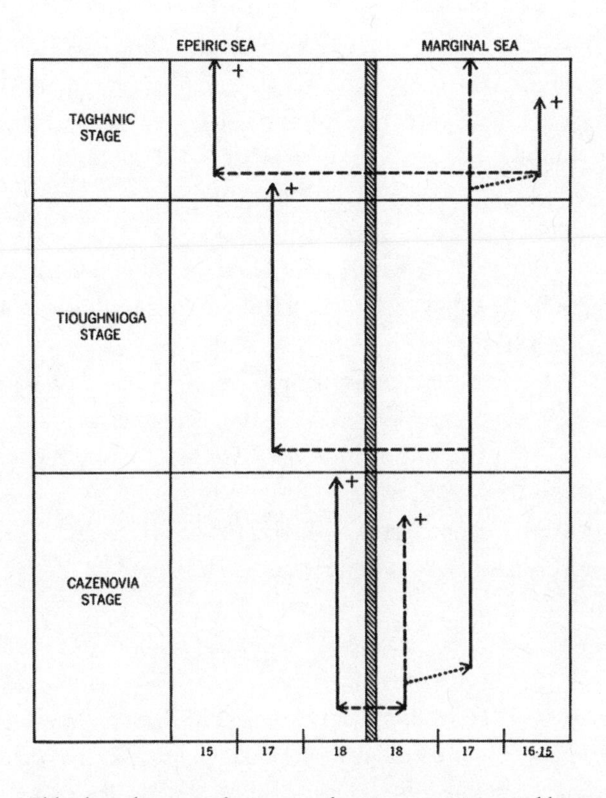

FIGURE 5.4 Eldredge's diagram of patterns of stasis, speciation, and biogeographic movements in the evolution of the Middle Devonian *Phacops rana* stock. (By permission, from "The Allopatric Model and Phylogeny in Paleozoic Invertebrates" [1971a:166, fig. 5])

In this paper, I was mostly concerned to establish that "morphological stability" within a species lineage is empirically documented, and thus generally "true." I go so far as to say this aspect, plus the geographic speciation explanatory model, probably holds true for most Paleozoic marine taxa, meaning not just my trilobites.

I wrote that "paleontologists have seldom emphasized this morphological stability" (1971a:160). I remember how thankful I was to find at least a few quotable examples that others besides myself had realized the emperor wears no clothes—people like H. J. MacGillavry (1968), who wrote that "many species do not show any evolutionary change at all." And the Finnish paleomammalogist Bjorn Kurten (1965), sounding very much like Moritz Wagner in the nineteenth century, when he wrote, "The situation suggests that new species arose comparatively rapidly, but once established, tended to continue without any change" (345).

Well, if morphological stability is the general rule and not the exception, how do we explain it? The best I could come up with was "stabilizing selection," but an entire far-flung species, living out a 5- to 10-million-year history? How does that work?

I basically punted, and did not produce, for that paper, a detailed model on how "stabilizing selection" could keep an entire far-flung species more or less constant for millions of years, let alone the other 100 or so other species living cheek by jowl with the focus species in those ecosystems of the deep past. The best I could do (and can do at this juncture in this narrative) is to quote from the summary of my paper:

> Emphasis on time and gradualistic transformation has led to the dominance of a strictly phyletic model of species transformation in paleontological thought. Even documented cases of lineage-splitting are often interpreted by recourse to a gradual (morphological) divergence model in which a strong element of phyletic thinking is incorporated.
>
> Allopatric speciation, not a strict alternative to gradual divergence, seems to fit the common pattern of non-intergrading species within a lineage as typically preserved in Paleozoic epeiric sediments. The majority of species preserved in epeiric sediments show no change in species-specific characters throughout the interval of their stratigraphic occurrence, and the phyletic model is inapplicable to most of these elements in the fossil record. Instead, change in, or development of, species-specific characters are envisioned as occurring relatively rapidly in peripheral isolates. Morphological stability of epeiric species is attributed to stabilizing selection. (1971a:166)

This model explaining the sudden replacement of one stable species by another has held up well. Whenever a species succumbs to extinction, it is often replaced by another, distinct species of the same lineage. The new species seems to come from the "source" area, in which conditions persist for longer periods of time. In terms of Paleozoic and Mesozoic epicontinental seas, the source areas typically lie in the deeper basins near the continental margins, while the regular patterns of abrupt replacement of stable species is mostly seen in the continental interiors where environmental conditions are less stable. It is fundamentally a pattern of speciation through isolation, later followed by migration of the descendant species into the territory vacated by the extinct ancestor. It is very much in the style of the young Charles Darwin.

I have but one regret about this paper: I cite neither Dobzhansky nor Mayr, and this after my own repeated assertions earlier in this present narrative of Darwin's (and many of his forbearers and contemporaries) failure to cite their own sometimes blatantly evident sources.

But here is the taxic view brought back—with teeth—into paleontology. There would soon be more to come.

PUNCTUATED EQUILIBRIA

So much for my worries over a fierce outcry over the publication of my paper "The Allopatric Model" (1971a). Instead, the silence was deafening, and I cannot recall, nor do I have a paper trail of, any reaction—positive, negative, or neutral—whatsoever. The paper rapidly sank without a trace. For those reasons alone; for the sake of the integration of the taxic perspective into what was soon to be called "paleobiology"; for the sake of the reintegration of paleontological data and perspectives into the mainstream of evolutionary biology; and (I readily admit) for the sake of my own career, it was fortunate indeed that by sometime in mid- to late 1970 I had joined forces with my friend and former fellow graduate student, Stephen Jay Gould (figure 5.5) in a project that led to a presentation at a symposium in the fall 1971 meetings of the Geological Society of America—and the eventual publication, in October 1972 of our paper "Punctuated Equilibria: An Alternative to Phyletic Gradualism."

Historian David Sepkoski (2013a), who has had access to all my files and recollections, as well as a good deal more material, has recently written what I consider to be a highly accurate, definitive account of

FIGURE 5.5 Stephen Jay Gould (*left*) and Niles Eldredge (*right*) with their mentor Norman D. Newell, in the late 1990s. (Courtesy of Gillian Newell)

the details of the history of punctuated equilibria and of my relationship with Steve Gould in the production of the paper. My plan is basically to stick to the established format of this narrative and analyze only original manuscripts that I find to be essential to unraveling the history of the taxic perspective in evolutionary biology, most certainly including my own papers. So some notes on the actual work with Steve Gould that led to the preparation of the manuscript (and despite the already available excellent account by Sepkoski) seem to be in order.

Steve was already at Harvard by then and very busy amplifying his already burgeoning reputation. He had gotten wind of paleontologist Thomas J. M. Schopf's plans to inject a renewed appreciation of theory into paleontological practice—initially via a symposium, followed by a book, both to be called "Models in Paleobiology." Unsurprisingly, Steve was eager to play a role in the project. He was disappointed, though, when the invitation came from Schopf that he was asked to handle the topic "Models in Speciation." According to Sepkoski (2013a:147), Steve wrote to Schopf in March 1970, saying that speciation was only number three on Schopf's list of topics that he (Steve) felt he could take on, and asking if he could handle, instead, "Models in Morphology" or "Models in Phylogeny"—if the already invited scientists (Michael Ghiselin and David M. Raup) for some reason dropped out and could not fulfill their assignments. I do remember the phone call from Steve, presumably sometime in the fall of 1970, when he told me of all of this, said that he was stuck with the topic of speciation, but that "for the life of me I can't think of a damn thing to say beyond what is in your manuscript"—that is, the draft of "The Allopatric Model" (1970a), which I had sent him—and asked me to join him as co-author of the speciation paper.

So we struck a deal: Steve was to give the paper at the GSA meeting, while I was to be senior author of the paper and write the first draft (Eldredge 1970b). That deal was in the end kept, despite a few bumps in the road: Steve's revision of my first draft came back to me listing Steve as first author and me as second author (Eldredge and Gould 1970; see also Sepkoski 2013a:174, fig. 5.3), along with a handwritten annotation: "We must discuss order. Very obedient Germanic secretary just put it this way." I wasn't pleased, but do not remember getting overly upset at that point: after all, a deal is a deal. But Steve did say "we must discuss order," and I had thought that we already had.

Sometime later (I believe it was in the spring of 1971), Steve paid a visit to the American Museum, and said something to the effect that "I [Steve] think that I should be the senior author," and I think I meekly said, "OK." But after dinner and a couple of Bergman movies that evening with our wives, looking across the Ninety-sixth Street subway platform at Steve and his wife while we waited to go north to take a bus home to New Jersey, I realized that I had made conceivably the mistake of a lifetime. I already knew that junior authors were routinely dismissed as also-rans. So when Steve, very fortunately, walked into my office the next day, I blurted out that I should be the senior author. And Steve immediately capitulated.

That this little drama was important to me and my career became all the more clear as the years went by. But equally was it to Steve, for although he was the junior author, he nonetheless, by dint of his writings, speeches, and sheer force of personality came to be identified with punctuated equilibria to the point of sometimes (even to this day) being regarded as sole author. Steve once said that people simply don't read very much, and the relatively modest sales of Schopf's book pretty much guaranteed that only a handful of people who went on to discuss punctuated equilibria (or "punk eek" as it is commonly called) had in fact ever read our paper. But the importance of punctuated equilibria to Steve's career is best seen in the myriad obituaries that followed his death in 2003, including the one that began on the front page of the *New York Times*: all of them mention our joint work (the authorship citations were nearly always correct) on punctuated equilibria, and, for the most part, punctuated equilibria constituted, in these obits, at least, Steve's main claim to fame as a scientist per se.

All of which is merely to say that academic life can be as bruising a sport as football. For the most part Steve and I were friends and good working partners. We remained so for the rest of Steve's life. Despite these bumps in the road, from then on it was pretty much Steve'n'me up against the world of doubters who immediately sprang up, saying variously that we were anti-Darwinians, saltationists, and even communist-inspired. No one at the time knew that the young Charles Darwin of the late 1830s would have found a lot to like in our paper.

My first draft was (no surprise) exceedingly prosaic, possessing none of the rhetorical frills and embellishments that Steve came to be famous for, but that were plainly in evidence in the final version of our paper (Eldredge and Gould 1972a). It was entitled "The Process of Speciation and Interpretation of the Fossil Record" (Eldredge 1970b). I sent it to Steve in December 1970.

"The Process of Speciation and Interpretation of the Fossil Record," 1970

My rough draft consisted of twenty pages of typewritten text, plus a bibliography and drafts of several figures. It forms the basis of pages 92 to 108 of the final published version, as rewritten for stylistic uniformity by Steve Gould. And though I had expected, upon rereading this draft for the first time in more than forty years, to find essentially just a rewrite of my "Allopatric Model" (1971a; still, at the time, in press), there was indeed something new in the argument.

Plus, there was an entirely novel conceptual addition thrown in as an addendum in the final pages of this draft, from December 1970, and eventually published pretty much verbatim in 1972. Entitled "some comments on trends," my brief discussion tackled the evident paradox posed by such phenomena as evolutionary trends. For if gradualism is ruled out as a dominant pattern of within-species evolutionary change, then how can we maintain that gradualism can be sufficiently long-lasting to account for the emergence of new species? And how can we take things further and proceed to cite long-term "orthoselection" as the major generator of directional change among species within clades? I toyed with the possibility that trends (such as brain size increase among hominid species through time) were largely fictitious, but rejected that easy out. Taking evolutionary trends as real phenomena, yet obviously not generated, as commonly supposed, by long-term "orthoselection," my solution was to see patterns of differential species survival within clades—an opening salvo in what a decade later emerged as the "species selection" debate. Indeed, my section on trends was the forerunner of much of the literature on hierarchy that began in earnest in the 1980s. Thus I will defer discussing this part of my draft/1972 "punctuated equilibria" until the beginning of the next chapter of this book.

The novel aspect of my argument (vis-à-vis "The Allopatric Model" [1971a]), when characterizing the nature of gradual change versus what one might expect to see in the fossil record if evolution was largely a matter of allopatric speciation, was largely rhetorical and philosophical. "Testability" had become the rage by then, and is still a major component of much methodological discourse in comparative biology. Which, of course, is all to the good. I argued that the outcomes of gradualism versus those of geographic speciation were bound to look very different in the fossil record, and that multiple instances would turn up familiar common patterns of geographic and temporal variation: of stasis and evolutionary change.

Fair enough. But I also argued that the anatomical features that are different between two closely related species evolve just before, during, and right after the onset of genetic isolation. So descendant species evolve in small populations in isolation ("peripheral isolates": Ernst Mayr's term, though I still did not cite his work!), *according to allopatric speciation theory*. This is very similar to Darwin's original view, when he saw, for instance, the differences among mockingbird species living on different Galápagos Islands within (human) sight of one another. I don't explicitly say so, but it is every bit as if genetic isolation freezes the action: not only are there two separate species when once there was but one, but the differences between them evolve quickly. Darwin saw that, saying that if adaptive change in isolation were the norm, there would be a ready understanding for the pronounced lack of change within lineages in the fossil record: most of the differences between closely related species arise in the isolated populations in the very process of speciation (in Darwin's terminology, the very process of the origin of the new species in isolation).

Again, fair enough. But in this rough draft of the punk eek paper, I went further. I insisted that, were this largely true (that is, that most differences between species arise in brief spurts of speciation), then it follows *as a prediction* that stability will thenceforth reign in both parent and daughter species. Newly separated species will not inevitably continue to diverge morphologically through time.

This time, for the life of me, I cannot divine my (1970s!) logic here. I don't see how stasis arises as a prediction from allopatric speciation theory, despite the fact that the theory is adequate to account for the morphological differences between closely related, adjacent species. After all, when allopatric speciation was reimagined and reconfigured (in short, reinvented), beginning with Theodosius Dobzhansky's work in the mid-1930s, and including Mayr's shortly after him, neither of these two thinkers felt the need to deny that two newly separated species would not, or could not, keep on diverging from one another—more or less in a gradual fashion.

And in 1839, the thirty-year-old Charles Darwin merely said that what we have come to call stasis (he said "uniformity") was understandable. Even more to the point, as far as my own efforts to read and understand myself in 1970, and as already made clear in the narrative in this chapter, I knew (as Darwin did) that stasis is an empirical reality. I already knew that stasis—relative non-change typically throughout the vast majority of a species' tenure on earth—was a phenomenon to be reckoned with in an evolutionary context. But I cannot see the logic behind my claim (in 1970, surviving into our punk eek paper [Eldredge and Gould 1972a]) that stasis arises as an expectation—literally, a predicted pattern—from the mere contemplation of the fact, and underlying explanatory theory, of allopatric speciation. We would predict stasis as an expectation of at least one aspect of evolutionary theory, but allopatric speciation is not that part of evolutionary theory that does so, despite the claims I was making in 1970 to 1972.

More on this as this chapter unfolds and concludes. For the moment, though, I will say that the only real other surprise to me as I reread "The Process of Speciation" (1970b) was that some of the language in my draft, especially some of the critical, punk eek terminology—in particular "phyletic gradualism" and "stasis"—are already there. Yet I persist in thinking that it was Steve Gould who coined those two phrases (along with "punctuated equilibria" itself—again, oddly changed by Gould later in his life to read "punctuated equilibrium"). And so I can only conclude that Steve and I must have been in active contact before, or while, I was writing "The Process of Speciation." Nothing happens in a vacuum, especially in the process of producing a joint manuscript.

For the rest of the draft, it was pretty much straightforward. I rendered Steve's Bermudan land snail doctoral dissertation analysis into an example of our theme, along with my *Phacops rana* example. Steve took this material and added his own spellbinding take on science, life, and evolution, including at least one original section at the paper's very end (though, naturally enough, at the time I didn't think much of it). The result was the paper that made our careers—Steve's as well as mine.

"Punctuated Equilibria: An Alternative to Phyletic Gradualism," 1972

Steve sent the next draft—the one with the authorship inverted (Eldredge and Gould 1970), and the version that was pretty close to what ended up being published—back to me sometime in December 1970 or January 1971. By the time we had further tweaked the manuscript and were in a position to submit it to Tom Schopf, Steve had departed for a sabbatical experience at Oxford University. Authors were instructed to submit multiple copies (I seem to remember it was as many as twelve) to facilitate pre-publication peer review. I remember being excited and probably a bit nervous as I assembled all the copies and sent them off to Schopf in February 1971. I was left in the United States to fend off the criticism that was sure to come in from Schopf and his phalanx of reviewers. The correspondence between Steve and me is pretty amusing. He instructed me to stand firm and not to water down our paper to placate the criticisms and editorial dictates that of course did materialize. I sent the final, post-pre-publication review version back to Schopf on June 21, 1971 (Eldredge 1971b).

Rereading the reviews (I have four in my files), plus Schopf's comments, I am struck with how mild they were. Tom basically did not like the paper, but on the whole, the reviewers did. What I find most interesting is that nearly all of them disputed what they said were our claims of originality. I stuck to our guns in reply:

> The only general criticism that bothered me at all was the assertion that we either state or imply that "we have discovered something new." We haven't, of course, nor do we claim so. We are merely pointing out the (largely unconscious, perhaps) sway that "phyletic gradualism" holds over our thinking—despite the fact that most of us have a coherent version of the allopatric model in our minds. The point is that on the whole we don't *use* it when confronted with species-level phenomena in the fossil record. Intellectual appreciation of modern neontological theory is one thing—but its application is quite another, and I believe that our paper pinpoints why this is so.
>
> I have made a thorough search of the ms. for hints that we feel that "we have discovered something new," or that we are the only paleontologists who have read Ernst Mayr." (Eldredge 1971b)

We also bent over backward (as I had originally in the first draft [Eldredge 1970b]) to insist that we, as paleontologists, could suggest "no new mechanisms"—meaning, of course, factors on the level, and akin to but somehow different from, mutation, natural selection, and the like. Of course not. But at the same time I wince a bit reading those words now, as we did suggest that there were explanations to be found for both stasis and the differential origin and survival of species within clades, as tentative and even ambiguous as those suggestions in fact were. In insisting that we could "suggest no new mechanisms," we were in fact denying that we were following George Gaylord Simpson's lead in insisting that the fossil record shows repeated patterns that require causal explanation, and that such explanations, involving the raw ingredients of selection, mutation, and the like, at the very least require novel combinations of such factors to explain patterns that had hitherto not been seen, or had been simply ignored. That's the core of novel theory, and there was that aplenty in our paper.

But of course the really big novelty was stasis. Not that we had discovered it, but starting with "The Allopatric Model" (Eldredge 1971a), at least, I had put in a hard search for predecessors who knew that the prevailing expectation of gradualism had created a sort of emperor's new clothes situation. Once again, I expected stasis to be the lightning rod provoking howls of outrage. But in the pre-publication reviews it seemed to pass largely unnoticed. It was only after our paper was published in October 1972 that we began to see the long-anticipated vehement attacks from many, starting mostly from paleontologists, but soon to embrace many "neontologists" who thought we were starkly anti-Darwinian, and who upheld the gradualism credo with ferocity.

Considering how radical was our claim that, after they first appear, species tend to change but little, often for millions of years, it is surprising to me to see how long it took us, in the development of the 1972 argument, to bring stasis into the picture. Like my 1971 paper, the argument does not begin with the empirical reality of stasis. It was as if we were still soft-pedaling the theme of stasis as Steve (despite his already developing a taste for novelty and provocation) amplified—and recast—my first draft. Consider our opening salvo of the paper, written by Steve—the closest thing to an "abstract," simply labeled "Statement":

> In this paper we shall argue:
>
> (1) The expectations of theory color perception to such a degree that new notions arise from facts collected under the influence of old pictures of the world. New pictures must cast their influence before facts can be seen in different perspective.
>
> (2) Paleontology's view of speciation has been dominated by the picture of "phyletic gradualism." It holds that new species arise from the slow and steady transformation of entire populations. Under its influence, we seek unbroken fossil series linking two forms by insensible

> gradation as the only complete mirror of Darwinian processes; we ascribe all breaks to imperfections in the fossil record.
>
> (3) The theory of allopatric (or geographic) speciation suggests a different interpretation of paleontological data. If new species arise very rapidly in small, peripherally isolated populations, then the great expectation of insensibly graded fossil sequences is a chimera. A new species does not evolve in the area of its ancestors; it does not arise from the slow transformation of all its forebears. Many breaks in the fossil record are real.
>
> (4) The history of life is more adequately represented by a picture of "punctuated equilibria" than by the notion of phyletic gradualism. The history of evolution is not one of stately unfolding, but a story of homeostatic equilibria, disturbed only "rarely" (that is, rather often in the fullness of time) by rapid and episodic events of speciation. (Eldredge and Gould 1972a:83–84)

This is basically a very good and eloquent summation of our thesis. But in maintaining that it is the importation of a novel picture—allopatric speciation—that causes us to rethink the fossil record, distorts in some respects the logical sequence of actual discovery.

Consider the case with the young Darwin, who saw that, if adaptive change in isolation was the predominant evolutionary mode, then the otherwise unwelcome and inconvenient fact of geological uniformity of species is rendered "intelligible." Much, much later, I had hit the stone wall of species intransigence—of stasis—that prompted my full embrace of "the allopatric model" to make sense of the *empirical* picture of spatiotemporal stasis plus change I encountered in my trilobite data in the latest 1960s.

Stasis was really the key to this, and that fact was not missed by many of our readers, however slowly we were to confront it in the 1972 paper. I saw that change hinged on geography and immediately latched onto allopatric speciation. What I did not do, in my 1971 paper, was pose an explanation for stasis beyond a vague invocation of "stabilizing selection."

In our 1972 paper, stasis does not come in, at least by its newly minted name, until after a disquisition on the nature and importance of theory in science. Next came a rather good characterization cum examples of phyletic gradualism, largely the work of Steve Gould. Stasis is characterized, but not named, on page 95, in much the way I had done in the first draft of the manuscript (Eldredge 1970b). As far as I can tell, the first use of the word "stasis" occurs in a sentence I had long forgotten, am not sure now what it means (in the context of the 1972 argument, at least), but that now I see as provocative, as it suggests the solution I have long since favored to the question: What are the causal underpinnings of stasis? "The key factor is adjustment to a heterogeneous array of microenvironments vs. a general pattern of stasis through time" (Eldredge and Gould 1972a:96).

I return to this statement later, in the next section when I review the post-1972 history of the empirical reality, and conceptual explanation, of the phenomenon of stasis. First, though, I close out this consideration of the 1972 paper with a few brief comments on the concluding sections of our paper—most especially Steve Gould's explicit attempt to grapple with the causal factors underlying stasis.

After the by-now familiar examples of Steve's Bermudan snails and my Devonian trilobites, we included a section called "Some Extrapolations to Macroevolution" (Eldredge and Gould 1972a:108–112), based, of course, on punctuated patterns and processes, rather than the "stately unfolding" (another of Steve Gould's phrases) supposedly yielded by phyletic gradualism. Steve contributed the first example on "Classes" of great number and low diversity, as exemplified by early phases of explosive echinoderm evolution in the Lower Paleozoic. The other example was my section on "trends," with which I will begin chapter 6.

But it is the "Conclusion: Evolution, Stately or Episodic" that warrants a few more words. It was, I believe, entirely written by Steve Gould. I admit I never liked it, but let it pass, as a conclusion to our paper. I think a bit more highly of it now.

Steve says that, given the recent (contemporary) attacks on gene flow as being the key to understanding species' internal cohesion, we must ask: What then provides that cohesion and stability of most species throughout the bulk of their histories? It is an attempt to confront the problem we otherwise continued to dodge throughout the paper: What causes stasis?

Steve says (technically, of course, *we* say): "The answer probably lies in a view of species and individuals as homeostatic systems—as amazingly well-buffered to resist change and maintain stability in the face of disturbing influences" (Eldredge and Gould 1972a:114). He goes on to elaborate a bit more, speculating that it is because new species habitually arise in small populations as peripheral isolates "that acquired its own powerful homeostatic system" (114).

I like this brief discussion a bit more now than I originally did simply because it was an attempt to grapple with a major issue: Why are species characteristically so stable? Why don't they just keep on evolving as Darwin concluded they must and surely do—and as Dobzhansky, Mayr (and, of course, Simpson) affirmed is generally the case?

It is true, of course, that Steve's formulation is hopelessly vague (though probably not much more so than my earlier handwaving at stabilizing selection). Attributing properties to species that have little or nothing to do with the interaction among component individuals (as Giambattista Brocchi, and many others after him, did in ascribing internally programmed longevities to entire species) is a dubious proposition at best. Then again, it is interesting that Steve made a very Brocchian analogy between species and individuals in this short passage. Both, Steve claims, are "homeostatic systems." Schopf wrote to me that this little concluding section left his students bewildered.

But at least Steve had asked the question.

STASIS: THE BIG GORILLA IN THE ROOM

This time, the reaction was swift and intense. Some paleontologists agreed with us, but many paleontologists and neontologists did not. Many of our sharpest critics wrote data-rich papers purporting to document examples of gradualism. This time it was stasis, with the implicit denial that phyletic gradualism has been an important mode of evolutionary change in the history of life at all, that proved to be the lightning rod. Our claim that the dominant pattern in the history of life was relatively rapid events of geographic speciation followed by much longer intervals of stability—"stasis"—was simply too much for many of our colleagues to swallow. Our thesis seemed to many biologists and paleontologists downright anti-Darwinian, downright heretical. And, again, it was stasis, not speciation per se, that rankled.

In 1976 Steve Gould decided that the time was ripe to clarify our own arguments, respond to our multifarious critics, and basically stand our punctuational intellectual ground. At first, I couldn't see the need, but of course he was right. Publishing in the Tom Schopf–inspired new journal *Paleobiology* immediately brought our ideas to an even wider forum than had been achieved with the initial chapter-in-a-book format.

This time around, after we discussed the general tone and tenor of the contents, Steve drafted the entire paper, which he initially called "Punctuated Equilibria: The Tempo of Evolution Reconsidered" (Gould and Eldredge 1977). I merely critiqued (followed up by a detailed letter of October 28, 1976), especially during one afternoon session in a hotel room while we were attending yet another Geological Society of America/Paleontological Society meeting. As far as I can recall, my major contribution to the entire enterprise was to insist that wherever Steve had written "tempo," I said he should write "and mode." Not only was that the title of George Simpson's (1944) early book, but, more to the point, I argued, Simpson had made a fundamental distinction between tempos (rates) of genetically based morphological evolution, and the modes of such change—by which he meant phyletic evolution, speciation, and his own concept of "quantum evolution."

I ended up getting my way on what to me was an important matter. In punctuated equilibria, we were basically rejecting the phyletic mode in favor of the alternative (geographic) speciation mode as the dominant locus, "mode" of evolutionary change. We changed the title to include "and mode" after "tempo," and said at the outset that our model of punctuated equilibria is "an hypothesis about mode." We made it clear that, in our view, speciation "is orders of magnitude more important than phyletic evolution as a mode of evolutionary change." And we said that "stasis is data." I now regard that introductory section, "Gradualism and Stasis," as a much needed, fairly concise, statement of what we at least had intended to say in "Punctuated Equilibria: An Alternative to Phyletic Gradualism" (Eldredge and Gould 1972a).

This little anecdote about my very slight contribution to the paper, stressing the importance of "mode," highlights the fact that Steve had seen our original paper as primarily a matter of tempo: slow, steady, gradual change versus rapid change, followed by much longer periods of morphological stasis, rather than speciation versus phyletic evolution modes. And this is interesting because it fits Steve's preference for seeing evolution primarily as a question of (genetically based) morphological change—linked, in his mind and psyche, with patterns and processes of developmental evolution. At heart, Steve was a transformationist in the purest sense. And it came as no surprise to me as I reread our 1977 paper that the most original thinking in the entire manuscript (again, entirely Steve's) concerned the relationship between the newly emerging molecular data (mostly involving proteins, as DNA and RNA sequencing still lay in the future) and patterns of morphological change. Steve posited the very interesting notion that punctuated equilibria might someday serve as a test of concepts of neutralism, the molecular clock, and similar notions that were beginning to be discussed intensely in the evolutionary literature. This turns out to have been correct, although it took something like thirty years before molecular biologists began seeing punctuational patterns in their data. More of this in the next chapter.

Some of the paleontologists who published purported examples of gradualism (see notes) have stuck by their guns over the years. Some, whether publicly or privately, have recanted and told one or the other of us that "you were right." Paleontologists began to publish hard, detailed looks at spatiotemporal patterns of within- and among-species patterns of morphological change. Among them was a very detailed multivariate multilineage study by Steven Stanley and X. Yang (1987) on "bivalve morphology over millions of years"; a careful multivariate statistical analysis by Bruce Lieberman, Carlton Brett, and Niles Eldredge (1995) on two brachiopod species-lineages from the Middle Devonian (a study that documented more morphological evolutionary "excursions" within ecological variants than the species-as-a-whole—a key to coming-to-grips with the causal basis of stasis); several chapters in a volume edited by Ross Nehm and Ann Budd (2008); and a marvelous paper by paleontologists Jeremy Jackson and Alan Cheetham (1999), who studied the evolution of the existing bryozoan species (defined morphologically and genetically) of the Caribbean, tracing the historical patterns of stasis and change of their well-preserved, complex skeletal characteristics using multivariate morphometric analysis down through the dense and rich fossil record for millions of years. Stasis was the overwhelming pattern they found.

Jackson and Cheetham's work is especially profound in the context of this present narrative, as it goes back to what I have shown was the original question in evolutionary biology: What is the nature and causal explanation of the origin of species of the modern biota?

By now, the majority of paleontologists consider stasis a common reality. Most, I believe, see it as the predominant pattern of the history of species. Because of such careful studies, few if any reject stasis outright, as multitudes did in the mid-1970s.

But consider a profound difference between the world of the 1820s and the modern day. Back then, especially in Edinburgh in the 1820s, conversant as they were with Georges Cuvier, Jean-Baptiste Lamarck, and Giambattista Brocchi, naturalists were equally at home with accounts of patterns of distributions and morphological variation within and among species of the living biota, and with paleontology. Most of the early "natural philosophers" interested in the origin of modern species keenly felt that such answers as might be forthcoming would best, or even only, come from a serious study of the fossil record. In short, everyone from Cuvier, Lamarck, and Brocchi, on up through Robert Grant, Robert Jameson, and John Herschel, knew all the work being done with living species and with extinct species as well.

All the work. Life was simpler then, and there may have been no one since Charles Darwin who has had such a mastery of everything relevant to the history of life. We've learned so much more. There have been multiple "speciation events" in biological, paleontological (geological), anthropological, psychological, and even sociological disciplines, all of which, at least in part, have a direct bearing on understanding evolutionary processes. In biology we've had successive revolutions in understanding the fundamentals of heredity. First, Gregor Mendel was rediscovered; then came population genetics; and now, of course, the successive waves of increasing sophistication in understanding the molecular basis of it all.

No one can hope to grasp all the basic fundamentals—data and insights—of all modern evolutionary biology and paleontology. And from the narrow standpoint of stasis, how can a paleontologist such as myself hope to convince an evolutionarily minded geneticist that stasis is a reality in a modern world in which I know as little of the details of molecular biology as biologists know about the details of the fossil record? It can be very frustrating for a paleontologist such as myself to talk to a biologist about stasis, expecting them to believe you when they can't see the evidence first-hand.

But ways to communicate across disciplines can be found. I participated, with nine colleagues, including five genetically imbued biologists and four other paleontologists, in the publication of a paper entitled "The Dynamics of Evolutionary Stasis" (Eldredge et al. 2005). The paper, published in a volume dedicated to the memory of Stephen Jay Gould, was the outcome of an ongoing workshop held at the National Center for Ecological Analysis and Synthesis (NCEAS), located in Santa Barbara, California, and sponsored by the National Science Foundation.

The workshop itself grew out of some post-conference conversations I'd had with evolutionary biologist John N. Thompson in Belo Horizonte, Brazil. John studies how plants and insects coevolve in nature and how bacteria and viruses coevolve in the lab. Year after year, he has studied how even the same plant and insect species differ in adaptations in population after population and how microbes placed in different environments in the lab evolve in different ways within just a few weeks.

In Brazil, John told me of the kinds of phenomena he routinely encountered with his evolutionary systems. In a nutshell, nearly constant evolutionary (genetic) change from field season to field season; and differences between his samples from different local populations, some of which changed more, and in different directions, than others. John Thompson has come to call this constant evolutionary change within populations of his study systems "relentless evolution."

Of course, I spoke to him about stasis—about the relative lack of accumulating morphological change throughout the often-long histories of entire species. How, we each wondered, could both be simultaneously true? John suggested we apply for funding to establish a study group of genetically minded evolutionary biologists and paleontologists to examine this very issue.

As we did. The very first item that arose, though, was the challenge to the paleontologists in the room to convince the biologists that stasis is indeed a reality—a common, even the dominant, pattern of (non-) change in the history of most species that have ever lived and have left behind an analyzable fossil record. Turns out that was not so very easy.

But the key turned out to be the presence of Jeremy Jackson, one of the paleontologists in the group, but one who had spent a great deal of time studying modern bryozoans, and someone who was up-to-date in the rapidly changing field of evolutionary genetics. It turns out that the living bryozoan species defined and recognized on the basis of morphological characters are (unsurprisingly) also genetically coherent and different from other bryozoan species, and that many of the living species appear to be preserved as lineages-in-stasis in the fossil record. In other words, bryozoan species defined on morphology are genetically discrete entities with long, well-preserved histories of stasis. Jackson had, as noted earlier, published with paleontologist Alan Cheetham the great study documenting stasis in the fossil record of many bryozoan species still extant in the Caribbean.

At one point at one of our NCEAS sessions, Jackson jumped into the conversation and simply said that when I published the original paper on punctuated equilibria with Steve Gould (Eldredge and Gould 1972a), he thought it was ridiculous (something like that; the actual expression might have been stronger), and that in any case our supporting examples were weak and really not well done at all. So, he said, he embarked on his collaboration with the paleontologist Alan Cheetham, whose extensive experience with the fossil record of Upper Tertiary bryozoans, coupled with Jackson's own work on the living representatives at the "tips" of these lineages, made for a compelling potential test case of punctuated equilibria. Jackson said he expected their study to falsify punctuated equilibria.

But, as I have already said, Jackson and Cheetham's (1999) study actually proved to be an iron-clad and utterly convincing case in favor of stasis as *the* pattern of the histories of Caribbean Neogene bryozoans. Jeremy's impromptu speech managed to command the attention and respect of the evolutionary biologists in the room. He was sufficiently convincing that the group was able to move on and consider what was causing stasis.

Prior to all this, several paleontologists, including myself and Bruce Lieberman (who had produced the brachiopod lineage study [Lieberman et al. 1995], and who was a member of the NCEAS group) had been considering the possible causes of stasis since at least the 1980s. Here I return to the somewhat cryptic remark made earlier in this chapter, in which I said that evolutionary theory could—and perhaps should—have actually predicted stasis to be the norm for the history of species. But such a prediction does not arise from the mere adoption of speciation as a model for understanding evolution in the fossil record, counter to the claim in the original punctuated equilibria paper. Instead, on rereading our paper, I am struck by that simple, somewhat enigmatic sentence: "The key factor is adjustment to a heterogeneous array of micro-environments vs. a general pattern of stasis through time" (Eldredge and Gould 1972a:96). Given what happened in the 1980s, 1990s, and early 2000s, it now seems to me that, whatever we meant by this sentence, and why we included it in our paper in the first place, the first phrase ("adjustment to a heterogeneous array of micro-environments") is not an alternative, but the actual explanation of the second phrase ("a general pattern of stasis through time").

The development of the ideas that I believe are the most fruitful in explaining stasis first arose from the extension of the metaphor of the adaptive landscape developed by Sewall Wright (1931, 1932). Wright's original imagery was devoted to a description of "more harmonious" gene combinations developed in some organisms over others within a population—or within an entire species. But even Wright himself extended the metaphor (as did Dobzhansky not long after) to embrace different populations (that is, geographically disjunct populations) within a species—and, ultimately, by Dobzhansky, to explain the structure and history of entire groups of related species, and indeed, entire families, orders, and so forth.

Given the ubiquity and extended use of Wright's adaptive peak imagery, it was only natural to start thinking of the quasi-independent histories of disjunct populations over the entire area occupied by far-ranging species. With their different mutational and selection histories, the latter occasioned by the very often quite different ecological settings in which these disparate local populations are living, it becomes challenging to imagine just how natural selection could in fact be expected to mold and shape a gradual evolutionary modification of an entire species in any one direction for any considerable length of time. Instead, the semi-autonomous evolutionary histories of these semi-isolated populations would be expected, instead, to cancel each other out—in terms, that is, of the long-range history of an entire species. That is, in fact, exactly what Lieberman found with his analysis of the two Devonian brachiopod species lineages.

In the NCEAS group discussions of stasis, another dramatic moment occurred one session, when the evolutionary biologists asked the paleontologists to remain silent ("shut up" would be more like it!) while they proceeded to list all the known causes of genetic change—and the causes that can counteract them. They filled three sections of a white board along an entire wall surface. One by one, they crossed out all the alternatives, as all known inhibitors of genetic change had countervailing causes that could override them. Except for one: complex population structures of far-flung species.

I can do no better than quote the conclusion of our paper, which reads in part: "Both theoretical and empirical studies of the past decade suggest that the complex pattern of selection imposed on geographically structured populations by heterogeneous environments and coevolution can paradoxically maintain stasis at the species level over long periods of time. By contrast, neither lack of genetic variation, nor genetic and developmental constraint is probably sufficient in and of itself to account for species-wide stasis" (Eldredge et al. 2005:142).

Surely, more data will undoubtedly arise regarding the underlying causation of prevailing patterns of stability within species. But the cooperation between evolutionary biologists and paleontologists on this particular study is a welcome reminder of what can be achieved if evolutionists of all stripes get together and achieve a common understanding.

STASIS, DARWIN, AND THE STRONG COROLLARY

If stasis is a truly common phenomenon throughout most of the histories of most sexually reproducing species that have ever existed, there is one ineluctable conclusion, however contrarian to received wisdom from the mature Charles Darwin on down to modern times: *most lasting morphological change in the evolution of life occurs in association with true splitting—speciation—events*. This is what I call the "strong corollary," if stasis is indeed the predominant pattern of species' histories.

As we saw in chapter 3, the young Darwin, in Notebooks B through E (1837–1839), was well aware of precisely this possibility. The way he put it was that *if most adaptive change occurs in conjunction with the origin of new species in isolation, then we needn't worry about the lack of change in species that even then was seen to be commonplace*. In his own words, Darwin (late 1838 or early 1839) said in Notebook E: "If separation in horizontal direction is far more important in making species, than time (as cause of change) which can hardly be believed, then, uniformity in geological formation intelligible" (135).

Translating this into the terminology of Niles Eldredge and Steven Jay Gould (1972a), Darwin's statement would read something like: "If separation in horizontal direction [allopatric speciation] is far more important in making species, than time [phyletic gradualism] (as cause of change) which can hardly be believed, then, uniformity in geological formation [stasis] intelligible." Or, more simply and directly, paraphrasing with the terms of punctuated equilibria: "If allopatric speciation is far more important in making species, than phyletic gradualism (as cause of change) which can hardly be believed, then, stasis intelligible."

The point is that stasis is a real phenomenon, known to Darwin, although ultimately brushed aside in favor of a view of gradual phyletic change for which he had no supporting evidence. He did so presumably because he could see no way for isolation to work sufficiently commonly and effectively on continental settings to account for the fact that most species occur on continents, not on the islands for which he had his own hard data on the origin of new species in obviously isolated settings. Darwin knew little of the nature and patterns of environmental changes on continents, starting with glaciations and associated with long-term climatic temperature and rainfall excursions (for a bit more on this, see chapter 6).

I will take this paraphrasing of this vital Darwinian sentence one step further. In my judgment, it does absolutely no harm to Darwin's meaning, in that marvelous sentence of Notebook E, if we rewrite it one more time, this time adjusting the position of the component concepts: "If stasis is real, thereby ruling out gradualism (which is easy to believe), then allopatric speciation is the locus of most adaptive change in the history of life" (135).

That is, in other words, and in my opinion, it seems implicit in what Darwin actually did write that he saw the possibility of what I am calling here the "strong corollary." And he saw this possibility as counter to the narrative that he ended up favoring in the (1839) single-sentence paragraph—as well, of course, in his later, published theory. Gradual change within lineages through geologic time, I believe Darwin saw and implicitly conceded in this passage of 1839, is absolutely antithetical to stasis, and if one is true the other cannot be—as the dominant explanation for the context of adaptive change in the history of life. If stasis is empirically established, gradualism cannot be true, and vice versa. And if stasis is empirically established as "true," then phenomena such as the production of new species on islands and everywhere else *must* be the time and place—the sorts of circumstances—in which the engine of adaptive change does most of its work.

In the late 1830s, the young Darwin would have liked the title of our paper, "Punctuated Equilibria: An Alternative to Phyletic Gradualism" (Eldredge and Gould 1972a), but would have insisted that we should have written "*The* Alternative," rather than "An Alternative to Phyletic Gradualism."

Again, as Moritz Wagner, Theodosius Dobzhansky, Ernst Mayr, and a host of talented evolutionary biologists who have so diligently studied the process (or processes) of speciation have made abundantly clear, speciation is by no means "instantaneous." Steve and I were initially accused by some critics as being "saltationists," so we took pains to make clear that the "rapid change" seen in the fossil record can take thousands of years. One catchphrase that I have been associated with is "5–50 thousand years"—based, somewhat sketchily, I admit, on estimates of minimal years for the deposition of sediments with apparent transitions to accumulate, like the transition from the 18 d-v to 17 d-v *Phacops* samples in that upstate New York cow pasture. "Five to fifty thousand years" is brief only with respect to the 5 to 10 million years of subsequent duration of many successful marine invertebrate species.

As to why adaptive change is so commonly and so directly associated with speciation events: many biologists (including myself) have thought that genetic isolation is the *sine qua non* for partitioning already existing within-species genetic variation. Such isolation, in addition, could also trigger bouts of rapid further adjustment to the different, or perhaps even novel, environmental circumstances that isolated populations find themselves in. It is as if isolated populations in marginal environments rapidly adapt under natural selection: a process that, in effect, redefines "marginal" to the new "optimal" environment—optimal, that is, to the new fledgling species. And that is, in fact, what the young Darwin himself thought when he developed this scenario of adaptive change—the only scenario other than the long-term accrual of adaptive modification of species that he, or for that matter I, see as possibilities.

What is it like to see yourself, now an aging paleontologist once thought to have been a radical thinker with new views who finds out, in the twilight of his career, that the young Charles Darwin essentially saw all the components of punctuated equilibria? I admit initially to being somewhat abashed when I stumbled on all this, beginning with the preparation of an exhibition and accompanying book on Darwin at the beginning of the twenty-first century.

But now I am more thrilled than anything else about it. Think of it: the young Darwin was such a careful and creative thinker that he essentially saw both punctuated equilibria and phyletic gradualism as alternative narratives of the story of adaptive evolutionary change through time. He got to a fork in the road, and took what has later turned out to be the wrong fork. Allopatric speciation lay in limbo, championed by very few biologists after Moritz Wagner, who was Darwin's far less well-known contemporary. It took Theodosius Dobzhansky, followed by Ernst Mayr, and eventually many others simply to resurrect the concept of allopatric speciation. Then we paleontologists came along and sealed the deal by resurrecting what Steve Gould famously called "paleontology's trade secret": stasis. These were the beginnings of the rebirth of the "taxic perspective" in evolutionary biology.

The "strong corollary" had immediate implications for many imaginative evolutionary biologists, neontologists and paleontologists alike: if stasis is empirically established as common and typical of the histories of most species in the evolutionary history of life, then most lasting adaptive change occurs in conjunction with the origin of species in isolation. In part, the strong corollary over the ensuing decades has led to an explosion of work and has pointed the way to work yet to be done. Ironically, it has led as well to still further resurrections of the sorts of themes pioneered, but long forgotten, in the work of Georges Cuvier, Jean-Baptiste Lamarck, Giambattista Brocchi, Alcide d'Orbigny—and, of course, the youthful Charles Darwin himself.

6

SPECIATION AND ADAPTATION

Large-Scale Patterns in the Evolution of Life, 1972–

Evolution, of course, implies that all organisms, living and extinct, are genealogically interconnected through a process of ancestry and descent. Life, in short, is monophyletic—a proposition repeatedly confirmed over the past two hundred years, and made all the more certain by the spectacular results of the application of molecular biology to evolutionary issues.

So if life is monophyletic, there must be one single evolutionary process that has generated life's history, whatever the relation between the process of adaptation and the origin of species may be. Whether one subscribes to a gradualist picture of the slow, steady, inexorable modification of existing species into the species of the modern world; or if instead one sees adaptive change largely associated with, and perhaps a function of, relatively brief events in isolation, and accordingly sees species as real and discrete in and of themselves; or if one supports some combination of views: regardless, that is, of how one thinks of the relationship of adaptation to the processes of speciation, all evolutionary biologists realize that the pattern of common descent among all known species bespeaks a fundamental sameness to the evolutionary process, at least insofar as metazoans, metaphytes, indeed sexually reproducing organisms in general, are concerned.

And that suggests to all evolutionists, regardless of their era, that the processes of selection-mediated adaptive change, and the processes of speciation lie at the heart of the entire evolutionary process, from the generation of patterns of genetic variation within species, up through and including the origins of major groups of animals and plants. Adaptation though natural selection has been seen by all at least since 1859 as integral to the evolution of everything from differentiated populations up through the origin and diversification of entire groups, like mollusks or mammals.

Once the importance of isolation was redeveloped, it naturally occurred to pioneers like Theodosius Dobzhansky and Ernst Mayr that species themselves might have some sort of role to play in evolution. Mayr, in particular, examined patterns of geographic replacement of closely related species. In *Animal Species and Evolution* (1963), Mayr examines phenomena such as adaptive radiation, suggesting that the evolution of larger-scale taxa can be viewed in terms of a sequence of speciation events coupled with adaptive change. Mayr (1963:587) wrote of species as potential evolutionary pioneers. Summing up his thoughts of the role of species in "transpecific evolution," Mayr wrote:

> The evolutionary significance of species is now quite clear. Although the evolutionist may speak of broad phenomena, such as trends, adaptations, specializations, regressions, they are really not separable from the progression of entities that display these trends, the species. The species are the real units of evolution, as the temporary incarnation of harmonious, well-integrated gene complexes. And speciation, the production of new gene complexes capable of ecological shifts, is the method by which evolution advances. Without speciation there would be no diversification of the organic world, no adaptive radiation, and very little evolutionary progress. The species, then, is the keystone of evolution. (621)

Strong words. Mayr's characterization of species as the "temporary incarnation" of "harmonious, well-integrated gene complexes" shows the (positive) grip of Sewall Wright's (1931, 1932) imagery on Mayr's thinking. Speaking as one who came along just after Mayr wrote these words (in 1963), I agreed with everything he says in this passage, with one demurral: Mayr calls species "temporary incarnation(s)" of "harmonious, well integrated gene complexes," implying more than a whiff of the traditional view of the necessarily transient nature of species. But when Stephen Jay Gould and I were contemplating our paper on "punctuated equilibria," I saw—and wrote in the first draft of our manuscript (Eldredge 1970b)—my first glimmers of the "strong corollary" and the challenge it posed for understanding the production of interspecific evolutionary patterns, including "trends." For if there were indeed apparent long-term *directional* patterns of among-species change within groups through geological time; and if stasis is a generally true characterization of the histories of component species within the group displaying the trend, then we can no longer simply cite direct extrapolation of natural selection into geological time to explain patterns of long-term adaptive change. In my cover letter to Steve when I sent him two copies of the first draft, I merely said: "I hope you find something of merit in my discussion of trends appended to the ms. Do with it as you see fit."

"TRENDS" IN "PUNCTUATED EQUILIBRIA": EARLY STAGES OF THE "SPECIES SELECTION" DEBATE

As it turns out, Steve left my discussion of trends in the first draft of our paper nearly intact in the published version (Eldredge and Gould 1972a:111–112). It was accompanied by a simple text diagram depicting differential production and survival of species within two subclades of a natural (monophyletic) group of species (figure 6.1). The species were all depicted as stable—that is, in "stasis."

After a brief flirtation with the idea that trends may in fact be nothing more than a selective rendering of species in the fossil record, we settled on a definition: "A trend is a direction which involves the *majority* of related lineages of a group (MacGillavry 1968:72)." Though we presented no empirical examples, whether from our own research or from that of others, I have long since thought that the net brain size increase in hominid evolution fits the bill rather well.

But if we take simple linear natural selection off the table, then what supplies the apparent directionality of adaptive morphological change in the generation of trends? I had read a provocative comment by Sewall Wright (yet another example of an analogy between distinct levels of

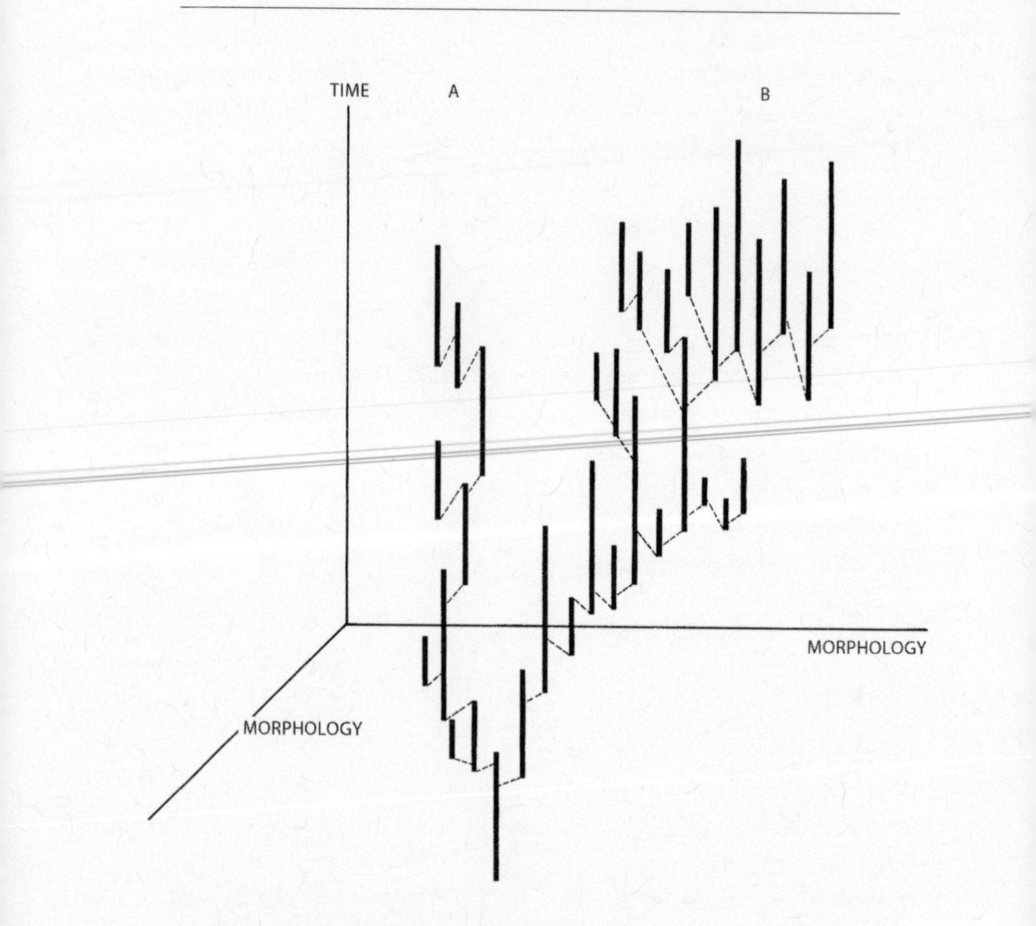

FIGURE 6.1 Niles Eldredge and Stephen Jay Gould's diagram of patterns of within- and among-species stasis and change—the early forerunner of the "species selection" literature. For the original caption, see the text. (From "Punctuated Equilibria: An Alternative to Phyletic Gradualism" [1972a:113, fig. 5-10])

biological organization) and used it here to generate a possible explanation for the phenomenon of trends in general:

> Sewall Wright (1967, p. 120) has suggested that, just as mutations are stochastic with respect to selection within a population, so might speciation be stochastic with respect to the origin of higher taxa. As a slight extension of that statement, we might claim that adaptations

to local conditions by peripheral isolates are stochastic with respect to long-term, net directional change (trends) within a higher taxon as a whole. We are left with a bit of a paradox: to picture speciation as an allopatric phenomenon, involving rapid differentiation within a general, long-term picture of stasis, is to deny the picture of directed gradualism in speciation. Yet, superficially at least, this directed gradualism is easier to reconcile with valid cases of long-term trends involving many species.

McGillivray's definition of a trend removes part of the problem by using the expression "majority of related lineages." This frees us from the constraint of reconciling *all* events of adaptation to local conditions in peripheral isolates, with long-term, net directional change.

A reconciliation of allopatric speciation with long-term trends can be formulated along the following lines: we envision multiple "explorations" or "experimentations" (see Schaeffer, 1965)—i.e., invasions, on a stochastic basis, of new environments by peripheral isolates. There is nothing inherently directional about these invasions. However, a subset of these new environments might, in the context of inherited genetic constitution in the ancestral components of a lineage, lead to new and improved efficiency. Improvement would be consistently greater within this hypothetical subset of local conditions that a population might invade. The overall effect would then be one of net, apparently directional change: but, as in the case of selection upon mutations, the initial variations would be stochastic with respect to this change [see figure 6.1]. (Eldredge and Gould 1972a:111–112)

This passage could have been more clearly written. But it was a speculation—one based on a very Giambattista Brocchian-styled analogy posed originally by Sewall Wright. And, I must say, I disagree with my co-author Steve Gould (2002:670), who wrote in his final manifesto on evolution: "When Niles Eldredge and I first formulated punctuated equilibrium [*sic*], I was most excited by the insight that trends would need to be reconceptualized as differential success of species, rather than anagenesis within lineages (a theme only dimly grasped in Eldredge and Gould, 1972, but fully developed in Gould and Eldredge, 1977)" (670). Au contraire. As this passage shows, we obviously grasped the essence of the problem: that stasis undermines the projection of within-species gradual change to explain among-species long-term trends. We made that quite clear.

What we were saying in this short passage on trends in our 1972 paper was simply that there may well be differential success (survival, in this context) of species that happened to invade certain kinds of environments over others—and that alone might suffice to produce what looks like a linear trend.

Fortunately, we said as much in the caption to the figure:

> Three-dimensional sketch contrasting a pattern of relative stability (A) with a trend (B), where speciation (dashed lines) is occurring in both major lineages. Morphological change is depicted here along horizontal axes, while the vertical axis is time. Though retrospective pattern of directional selection might be fitted as a straight line in (B), the actual pattern is stasis within species, and differential success of species exhibiting morphological change in a particular direction. (Eldredge and Gould 1972a:113)

There it was, fortunately: the phrase "differential success of species." In the somewhat overblown, if not wholly false, modesty we adopted throughout this paper, we went on to say, as a sort of appendix to the text just quoted: "We postulate no 'new' type of selection. We simply state a view of long-term, superficially 'directed' phenomena that is in accord with the theory of allopatric speciation, and also avoids the largely untestable concept of orthoselection" (112).

But the cat was out of the bag: for the first time, I believe, since Dobzhansky (1937) had effectively reinstated the discreteness of species and their origins in small isolated populations, with the addition of stasis it was no big imaginative leap to begin to think of species as discrete entities, with their own births, histories, and deaths. Brocchi, if not wholly reinvented, close to becoming so. And if species are discrete in space and in time, why not think of them, in the context of monophyletic groups, as entities that can indeed have differential fates—some more successful than others?

In retrospect, it was a mistake not to take the opportunity and actually indeed name a "new" type of selection. A few years later, paleontologist

Steven Stanley (1975) did just that. He dubbed the pattern and presumed underlying causal explanation "species selection." There was more to come.

ELISABETH S. VRBA: SPECIES SORTING VERSUS SPECIES SELECTION, 1980–

No one contributed more to the taxic perspective and macroevolutionary theory in the 1980s and 1990s than paleontologist Elisabeth S. Vrba (figure 6.2). Her contributions were many, varied, and significant, including an important clarification and extension of concepts surrounding "species selection." In addition, she published novel insights and ideas on many of the subjects treated throughout

FIGURE 6.2 Elisabeth Vrba at the Kromdraii fossil site in the Transvaal, South Africa. (Courtesy of Michelle Eldredge; by permission of Elisabeth Vrba)

the remainder of this chapter, including hierarchy theory; the "species as individuals" debate; and possibly the most significant of all, the all-important pattern of species level, ecosystem-wide, cross-genealogical extinctions and consequent bursts of speciation—her "turnover pulses."

Yet another noteworthy contribution, not further discussed herein, as it is not explicitly a contribution to taxic evolutionary theory, is her concept of "exaptation." The term is usually associated with Stephen Jay Gould, but only because he managed to appropriate senior authorship of their paper announcing the concept of exaptation (Gould and Vrba 1982).

But first things first. Sometime in the late 1970s, Elisabeth mailed both me and, I believe, Steve Gould a copy of her manuscript "Evolution, Species and Fossils: How Does Life Evolve?" (1980). Elisabeth was raised and educated in southern Africa, and the package containing her manuscript was posted from the Transvaal Museum in Pretoria, where she was working as a paleontologist (specializing on several groups of antelope, including impalas and wildebeests) and as assistant director. That paper, which became informally known as the "Green Monster," owing to the dominant color of the reprint cover, and perhaps especially to the length and density of the exposition (it was twenty-three pages long in final printed form), took a determinably taxic stance from its inception. I remember being very happy when I saw Vrba's contrasting images and discussion of gradualism versus a taxic, punctuational take on the history of life—with a decided preference for the latter.

But I think the paper is best remembered for its analysis of the varied, and up to that point rather vague and confused discussion of what had come to be called "species selection." In it Vrba makes two particularly cogent additions to the very concept of species selection. To properly be called a form of selection, truly analogous to the lower level of organismic natural selection (levels again—more shades of the long-forgotten Giambattista Brocchi), focus should—must—be on truly "species-level" properties. Moreover, and again to be a true analogy with natural selection, species selection should be seen as more than just differential deaths of species within a clade. After all, biologists had long since come to measure the effects of natural selection in terms of "relative fitness," meaning relative reproductive success, among organisms within a deme, or "population," or even an entire species. In other words, for the analogy to be complete, differential rates of speciation—births of new species—should be added to the very definition of "species selection."

Vrba leaves no doubt that there are such patterns of differential births and deaths of species—exemplified very clearly in the contrasting patterns of births and deaths of species within two sister clades (taxa) of African antelope drawn from her own data. But she casts considerable doubt on the very existence of truly species-level ("emergent") properties. She suggests, instead, that down-to-earth organismic properties (morphological, behavioral, physiological) may well be all that is required to produce contrasting histories of births and deaths of species among related clades. Her own antelope example had to do with relative "niche width" of species. Impalas have low rates of speciation and extinction and are eurytopic, meaning they occupy a wider variety of environments than the more narrowly niched wildebeests and kin. The latter group has far greater species diversity and higher rates of species extinction. True both in the past and present, since part of the mate recognition system of these antelopes resides in the shape of their horns, commonly preserved in the fossil record. I had made a similar argument involving the effect of niche breadth on rates of evolution in a paper outlining the differences between the "taxic" and "transformational" "alternative approaches" to thinking about evolution (Eldredge 1979).

But niche breadth arises from the adaptive phenotypes (again, the morphology, behavior, and physiologies) of organisms. And just because all impalas not only look pretty much alike but behave pretty much alike does not thereby imply that the adaptive complex of modern impalas is thereby a property of the species. In other words, organismal characteristics that are species-wide are nonetheless just that. They are organismal, rather than truly emergent species properties. Vrba said that patterns of differential species births and deaths rising purely from organismal characteristics could not properly be called "species selection." Instead, "species sorting" is a more appropriate term. Because macroevolutionary patterns can arise as side effects of differing organismal adaptations, Vrba (1984) dubbed this "effect macroevolution":

> If one can explain sorting among species solely by comparison of characters and dynamics at the level of organisms and genomes (the effect hypothesis), then there is no need to invoke species selection. The argument amounts to a plea for recognition that not all non-random sorting among species need be caused by selection (see discussion in Vrba and Eldredge, 1984) and for restriction of a hypothesis of species selection to a hypothesis of species aptation. (322)

I remember a lively discussion with Vrba sometime in the 1980s (she had moved to Yale University in the 1980s). Her work had convinced me that "species selection" was a misnomer, as it was next to impossible to imagine credible "species aptations," meaning true examples of emergent species-level properties. The one example I could accept (then and now) is the Species Mate Recognition System (SMRS; Patterson 1985 [see chapter 4]) in sexually reproducing species, as it is an adaptive system involving males and females that enables continued reproduction and, as a side effect, precludes the possibility of successful mating with organisms from a closely related species. Vrba disagreed with me, saying she hadn't ruled out species selection altogether. Instead, she had just restricted its use and (my words) raised the bar considerably for actual demonstration of an example of species selection.

And there the matter seems to stand: some (like myself) follow Vrba's argument on the necessity of emergent properties (characters of species) before invocation of "species selection." Others (for example, Gould and Lloyd 1999) have a looser criterion involving "emergent fitness" at the species level, eliminating the need to find what well may not exist (beyond the SMRS): actual examples of true, emergent species-level properties.

So there was some give-and-take, some disagreement, among Vrba, Gould, and me on the subject of species selection. But such is inevitable in close working relationships: it is part of the fun, actually. I have always said that the best part of working with Steve Gould was finding the places where we disagreed. Much the same is true about working with Elisabeth Vrba, who is a most careful and thorough, as well as highly creative, thinker on evolutionary issues. Whatever differences may have developed and persisted among us three, it is worth quoting Steve's dedication to his last book, *The Structure of Evolutionary Theory* (2002):

> FOR NILES ELDREDGE AND ELISABETH VRBA
>
> May we always be the Three Musketeers
> Prevailing with panache
> From our manic and scrappy inception at Dijon
> To our nonsatanic and happy reception at Doomsday
> All For One and One For All

For that is how we thought of ourselves—as the "Three Musketeers." We banded together to help develop these new lines of thinking, and also to stave off the strong resistance to our way of looking at things frequently evident at meetings. We became the Three Musketeers in a meeting in Dijon, France, where the going had been particularly rough. That was also a lot of fun.

In any case, and strictly from my perspective, any higher-level analogue to natural selection must conform to criteria set forth by philosopher David Hull (1980): for an entity to be selected, it needs to be a "replicator," as well as an "interactor." Organisms are clearly both. They reproduce (replicative fidelity supplied by the genome) and they "interact" (that is, organisms are matter-energy transfer entities, with active roles to play in the economy of nature). Some of the most important work done in the 1980s involved clarification of the ontological status of biological entities, especially species. I'll take this up in the next section on hierarchy theory. But suffice it to say that species can reasonably be seen as "replicators." Species are "packages of genetic information." Also, species make more species: new species arise from old species via speciation. But species are *not* interactors in an ecological sense. More on this directly.

So what has been gained by the gallons of ink spilled over "species selection"? What we do have is a return to the realization that species not only replace one another in time and in space, but also that some groups show higher rates of speciation often correlated with higher rates of extinction. The result is recognizable, repeated patterns in the history of groups, starting with trends. Indeed, Mayr (1963) singled out "trends," as did Eldredge and Gould (1972a), and later Vrba (1980) as an obvious "macroevolutionary" pattern arising among many species within a clade.

As far as I am aware, "trends" represent the most compelling application of the notion of differential species production and survival. Why? Because of the paradox between stasis and the long-held belief that any apparent directionality among species is due to simple natural selection writ large: as "orthoselection" (a neologism invented originally, I believe, to avoid the teleology of "orthogenesis"). Beyond trends, though, "species selection" per se seems to have done very little meaningful "work" in evolutionary biological discourse.

But the very valuable residue of all this discussion of among-species patterns of origination and extinction must not be overlooked. The rise and fall of monophyletic taxa from the level of genera through phyla and kingdoms has become, once again, fair game for empirical research. Jack Sepkoski's (1978) monumental analyses of large-scale patterns in the fossil record drawn from compendia of published sources; Bruce Lieberman's (2000, 2001, 2003) investigations of the rates of speciation and extinction among the earliest trilobites at the "Cambrian explosion"; Lynn Margulis's (1974) demonstration, based on anatomical and molecular evidence, of the origin of the Eukaryota; and recent (as of this writing) work on the origin and diversification of placental mammals; and so forth. All ultimately stem from the quantitative work of the earliest evolutionary paleontologists, especially Jean-Baptiste Lamarck, Giambattista Brocchi and Robert Jameson; and, at least in terms of the understanding of the origin of major taxa of the modern biota, stemming as well from Ernst Haeckel and the embryologists who sought to flesh out the reality of Charles Darwin's seminal vision of evolution as finally published in 1859.

In short, none of this explosion of modern work was wholly new under the sun. And still less did it all derive from the resurrection of the taxic perspective as embodied in punctuated equilibria and its application (via our brief treatment of the apparent paradox of evolutionary trends) to among-species, higher-level evolutionary phenomena. But I do think we did give a boost toward thinking above the Darwinian-restricted level of natural selection within species, while establishing a counterfoil to the focus on the gene level to the near exclusion of everything else, as usefully developed in the 1970s by Richard Dawkins (1976).

For if it is true that species are usefully construed as "syngameons" (as Dobzhansky [1935] took from J. P. Lotsy), and I believe it is, then the differential production and fates of species are all about the differential production and survival of discrete packages of genetic information. From that perspective, what could be more important, beyond basic genetic processes and natural selection itself, in understanding the evolutionary history of life and the processes that have shaped that history than the phenomena of speciation and extinction—and the differential births and deaths of species?

SHADES OF BROCCHI: ARE SPECIES "INDIVIDUALS"?

Two years after we published our paper on punctuated equilibria (Eldredge and Gould 1972a), biologist Michael Ghiselin published a paper with the arresting title "A Radical Solution to the Species Problem" (1974; see also Hull 1976, 1978). Ghiselin's paper had nothing explicitly to do with our ideas on speciation plus stasis, let alone our thoughts on the concentration of lasting evolutionary change in speciation events, but many of us immediately saw the latent connections.

In this brief paper, Ghiselin argued that traditionally species have been seen as "classes," with members (organisms) and definitions: to be a member of a species, an organism had to possess the defining properties of that species. Many systematists even in post-Darwinian years more or less understood species in this way. Hypothetically, if organisms were to be found in far-flung regions well away from where a species is regularly known to occur, they would nonetheless have to be considered as members in good standing of that species if they possessed the properties of that species. Thus even if a butterfly seemingly identical to a species of a family endemic to Africa were to be found in, say, the Amazon rainforest, there would be no question that it would belong to the "same" species (though nowadays DNA sequencing would quickly resolve the issue). As an example from the physical realm, consider gold—a category ("class") of metallic elements. Every atom of gold is an individual, but, under the right circumstances, atoms of gold can be formed at any time anywhere in the universe. If an atom has an atomic number of 79 and an atomic weight of 196.966569, it is by definition an atom of gold.

Not so for biological species, Ghiselin argued. Ghiselin saw species as reproductive entities (J. P. Lotsy's/Theodosius Dobzhansky's "syngamea," in fact, plus a concept of reproductive isolation), and as the largest-scale units with what he called "reproductive competition" among its component organisms ("parts"). Ghiselin repeats the by-then common assertion that Darwin for the most part paid insufficient attention to the origin of species, seeing them instead more like boundless classes, given their constant internal evolutionary makeovers. He was alluding here to the published work of the older Darwin, and not, of course, to what the younger Darwin thought about things.

Darwin to the contrary, Ghiselin asserts, species are indeed "real." They have names. They have origins (and histories and deaths, though Ghiselin does not elaborate on these latter two features of species). And to illustrate his proposition that species are "individuals," he makes analogies with economic entities, such as automotive firms.

Ghiselin (1974) ends his paper with a call for more careful consideration of the fundamental natures of biological entities ("ontology")—which alone, in itself, was a stirring call for action that, I think, underlay much of the work in evolutionary biological hierarchy theory that was to follow, predominantly in the 1980s:

> The species problem has thus involved difficulties in understanding the ontological status of fundamental biological units, and failure to interrelate levels of integration in the appropriate manner. The reason why

> the economic analogies have so much heuristical value is that we find it much easier to think of firms as individuals. For this very reason a host of biological problems are more readily soluble if one learns to think like an economist. In addition we should note that we often fail to solve our problems because we cannot even identify them. Under such circumstances, conceptual investigations do more than just help. They are the only way out. (543)

These were stirring words, and some of us went on to take them as a call to arms. The Supreme Court decision in *Citizens United v. Federal Election Commission* (2010), affirming the status of corporations as "individuals," and extending that concept into the realm of campaign financing, one assumes, had nothing to do with Ghiselin's argument.

Ghiselin noted that his thesis would perhaps be primarily of interest to philosophers. And, in fact, the most immediate reaction came from David Hull, a philosopher of science, who very much liked Ghiselin's core idea of species as individuals. Hull (1976, 1978) developed various aspects of Ghiselin's idea still further over the years. It must be said, though, that Ghiselin's distinction between "classes" and "individuals" did not win universal approbation among philosophers; indeed the matter is still being debated in some philosophical quarters.

But it is the core biological implication of "species as individuals" that matters to this narrative. I liked the idea very much, as of course punctuated equilibria painted a picture of species with discrete births and eventual deaths. Species are also "more-makers" on their own account, meaning that species from time to time give rise ("birth") to other species, as organisms themselves do (albeit speciation is more like asexual fission than the process of sexual reproduction among complex organisms). I liked the idea still more because of *stasis*: species are discrete from birth and remain so, consistent throughout their histories as recognizable entities in their own right. They do not, upon speciation, resume an imagined course of continued gradual change, thus adopting a sort of class-like status of open-ended non-existence.

Philosopher Marjorie Grene first caught my attention when I was a graduate student, with her paper "Two Evolutionary Theories" (1958).

There she compared the evolutionary theoretical positions of two paleontologists: the American vertebrate paleontologist George Gaylord Simpson and the German invertebrate paleontologist Otto Schindewolf, whose saltationist views Simpson staunchly opposed (despised would perhaps be more like it). Grene astonished me by concluding that Schindewolf's work was the better science, not least because it seemed to rely more closely on actual data. (I remember turning to the published rebuttals of her paper, because as a mere student I myself felt inadequate to refute her analysis. One of the rebuttals was by Walter Bock, a professor of mine at Columbia, and himself a distinguished evolutionary biologist. I was further dismayed when I found myself unconvinced by Bock's comments.)

Not that Grene was a "saltationist." Like any good philosopher, her interest was in the formal structure and logic of arguments and propositions. But she did go on and do what I view as important biological "work" in the conceptual manner that Ghiselin (1974) was calling for, as the work on "species as individuals" and the issues of "integration of levels" developed in scope and intensity in the work on biological hierarchies in the 1980s.

I mention Grene here because of her interest in the "species as individuals" proposition, which for the most part appealed to her. But I was once again surprised by her, this time when she assured me that both Ghiselin and Hull were attracted to the idea of species as individuals not least because species *do* continue to change gradually through time. And this would be a direct analogy between organismic-individuals and species-individuals: all organisms change, radically, from a fertilized egg through embryonic changes to a fully formed individual at birth. And organisms change more slowly as growth, sexual maturity, and aging proceed apace. Yet as Ghiselin (1974) says tartly, we remain ourselves—an individual organism—despite all such change. Species, on the other hand, are bound to change slowly, evolutionarily through geological time.

Grene was saying to me that in making the analogy explicit between organisms and species as individuals, Ghiselin and Hull were taking into account similar patterns of change through time in both organisms and species, and that such change does not preclude their being conceptualized as "individuals." In other words, stasis has nothing to do with the proposition.

But I liked linking stasis up with the "species as individuals" debate, and I continue to do so. As a paleontologist looking at the fossil record, I see species stand out as discrete, with very little ambiguity engendered by variation in space or in time. Damn right they are individuals! It is only a matter of figuring out just what sorts of individuals they are, and how close the actual analogy between species and individual organisms really is. For it was not Ghiselin's original intent, at least in his seminal paper, to make the actual analogy between species and organisms, except to postulate that they both have the ontological status of "individuals," rather than "classes."

Once analogies are sought between individual organisms and individual species, we are, at long last, catching up with Giambattista Brocchi, who initially made that analogy in 1814. It is marvelous, in a very satisfying way, to measure conceptual progress by the degree that we can see ourselves catching back up with the wisdom of our elders. By the end of this narrative, we'll be back in time to even before the powerful analogy that Brocchi made. But first we must take a harder look at levels, the ontological status of biological entities, and the degree to which the analogy between species and organisms as individuals holds ontological water.

THE BEGINNINGS OF MODERN HIERARCHY THEORY, 1951–1980

In some ways biologists have always been comfortable with hierarchies—the archetypal example being the Linnaean hierarchy as codified no later than 1758. Organisms are parts of species, species parts of genera, and so on up the scale. The categories ("species," "genera," "families," and on up) are arranged in a linear array of greater actual or potential inclusivity (there are some monotypic higher taxa known, such as aardvarks). But the taxa themselves, the actual species, actual genera, actual families, occur as nested sets. *Homo sapiens* is a species within the genus *Homo*, and *Homo* is a genus within Family Hominidae. As a vivid example of the sort of ontological clarification of biological entities through the "conceptual investigation" that Michael Ghiselin (1974) was calling for, consider that it wasn't much more than a decade before Ghiselin's paper that George Gaylord Simpson (1963) published his penetrating analysis, with his conclusion that there is a fundamental distinction between "species, genera, classes" (as categories of the Linnaean hierarchy) and examples of actual species, genera, classes, and so on: actual taxa such as *Homo sapiens*, Hominidae, and the like. Taxa are at least arguably phylogenetically historical "individuals," whereas Linnaean categories are "classes."

But in terms of understanding how the evolutionary process works, there has been a much less coherent history to thinking in terms of biological levels. This is so both in the organization of biological entities, and perhaps especially how individuals within such levels interact, especially between higher and lower levels. Indeed, it wasn't until the pioneers of modern speciation theory who brought back the taxic perspective to evolutionary theory started writing, that the possibility of seeing the evolutionary process as the totality of processes at discrete levels of biological organization was discussed. Perhaps unsurprisingly, an early discussion of the importance of hierarchically arrayed levels of biological organization came from Theodosius Dobzhansky himself, my candidate for the one who got the modern era of taxic thinking via speciation theory going. I would also point out that Dobzhansky's (1935) laconic remark is another example of precisely the sort of conceptual analysis that Ghiselin (1974) was calling for. Dobzhansky says we must ask: "whether the species is a purely artificial device employed for making the bewildering diversity of living beings intelligible, or corresponds to something tangible in the outside world" (345). In other words, is the species "a part of the 'order of nature,' or [merely] a part of the order-loving mind?"

But as an example of pure hierarchical thinking in evolutionary theory, consider Dobzhansky's words in the first edition of *Genetics and the Origin of Species* (1937):

> Mutations and chromosomal changes arise in every sufficiently studied organism with a certain finite frequency, and thus constantly and

> unremittingly supply the raw materials for evolution. But evolution involves something more than origin of mutations. Mutations and chromosomal changes are only the first stage, or level, of the evolutionary process, governed entirely by the laws of physiology of individuals. Once produced, mutations are injected in the genetic composition of the population, where their further fate is determined by the dynamic regularities of the physiology of populations. A mutation may be lost or increased in frequency in generations immediately following its origin, and this (in the case of recessive mutations) without regard to the beneficial or deleterious effects of the mutation. The influences of selection, migration, and geographical isolation then mold the genetic structure of populations into new shapes, in conformity with the secular environment and the ecology, especially the breeding habits, of the species. This is the second level of the evolutionary process, on which the impact of the environment produces historical changes in the living population.
>
> Finally, the third level is a realm of fixation of the diversity already attained on the preceding two levels. Races and species as discrete arrays of individuals may exist only so long as the genetic structures of their populations are preserved distinct by some mechanisms which prevent their interbreeding. . . . .The origin and function of the isolating mechanisms constitute one of the most important problems of the genetics of populations. (13–14)

In other words, there are three ontologically distinct levels of biological organization (individuals, populations, species) that each play a distinct role in the evolutionary process. All three are necessary for the generation, fate, and conservation of genetic change—Dobzhansky's not unrealistic definition of evolution itself. Interestingly, Dobzhansky saw that at the lowest level, discontinuity (the production of discrete mutations) is the predominant theme. At the next higher level of populations, gene frequencies are the coin of the realm, and thus continuity is the dominant theme. But at the third and highest level, that of the origin of discrete, reproductively isolated species, discontinuity is once again the important theme. That is where existing variation is "fixed," or conserved. And, I might add, in concert with

Dobzhansky's own rhetoric with mutations at the lower level, "injected" at this higher level into the mainstream of phylogenetic history, there to meet their own differential fates.

As we have already seen in this chapter, Ernst Mayr (1963) implicitly spoke about levels when he stressed the role of species in generating large-scale patterns of "transpecific evolution." And Niles Eldredge and Stephen Jay Gould (1972a), riffing off a similarly hierarchically imbued analogy from Sewall Wright (1967), started a discussion of the differential fates of species as a higher-level way of understanding how phylogenetic trends might be generated, given the *prima facie* falsification posed by stasis of the simple extrapolation of truly long-term patterns of selection-generated change through time.

These were just some of the early mileposts of what became, in the late 1970s and throughout the 1980s and still ongoing, conceptual analyses and ontological clarifications that soon came to be known as "hierarchy theory." One of the dominant themes to emerge early on was the relationship between ecological and evolutionary processes.

CONNECTIONS BETWEEN ECOLOGY AND EVOLUTION

The history of taxic thinking in evolutionary theory, as revealed so far in this narrative, has had little to say on the subject of ecology. Even at the level of within-species evolutionary change, much of the work of population geneticists, for example, has been concerned with the fates of genes in populations. And certainly, when a hierarchical perspective has been mooted, such as Dobzhansky's early foray, distinguishing between the organismic, population, and among-species levels, his discussion was all about the generation, modification, and, ultimately, conservation of genetic variation. Most evolutionary biologists most of the time, myself included, are comfortable with a definition that sees evolution fundamentally as a matter of changes (or lack thereof) of genetic information through time.

And yet everyone "knows" that evolution is deeply connected to, and even deeply dependent on, the "environment." After all, Darwin's mature evolutionary theory was essentially a theory of adaptation through natural selection. And a moment's reflection reveals the obvious: in addition to adaptations that are strictly concerned with reproduction, the vast majority of adaptations—phenotypic, behavioral, and physiological functions of individual organisms—are concerned with the somatic side of life. Organisms are matter-energy transfer machines. Organisms spend the greatest amount of time and energy simply on the basic functions of living. They are procuring or manufacturing renewed energy sources and staving off ultimately inevitable death from predators, disease, or simple starvation.

In his riveting account of the relationship between the study of evolution and the study of ecology, evolutionary biologist John N. Thompson (1994:chaps. 2, 3) makes much the same point. The early "naturalists," starting with Charles Darwin (with adumbrations of earlier scientists, including, I was surprised to learn, Joseph Priestly, better remembered for his chemistry), used the word "adaptation" freely. The search was on for a natural causal explanation for the sometimes remarkable match between organism and environment.

Thompson makes a compelling case that Darwin's own studies of species interactions show how ecological factors played a major role in the early stages of Darwinian evolutionary biology. This is especially true of Darwin's amazingly detailed and careful studies on pollination, and also his work on specialization—both quickly followed by many naturalists in the field and lab.

And yet, Thompson points out, with the advent of genetics in the early twentieth century, the original environmental perspective of the early evolutionarily minded naturalists ebbed. Ecology developed its own theories and perspectives. Relatively few ecologists, excepting, most notably, those of the "Chicago School" of the earlier decades of the twentieth century, were concerned at all about evolution. And the reverse, I think it is fair to say, was also the case. Evolutionary biologists might have occasionally alluded to the importance of ecology (as Dobzhansky did in the "levels" quote earlier in this chapter), but that was usually as far as it went. Ecologists were more concerned with how things work in biological nature than how they came to be the way we see them today. Evolutionists were interested in the intricate details of how things worked, but as a means to understand how adaptations—especially extreme adaptations—could have arisen.

Thompson closes his historical narrative with an account of how ecology and evolution began to come together, starting roughly in the 1960s. And there is no doubt this is true, with recent developments in "ecological speciation" reaching the level of what I have been calling "taxic" thinking.

Yet the evolutionary biological perspective on selection and adaptation has remained, of necessity, focused on within-species structures and processes. In co-evolutionary phenomena, the evolutionary play is greatly expanded, of course, as in the interplay between two or even more (unrelated) species, such as in insect pollination of plants.

I also think it fair to say that larger-scale systems and processes that had been the staple of ecological theory for much of the twentieth century—especially ecosystems of varying spatial magnitude—would seem to have not been explicitly or successfully connected with larger-scale evolutionary events and processes in evolutionary theory.

As ever, there are exceptions to these historical generalizations. One such attempt to link evolution and ecology on a progressive from-species-through-higher-taxa scale was made (by now, unsurprisingly) by Theodosius Dobzhansky, in the third edition of *Genetics and the Origin of Species* (1951). In this striking passage, Dobzhansky expands on the metaphor/analytical rubric of Sewall Wright's (1931, 1932) concept of "adaptive peaks." Wright, starting with the simple claim that some gene combinations are "more harmonious" than others, depicted the more harmonious gene combinations as occupying "peaks" in an "adaptive landscape" that looked like a topographic map. Less harmonious gene combinations would be located down the slopes of the peaks—the least harmonious occupying the valley floors. The process of evolution, Wright said, is geared to maximize the number of individuals within a population on the peaks.

Wright himself was the first to extend the metaphor of adaptive peaks, seeing populations, indeed even species, as occupying different peaks in the adaptive landscape. Others took this gambit and ran with it: George Gaylord Simpson's (1944) initial theory of "Quantum Evolution" was nothing less than a model to explain how species come to occupy new peaks in the realizable ecological adaptive field. Simpson thought that a portion of a species would have to go through an "inadaptive" phase to get off the presently occupied peak, pass through the valley (I cannot write these words without adding "of the shadow of death"), through a "pre-adaptive phase," beginning to climb the next peak to the point at which selection takes over (the "selective phase"), to complete the transition and form a new taxon (species, to be sure, but his theory was mainly about major adaptive change exemplified by higher taxa) on a newly occupied adaptive peak. Very imaginative imagery—though, as noted in chapter 5, it was Wright himself who basically shot down this early expression of "Quantum Evolution."

Dobzhansky's (1951) gambit utilized adaptive peaks in a somewhat different way:

> The enormous diversity of organisms may be envisaged as correlated with the immense variety of environments and of ecological niches which exist on earth. But the variety of ecological niches is not only immense, it is also discontinuous. One species of insect may feed on, for example, oak leaves, and another species on pine needles; an insect that would require food intermediate between oak and pine would probably starve to death. Hence, the living world is not a formless mass of randomly combining genes and traits, but a great array of families of related gene combinations, which are clustered on a large but finite number of adaptive peaks. Each living species may be thought of as occupying one of the available peaks in the field of gene combinations. The adaptive valleys are deserted and empty.
>
> Furthermore, the adaptive peaks and valleys are not interspersed at random. "Adjacent" adaptive peaks are arranged in groups, which may be likened to mountain ranges in which the separate pinnacles are divided by relatively shallow notches. Thus the ecological niche of the species "lion" is relatively much closer to those occupied by tiger, puma, and leopard than to those occupied by wolf, coyote and jackal. The feline adaptive peaks form a group different from the group of the canine "peaks." But the feline, canine, ursine, musteline, and certain other groups of peaks form together the adaptive "range" of carnivores, which is separated by deep adaptive valleys from the "ranges" of rodents, bats, ungulates, primates, and others. In turn, these "ranges" are again members of the adaptive system of mammals, which are ecologically and biologically segregated, as a group, from the adaptive systems of birds, reptiles, etc. The hierarchic

> nature of the biological classification reflects the objectively ascertainable discontinuity of ways and means by which organisms that inhabit the world derive their livelihood from the environment. (9–10)

Dobzhansky's vivid imagery is not detailed in the way of Simpson's (1944) models of evolution utilizing the imagery of "adaptive peaks," and alternatively, "adaptive zones," proposing evolutionary processes that presumably underlie diversification and movement from peak to peak. Nor is it quite as poetic as Robert Grant's (1829) characterization of species and individual organisms as fleeting entities with, in the case of organisms, characteristic roles to play in the environment.

But I have always found it compelling, even though I have picked arguments with it in the several times in my career that I have focused on this passage. Dobzhansky, after all, is attempting to paint a picture of evolution. He links the evolution of diversity with the environment in which species and their descendants and eventually collateral kin are living. Above all else, it is an image of the importance of adaptation through natural selection to new "niches." And at the very least, it does much to address the criticism that evolutionary biologists, entranced with genes, pay scant heed to the environment, or indeed to ecological matters in general. Dobzhansky was, after all, initially trained as a naturalist/systematist.

But it seemed to me in the 1970s and early 1980s that there was something fundamentally awry with the imagery Dobzhansky developed in this passage. It was very much as if the sort of ontological conceptual analysis Michael Ghiselin (1974) was calling for was crying out to be applied here: What is a niche? What is a species? And, most importantly, do species really have niches? For if they don't, the entire extended analogy falls apart.

HIERARCHY THEORY: THE ECOLOGICAL AND EVOLUTIONARY HIERARCHIES, 1980–

The subject of hierarchy theory, like virtually everything else in evolutionary biology since World War II, is enormously complex, multifarious, and impossible to cover in any detail. I will focus on my preferred views on the ontological clarifications of various sorts of biological entities and systems, including how they are ordered in the natural world and how their components interact to give us what we call evolution. I do this not so much because I am wedded to these views (I am!), but to provide an illustration of the sort of enterprise under the taxic perspective that thinking hierarchically typically produces. I did not develop this on my own, but with colleagues (especially Stanley Salthe [Eldredge and Salthe 1984], Elisabeth Vrba [Vrba and Eldredge 1984], and Marjorie Grene [Eldredge and Grene 1992]). Nor did we agree with one another on all points. Others have developed alternate hierarchical schemes.

I certainly do not pretend that this work met with anything like universal approbation. Such would be unlikely in any case, but perhaps the more so since examining the disconnect between the worlds of ecology and evolution and then setting about to see how they are in fact connected is an intrinsically perilous enterprise. Stephen Jay Gould, for one, utterly rejected one half of the resultant picture, seeing no need for an "ecological hierarchy," at least in conjunction with a formal evolutionary theory, on the grounds that such a move would only make things too complex. Gould (2002:642n) managed to dismiss the entire ecological hierarchy in a single footnote—albeit a rather long one.

At first sailing solo, I attempted in 1982 to cram the Theodosius Dobzhansky–style hierarchy of genes-organisms-populations-species-monophyletic higher taxa, together with regional biotas (larger-scale ecosystems), and, at the highest rung of the ladder of biological system scale, the entire "biosphere." In other words, genes-organisms-populations-species-monophyletic higher taxa PLUS ecosystems-biosphere. Somehow, I saw monophyletic higher taxa as parts of regional ecosystems, which I should have seen from the get-go was a non-starter. It was apples and oranges, this strange, single hierarchy of biological entities mixing ecological and evolutionary components in part/whole relationships.

But it was not until I encountered an earlier paper by biologist Stanley N. Salthe (1975) that I began to see that however ecology and evolution may be connected, the connection could not be envisaged as a nested series of biological systems that included ecological as well as the traditional "genealogical" entities of evolutionary discourse, such as those that were specified by Dobzhansky. An evolutionary hierarchy had to be something like genes-organisms-demes-species-monophyletic taxa. "Demes" are Sewall Wright's term for "populations," in a genetics and evolutionary sense.

Salthe's paper had the beguiling—but also bewildering—title "Problems of Macroevolution (Molecular Evolution, Phenotype Definition, Canalization) as Seen from a Hierarchical Viewpoint." I knew Stan from casual encounters at the American Museum over the years; Stan was working at Brooklyn College in the CUNY system, and the museum's monthly "Systematics Discussion Group" was a great drawing card for the evolutionary biologists in the greater New York area to get to know one another.

But I think it was at another meeting of yet another professional society in yet another city where I had a chance to sit down with Stan and discuss what I was seeing as a major disconnect between how he and I saw the world. I simply said to him something like "I agree that the biological world is hierarchically constructed, and that hierarchical structure has a lot to do with evolution," but that "your idea of what that hierarchy looks like is profoundly different from how I see the world." Stan, with his emphasis on the organismic phenotype, was profoundly ecological in his perspective. The version of hierarchy I was toying with, based on Dobzhansky, had more to do with gene-based information systems—the traditional purview of evolutionary theory "properly" (or so I thought) construed.

The hierarchy that Salthe (1975) published was also linear, consisting of a list of entities made up of the next lower component entities. He called it an "incomplete hierarchy of natural phenomena as viewed by biologists": molecular-cellular-organismic-population-community, with the "nuclear" and "ecosystems" being the next lower and higher levels, respectively, though not explicitly addressed in his paper.

Salthe's view was profoundly ecological. It hardly intersected at all with the more traditional and conventional hierarchies of most evolutionary biologists (again, like Dobzhansky) who had broached the subject. We decided to get together once a week to see if we could bridge the gap between our perspectives.

At first we got nowhere. We were just spinning our conceptual wheels trying to conjure up a single, linear hierarchy of nested elements that had both sets of components in them. It finally dawned on us that it just couldn't work. And the reason was, retrospectively, obvious. We were dealing with two sets of biological levels, processes, and entities that were fundamentally different in their basic natures. We were staring at an ontological disconnect.

So, in the end, we simply lined up versions of the gene-based ("genealogical" or "evolutionary") hierarchy side by side with a simplified version of ecologically based hierarchy ("ecological" or "economic") that Stan had originally delineated and discussed, as seen in the accompanying table (Eldredge and Salthe 1984). We saw the disconnect we were trying to bridge as flowing, at base, from August Weismann's late-nineteenth-century distinction between the germ line and the soma: the germ line specifies the soma, but the soma does not affect the germ line—in terms of the physiological lives of organisms. Weismann's insight returned as the so-called central dogma, in the 1980s, with the advent of molecular biology.

The key elements are common to all organisms: organisms carry genes that, in the germ line, are utilized and passed along in reproduction; in the soma, genes are used for the elaboration of gene products

Genealogical hierarchy	Ecological hierarchy
Codons	Enzymes
Genes	Cells
Organisms	Organisms
Demes	Populations
Species	Local ecosystems
Monophyletic taxa	Biotic regions
(Special case: all life)	Entire biosphere

Source: From "Hierarchy and Evolution" (Eldredge and Salthe 1984:187).

essential to survival, including the ongoing "reproductive" process of generating new cells of all the myriad different types in a complex organism's body. In terminology proposed by Hull (1980), most biological entities are either "interactors" or "replicators"—the major exception being organisms, which are both. Once again, the organismic soma is, in essence, a matter-energy transfer "machine," while the germ line is a "replicator" par excellence, replicative fidelity being supplied by the component genes. This line of thinking has clear conceptual links with the parallel, independent work by Daniel Brooks and E. O. Wiley (1986) on thermodynamics and evolution, in which information theory, derived initially from thermodynamics, applies to the nature, transmission, and modification of heritable information ensconced in the germ line genome. And (non-equilibrium) thermodynamics has equally clear links with the fundamental matter-energy transfer natures and behaviors of organisms and the other components of the ecological hierarchy.

And that is why and how natural selection works: within populations ("demes" in a reproductive sense), ongoing reproductive behavior by organisms typically produces more offspring than can possibly survive and themselves reproduce. That's the population-level statement of Weismann's doctrine. But the causal vector is then reversed. The relative economic success of organisms, as members of local populations integrated into local ecosystems, determines (on a statistical basis) what organisms will survive and reproduce. This biases the representation of genes in the next generation. At the population level, then, relative success of the soma will indeed affect the distribution of genes in the next generation. The one exception to this is Darwin's original concept of "sexual selection," in which relative reproductive success arises solely from the relative capacity of organisms to reproduce, dependent solely on their reproductive adaptations.

From there it was a relatively small leap to see that all members of the ecological hierarchy are matter-energy transfer systems composed of next-lower-level entities that are themselves matter-energy transfer systems. Unlike reproduction (at least in higher organisms), matter-energy transfer goes on constantly on a moment-by-moment basis. Somatic cellular processes, dependent on somatic genes, comprise the organismic soma—a matter-energy machine par excellence. Organisms are members of local interactive economic systems. They are local populations ("avatars") integrated into local ecosystems. It is at this level that we felt the term "niche" makes the most sense. Local ecosystems are linked laterally as matter-energy flows between usefully distinguished local systems, such as at the margins of ponds, forming contiguous networks of increasingly larger ecosystems, and eventually forming the entire global biosphere.

Likewise, genealogical systems are all packages of genetic information that are composed of next-lower-level entities that "reproduce." In other words, genealogical entities are all "more-makers." Genes "duplicate," organisms "reproduce," "demes" can divide or merge, and species also make more of themselves. Thus we preferred "more-making" over "reproduction" as the general process that characterizes genealogical entities of these various levels.

However, the "more-making" buck stops with species. Genera and higher taxa do not make more of themselves through an analogous natural process. Instead, higher taxa, which are indeed collectivities of genetic information, are simply skeins of phylogenetically interrelated species.

Nor are genealogical entities parts of matter-energy transfer systems—again with the major exception of individual organisms. Perhaps the most important conclusion of all this work in hierarchy theory in the 1980s and 1990s is that species are *not* economic entities. Entire species are not parts of local ecosystems, or indeed ecosystems of any size, nor do they have niches. Species do not play coherent direct economic roles in the natural world—not, as Robert Grant said, the way organisms or, as we maintained, local populations within local systems can be seen to do.

The best way to view this important point is to imagine an ecosystem of any size in which all species present as local avataral populations have exactly the same total geographic distributions. This is never the case. Leopards, for example, range from Africa's Cape province through Asia—eating, among other things, antelopes in Africa, and deer (which do not occur in Africa) as a main staple of their Asian diet.

Species are strictly genealogical, evolutionary units. They cannot meaningfully be seen as economic entities with specifiable roles in ecosystems, except in the rare circumstances in which a species is the size of a single population, as when they arise in isolation, or are perhaps dwindling toward extinction.

And this, of course, implies that species cannot and do not occupy "adaptive peaks," as suggested by Sewall Wright, or Simpson's "adaptive zones." Species cannot be niche-occupying parts of higher taxa that have their equivalent of many adjacent adaptive peaks, occupying entire mountain ranges of adaptation, in the otherwise beguiling imagery of Theodosius Dobzhansky.

I have presented only the barest sketch of what constituted at least a major part of "hierarchy theory" as it was developed throughout the 1980s and 1990s. I believe it was a useful contribution to clarify the ontology of the various sorts of biological entities and systems conventionally seen to have something in general to do with the evolutionary process. But it was, of necessity, rather dry and analytical. And it must be said that practicing evolutionary biologists and paleontologists in general paid hierarchy theory scant heed. Indeed, at least one evolutionary journal rejected my paper with Salthe by return mail. The editor (with whom I had a friendly relationship) remarked tersely that his was a journal of data analysis and theory—not philosophy. Nearly that same day I received an invitation from none other than Richard Dawkins to submit a paper to his new journal. I wrote him back, saying I had a freshly rejected manuscript, but that he would not like it, as it was a lengthy disquisition on hierarchies in evolutionary theory. Richard wrote right back, in effect asking me why I would simply assume he wouldn't like it. And so, in due course, I remain very glad to say, it appeared in volume 1 of *Oxford Reviews in Evolutionary Biology* in 1984.

I conclude with a quote from Marjorie Grene (1987)—in itself a conclusion of an important paper: "All evolution must occur through the interaction of the economic with the genealogical hierarchy. It is the investigation of such causal relations that current hierarchy theory should help to initiate" (509).

I agree. But still, the subject of hierarchy theory itself is indeed a bit dry and recondite, I must admit. It needed to be fleshed out with direct reference to the real world—to which we now turn.

ECOLOGICAL SUCCESSION, TURNOVERS, MASS EXTINCTIONS—AND THE SLOSHING BUCKET

A Moving Sandbar: Local Extinction and Recruitment Renewal

There were, of course, a number of compelling papers in Tom Schopf's compendium *Models in Paleobiology* (1972) other than our own "Punctuated Equilibria" (Eldredge and Gould 1972a). Of these, I was most struck by Ralph Gordon Johnson's "Conceptual Models of Benthic Marine Communities." Johnson, as it happens, was the editor of the journal *Evolution* who had shepherded "The Allopatric Model and Phylogeny in Paleozoic Invertebrates" (Eldredge 1971a), forerunner to the 1972 "punk eek" paper, through to final publication. Like many invertebrate paleontologists back then, Johnson was keenly aware of the insights to be gained from studying the modern biota, as well as the implications of the fossil record for understanding the recent. In his case, Johnson's interest lay in community-level processes in marine benthic invertebrates—mostly clams, snails, polychaete worms, and other invertebrates living on or in the sea floor.

Johnson's (1972) paper followed up on an earlier study conducted over a five-year period in the shallow marine waters of California's Tomales Bay. The picture he painted that I found most arresting was the slow movement over that five-year period of a "wave of sand"—in more prosaic terms, a sandbar. The sand slowly progressed over the community of species-populations lying in its path. Many of the hapless who could not move were killed, and the benthic community was severely downgraded in terms of the diversity and sheer numbers of organisms.

Death, destruction, and degradation lay in the face of the inexorable movement of the sandbar. But what happened as the sandbar passed, revealing the familiar sediments of the old sea bottom? Johnson was able to document in detail the renewal of the invertebrate communities as the barren floor was repopulated by individuals of all (or at least most) of the species that had originally lived there. This happened with some coherent order, so Johnson was able to speak convincingly of true "ecological succession."

I mention Johnson's work here because in the mid-1980s or thereabouts, thinking about the ontological nature and functions of genealogical and ecological entities, I remembered that wave of sand and what happened to the benthic communities before and after the sand encroached. Here was a role clearly played by local populations—call them "demes," call them "avatars"—in addition to the interplay that is natural selection itself. For the re-established communities, at least on the phenotypic level (if perhaps not exactly on the genetic level) were nearly exactly the same as the destroyed preceding communities. In other words, larval dispersal and perhaps active mobility brought recruits in from nearby, undisturbed communities.

It's so simple as to not be remarkable at all. But it does highlight the nature of demes as genetic reservoirs. And though someone like myself is perhaps overly fond of saying that nothing much happens in terms of lasting evolutionary change unless and until it is generated and/or conserved in the speciation process; and granting that the sampling from even closely adjacent populations will likely, on chance alone, present a slightly different array of genetic information than what was there before the sandbar knocked them off, nonetheless the net effect of that passing sandbar was little if any detectable change, whether measured in terms of community structure and composition, or, even more, in terms of palpable evolutionary change in morphology. John Thompson's (2013) "relentless evolution" is very real, made all the more so by recent techniques that have been developed that allow the analysis of changes in genetic composition within and among populations. But according to Thompson, such changes are far more concerned with survival of the local population than with generation of lasting, long-term evolutionary change.

Thus I came away from Johnson's elegant study of this five-year sandbar movement struck with a lasting impression of the "function" of populations, species, and, for that matter, larger-scale monophyletic taxa: they don't *do* anything except act as reservoirs of genetic information. And that is huge.

Shades of d'Orbigny: "Turnovers" and "Coordinated Stasis": Extinction, Speciation, and the Generation and Conservation of Adaptive Change at the Regional Ecosystem Level

Somewhere between local disturbance and death of individual organisms, on the one hand, and the wholesale carnage of global mass extinctions, on the other, lies the regional occurrence of sufficient environmental change to trigger the extinction of many, if not all, of the species of, say, half a continent's geographic area. The abrupt disruption of such regional ecosystems and the concomitant extinction of many entire species is in fact a watershed. As long as species survive, samples of their genetic information survive and no radical changes can be expected. But when a number of species living cheek-by-jowl are driven to extinction in a brief span of time, that's when speciation takes over and becomes rapid and rampant. Evolution doesn't slow down until new ecological systems are restored in different, if often reminiscent, form to what had preceded them.

This, I believe, is where most lasting adaptive change actually occurs in the evolutionary history of life. The view started truly being articulated in the work of paleontologist Carlton Brett and colleagues (for example, Brett and Baird 1995), who called the pattern "coordinated stasis"; and, I think especially, in the work of another paleontologist, Elisabeth S. Vrba (1985), who managed to link her work more closely to existing evolutionary theory. She called her pattern "turnover pulses," or "the turnover pulse hypothesis," which is underlain by regional extinctions and speciation (see also Van Dam et al. 2006).

But the stunning thing to me, as we turn to this pattern in a bit more detail, is that we find ourselves confronting Alcide d'Orbigny's (1842) *étages*, the same word as the English "stages" in a geological sense. These are blocks of rock defined and recognized by their faunal content. In a typical geological stage, most of the species within the interval seem to originate at the lowest levels near the very beginning; and many, if not all, species succumb to extinction at or near the very end of the interval. Though d'Orbigny did not stress this, it has become apparent to most paleontologists confronting "turnover pulses"/"coordinated stasis," that the individual species within each interval—each geological stage—are in fact for the greatest part in stasis.

To be sure, some of the newly appearing species are migrants in from source areas. And, indeed, within these intervals, there still seems to be some degree of speciation, as well as the pattern of extinction of a species, followed by the arrival of an already-evolved replacement species from an outlying source area. But the bulk of the appearances of what seem to be newly evolved species happens in waves of extinction and speciation that define the upper and lower boundaries of d'Orbigny's *étages*.

D'Orbigny thought *étages* were global. They are not. Instead, they are regional, reflecting the common situation of biogeographic differentiation so familiar to the early naturalists of the late eighteenth and early nineteenth centuries. And geologically speaking, they take on average some 5 to 10 million years to run their course in the marine realm—less, averaging only a few millions of years, in the terrestrial realm.

So these events of regional extinction and evolutionary reproliferation of species have been common given the nearly 600 million years of complex multicellular life. And as I say, this is precisely the locus for most lasting adaptive change in the history of life. Adaptive change is tied up in speciation events occurring in the aftermath of regional extinction of many phylogenetically unrelated species.

The work of Brett and colleagues in the 1980s and later was focused on a succession of such stage-level events involving marine systems of what is now the Appalachian Mountains and adjacent areas during the mid-Paleozoic, including the Hamilton Group of Middle Devonian rocks that provided my original example of punctuated equilibria involving the *Phacops rana* trilobite species lineage (Eldredge 1972). Brett and colleagues showed that many of the species of marine invertebrates (such as brachiopods) of the mid-Paleozoic that became extinct near the end of each interval of "coordinated stasis" were replaced by closely related species within the same genus. Since their work appeared, many other marine examples up and down the geological column have been published, including the bryozoans and other invertebrates from the Caribbean Tertiary mentioned in the previous chapter as a pivotal confirmation of the mere existence of within-species stasis through time.

In dubbing the pattern "coordinated stasis," Brett chose to emphasize the parallel patterns of species-level stasis within the intervals demarcated by abrupt extinction and consequent evolution of species. In contrast, Vrba's "turnover pulses" emphasized the extinctions/evolutionary rebounds, not the stability of the component species in the longer periods of relative environmental quiescence. Yet most would agree that the two patterns are fundamentally the same, both fitting comfortably into the sort of thing d'Orbigny had in mind with his concept of *étages*.

Vrba illustrated her notion of "turnover pulses" with a detailed analysis of the climate change-induced extinction of many terrestrial, predominantly mammalian, species of the late Pliocene of eastern and southern Africa. Admirably, she was able to link her data up with familiar elements of evolutionary theory, especially speciation theory. The raw notion that new species would as a matter of course be formed to "take the place" of closely related predecessors that were succumbing to extinction (as Brett and colleagues were seeing in their Paleozoic examples) at first glance seemed to conjure up images of evolution rushing in to fill "vacant niches"—an image that I, for one, always had difficulty imagining in any satisfying detail. Species, in my book, simply do not "have" niches.

But Vrba made it easy. According to her model, the environmental changes that both caused the extinctions and triggered the origin of new species was a period of cooling that resulted in the transformation of the dominant tropical rainforests of eastern and southern Africa into vast swathes of grasslands. In such an ecological transformation, there would be pockets of residual habitat interspersed with the new dominant habitats, forming a mosaic of habitats, and possibly ecotonal situations as well. Such radical habitat transformation would trigger the extinction of large numbers of species. But the resultant habitat mosaic would also set the stage for relict populations in isolation to undergo rapid adaptive change. And some might emerge as successful new species in the transformed landscape. It would seem likely that specific speciation models, such as "ecological speciation," will prove fruitful in understanding the sorts of evolutionary dynamics that can occur during these periods of regional environmental stress, extinction, species survival (some of course do survive), and the origin of new species bearing the stamp of adaptive change.

So here is a new take on an old evolutionary image: the origin of species of the modern biota is to be seen as a series of replacements of pre-existing and now-extinct ancestral species by new derivative species produced by natural, non-miraculous causes. The phenomena of turnover pulses/coordinated stasis are deeply cross-genealogical, involving tens, even hundreds of species from different lineages virtually simultaneously. If evolution does boil down to ancestral species giving rise to descendants within a single lineage (or a splitting event, in which a descendant species lives alongside the ancestral species), nonetheless this does not happen in a vacuum. Arguably most speciation events occur in the throes of profound environmental disruption that kills as many species off as it produces new ones to "take their place." The newer species look a bit different and are presumably somewhat differently adapted from their now-extinct progenitors. This is perhaps a difficult proposition to establish with certainty with Paleozoic marine invertebrates, but a lot more plausibly established with examples like Vrba's antelopes and other associated mammals of the African Pliocene including hominids; the first species of the genus *Homo* appeared in the aftermath of the turnover pulse that Vrba has detailed.

I have never seen Olduvai Gorge or any of the other classic East African vertebrate paleontological localities. My experience is rooted and pretty much restricted to the turnover patterns of the marine Paleozoic, predominantly in the eastern half of North America.

But I did have the marvelous experience of seeing some of the great hominid sites in South Africa with Elisabeth Vrba. In East Africa, the fossils occur in layered sediments with occasional lava flows interspersed, providing a welcome source of radiometric dates. In contrast, the South African localities are limestone cave deposits. There the very hard rock resembles layers of candle wax drippings, and the sequence of deposits is often more difficult to decipher than the simpler "layer cake" structure found in East Africa. Fossil bones and admixtures of other jumbled rocks ("breccias") occur in this dense, tough, carbonate version of candle drippings.

Though some of the South African sites are still caves, others are simple surface exposures of what were originally cave deposits, like Elisabeth's own research site, Kromdraii in the Transvaal (see Figure 6.2). It was a place where the Scottish-born physician/paleontologist Robert Broom had earlier worked. Kromdraii had produced some of the earliest known specimens of robust australopithecines—early members of our own hominid lineage.

Kromdraii lies exposed in an agricultural field. It is an open pit, with spitting cobras and other African wildlife occasionally showing up to distract a paleontologist. The fossils are rare and difficult to extract; in fact, they must be removed within entire chunks of limestone to await eventual escape from their rocky encasements back in the lab.

It's a tough place to work. But what a story has emerged from Kromdraii and nearby caves such as Swartkrans and Sterkfontein. The dry plains that dominate the scene these days lie in sharp contrast to the wet riverine woodlands that predominated when most of those caves filled up with the flotsam and jetsam of life. The bodies of the australopithecines may well have dropped from trees where leopards stashed them. Other bones were either brought in by carnivores or by accidental death. Fig trees tend to stand sentinel at the mouths of caves across southern Africa. And the largest leopard I have ever seen stuck his head out of an opening to Drotsky's Cave near the Botswana–Namibian border.

Once you see the modern scene, including the abundant African fauna still surviving in places in southern Africa, and then compare it with the environment reconstructed from these fossil deposits, a vivid picture emerges of climate change and its effects on animal and plant life both ecologically and evolutionarily. I have explored riverine forests in southern Africa, primarily in northern Botswana in the Okavango Delta, far away from these South African caves and their treasures. The Okavango Delta is a remnant of much more widespread wet woodlands, riverine forest African habitats that were much more prevalent 3 million years ago.

So here, finally, in flesh and blood, is the nexus of ecology and evolution at the level of the origin of new species. Natural selection effects the change of adaptations, and the evidence has been mounting that most such change occurs in conjunction with speciation in isolation, as Darwin suspected but ultimately didn't pursue. But turnover pulses/coordinated stasis show the direct impact of environmental change. And they suggest how adaptive change can be generated in independent phylogenetic lineages in more or less simultaneous bursts of extinction and evolution of species.

Mass Extinctions and Macroevolutionary Rebounds in Higher Taxa

On the other extreme of spatiotemporal scale, consider the five or six (depending on how you count) global mass extinctions that have severely "impacted" life (literally—at least some of these seem to have been caused by extraterrestrial impact events) over the past 500 million years or so.

Our mentor in the Columbia/American Museum graduate paleontological studies program, Norman D. Newell (see figure 5.5), was virtually the only scientist of note who paid any attention to mass extinctions whatsoever, at least in the decades of the mid-twentieth century. Newell was a specialist in benthic marine invertebrates (mainly bivalved mollusks) of the Permian Period, the final geological period of the Paleozoic Era. He referred to events such as the Permo-Triassic mass extinction, which killed off perhaps as many as 90 percent or more of the species on the face of the earth, as "crises in the history of life." Stephen Jay Gould, Bud Rollins, and I, along with our fellow graduate students, often complained that, with Newell's emphasis on death and destruction, there was little time to "accentuate the positive," and talk about evolution—until the light finally dawned on us that mass extinctions were fundamental game-changers in the history of life.

But then the work of Luis Alvarez's team (1980) on the Cretaceous-Tertiary boundary (65 to 66 million years ago) focused attention on the most recent mass extinction. (So far! We seem to be in the midst of another, this time anthropogenic, extinction event right now.) This "K-T" mass extinction was when the dinosaurs finally succumbed, as did a host of marine invertebrate groups. Their work excited a lot of attention in the early 1980s. The geological community got on board, and debate raged over the causes of the mass extinction. The Alvarez team thought that the iridium "spike" it discovered right at the boundary level, first in Gubbio, Italy, and then elsewhere, was indicative of cometary (or at least extraterrestrial) impact with the earth. The impact(s) hit with great force, and the ensuing environmental chaos globally reached where many species, including many higher taxa such as ornithischian and saurischian dinosaurs, and the marine ammonites, fell victim to extinction. It was the Tomales Bay sandbar that degraded a local system and destroyed local populations of various invertebrate species on a mega-scale.

The point here is not so much the details of this particular extinction scenario, than it is that the non-bird dinosaurs are simply not here now; that, despite the fact that their diversity was dropping as the Cretaceous

Period wore down, nonetheless their total, final demise was abrupt; and that mammals, present in the form of various non-surviving Mesozoic groups, along with the root stock that eventually radiated into the modern mammals of the Tertiary only took their present forms and shapes in an evolutionary sense after the dinosaurs were out of the ecological picture. Mammals radiated and took the place of dinosaurs as the dominant tetrapods of terrestrial environments. In a very real sense, the mammalian radiation that came after the demise of the dinosaurs was very much like reinventing the ecological wheel.

Imagine, if you will, the rather stunning implications that life today, as diverse at the species level and more so than ever it has been in the past, was derived from as little as 10 percent (or even less) of the relict genetic information that managed to squeak through that greatest of all mass extinctions, Newell's (1967) Permo-Triassic "crisis in the history of life." Not only is extinction not off the subject of evolution, as we graduate students originally thought in the 1960s. The picture we have now is that, without ecological disruption, little conspicuous adaptive morphological change is generated and conserved. Mass extinctions make this point all the more clear. It is massive changes in the physical environment, intruding far beyond the normal adaptive capacities of organisms to survive, that drive species and even higher taxa to extinction. And only when the dust settles do we see the rapid reproliferation of species, including the origins of major new taxa, as the ecological wheel returns to something resembling normalcy and stability.

The intimate relationship between extinction and evolution was an important component of the fledgling science of evolutionary patterns and causes in the first four decades of the nineteenth century. It was yet another virtually forgotten line of thinking that needed resurrection in the twentieth century.

THE SLOSHING BUCKET

So there we have it: minor, local disturbance kills off local populations, as when Ralph Johnson's (1972) "wave of sand" passed over the sea floor in Tomales Bay. It killed many organisms in different species.

But it drove nothing—certainly no entire species—extinct. And the aftermath was just as mild and, superficially at least, unremarkable. The communities were rebuilt by larval and adult recruitment from nearby demes in nearby communities. Chaotic, but morphologically slight or undetectable, genetic change, one now expects, must have been there. But on the surface of it all, the "same" species came back and the "same" ecological communities were soon back in place.

The intermediate, regional level, at which extinction and generation of species occur in "turnover" events, is where the irrevocable moment occurs. Once a species is gone, it can never come back. It may mark the end of an entire lineage or a replacement of the old by some modified version of the "new." This is where real change occurs most of the time in evolution.

Likewise the extreme other side of things: near total, global ecological devastation, killing off not just individuals but entire species and entire large-scale higher-ranked taxa, with a proportionally large evolutionary reaction as new taxa quickly appear to "take the place" (as the Edinburghians would have put it) of the taxa driven to extinction.

But the basic idea runs through these different scalar examples. Each time, life resumes some semblance of "normalcy," either based on the genetic information that survives to supply the raw materials for evolution or simply and easily recruited from next door.

This strikes me as a sort of graphic realization of the abstract ontological picture some of us were developing of the hierarchical structure of both genetic (genealogical) and ecological (economic) biological systems, and the potential interactions among them. Almost on a whim, in the late 1990s, I took a shot at literally fleshing out the hierarchies based on the sorts of examples just mentioned: phenomena ranging from disturbance of local ecosystems, causing deaths or displacement of individuals and subsequent recruitment from outlying demes, so that the ecosystems would be restored to more or less their "normal" form; through "turnover pulses"/"coordinated stasis," with extinction of entire species, and the evolution of new, replacement species in some instances; on up through global mass extinctions of entire higher taxa—sparking, in many instances, the origins and diversification of new, higher taxa.

The core idea of the "sloshing bucket" is simple (Eldredge 2003). The greater the scope and severity of environmental change, the more extinction occurs and the greater its reach globally and in terms of the size of the taxa driven to extinction. Below the species level, little or no lasting palpable evolution at the adaptive phenotypic level occurs. The key level is the species. If species are driven to extinction, their genetic information is lost, unless it lives partially on in a new, adaptively modified genetic system in the form of newly evolved replacement species.

Indeed, even after global mass extinctions and the eventual origin and diversification of truly new higher taxa, all the evolutionary action starts with speciation events. In the origin of new, higher taxa, presumably a series of speciation events are involved.

The "sloshing bucket" imagery is by no means as poetically expressed as Robert Grant's (1829) beautiful passage resonating the Giambattista Brocchian imagery of individuals and species and their roles in nature. And it is not as compelling as Theodosius Dobzhansky's, I think, ontologically shaky but otherwise almost lyrical imagery of species and higher taxa evolving on, and continuing to occupy, adaptive (mountain) ranges. Nonetheless this simple model of the "sloshing bucket," a meld of experiential data draped over the conceptual framework of the twin ecological and genealogical hierarchies, has been a lot easier to communicate than the bare bones of hierarchy theory itself.

Darwin possessed a copy of his old South American rival Alcide d'Orbigny's (1842) monograph of French Jurassic ammonoids, in which d'Orbigny first developed his concept of *étages*. It was Darwin's reading of this copy (though not annotated), that, I firmly believe, drove him to write the footnote to the "Fair Copy" of Essay (1844a): "Better begin with this. If species, really, after catastrophes, created in showers over world, my theory false." D'Orbigny's *étages* were not global, as he had originally thought, but they were nonetheless very real, if regional. D'Orbigny's *étages* are known to be exactly as originally described. And they conform precisely to the sorts of patterns developed by Vrba (1985; "turnover pulses") and Brett and Baird (1995; "coordinated stasis").

Even if restricted regionally, though, d'Orbigny's observations and generalizations would have been deeply problematic to the Darwin of the early 1840s, as they stood firmly against his view of the gradual transformation of individual species lineages through time. Had Darwin chosen that other fork in the road, had he accepted that most, if not all, adaptive change occurs in conjunction with the origin of species in isolation, and had he taken d'Orbigny's empirical *étages* seriously (after all, he did know about stasis, which was a start), he would himself have grasped the picture presented by biologists/paleontologists in the 1980s as they formulated their notions of "coordinated stasis" and "turnover pulses."

Though lacking little, if any, empirical support, Darwin nonetheless adopted the picture of gradual adaptive transformation of species through time at least in part because he could not see how isolation can arise on broad continental areas—save for large-scale geographic barriers such as mountains, and to a lesser degree, rivers. He was not aware of the effects of climate change on terrestrial habitats, which, if sufficiently pronounced, can produce the sorts of habitat mosaics and resultant effects that would promote extinction and speciation events more or less simultaneously, as proposed by Elisabeth Vrba and others (Jürgen Haffer's [1974] linkage of speciation patterns in Amazon bird species with climate-induced habitat changes is very much the same sort of thing). Vrba's data was largely vertical with a geographic component. She had a good fossil record in eastern and southern Africa to work with. Haffer's deductions depended to a large degree on the geographic distributions of still-living species.

MOLECULAR BIOLOGY AND THE FUTURE OF THE TAXIC PERSPECTIVE

So far, I have had little to say about the molecular revolution and how it has affected evolutionary theory since its advent in the 1950s. Although much of the early work amounted to the exciting description of the structure and function of the hereditary molecules themselves, inevitably there were those adventurous investigators eager to link the new genetics with evolutionary theory. Many of the arguments at the Chicago meeting on macroevolution held in 1980 involved molecular biologists engaging, sometimes rather heatedly, with population geneticists—who, at that time, seemed perched alone in the catbird seat of prevailing evolutionary theory.

And once, at about the same time, a molecular biologist came up to me with a gleam in his eye and proclaimed that he had a molecular explanation for punctuated equilibria. Gone, in one fell swoop, was any need for models of speciation in isolation!

One of the greatest obstacles to achieving a fully satisfactory and "complete" evolutionary theory lies in the extraordinary number of biological systems that seem to be involved in this thing we call "evolution." Indeed, one of the simplest ways to open a discussion of the ontological status of any biological system, including its putative "reality," is simply to ask if there is a separate field of study within biology dedicated to it: fields like cytology, histology, developmental biology, ecology, population genetics, along with the finite but long list of other subdisciplines, suggest that there are indeed discrete levels and kinds of biological systems.

As Theodosius Dobzhansky was perhaps the first of the "modern" era to make clear, recognizing the ontological status of discrete biological systems nested within one another achieves at least two things: it teases these levels, with their disparate properties and processes, apart from one another, the better to study their internal dynamics and structure. But, in so doing, it also invites further analysis, as Dobzhansky (1937) did, of how these various levels and subject matters do interconnect in dynamic ways in the evolutionary process.

As I have already remarked earlier in these pages, the day has long since gone when someone like Charles Darwin could be reasonably conversant with all the levels, all the fields of study, to forge a coherent, single theory of the evolutionary process. Paleontologists and geneticists, traditionally, do not have more than a nodding acquaintance (oftentimes not even that) with each other's data—witness the anecdote in chapter 5 about some paleontologists trying to convince some evolutionary geneticists that species-level stasis is a real and common phenomenon that needs to be taken seriously in evolutionary theory.

Matters became even more complicated for nearly all of us when molecular biology matured in the 1980s and 1990s—and, among other places, began to make a significant impact in the study of biological systematics. At least one graying paleontologist found himself too old to become thoroughly conversant and even comfortable with molecules. Still, I made sure my students were au courant, and I provided the space and staff openings for the very first molecular lab at the American Museum of Natural History; after all, I could see the handwriting on the wall.

One of the earliest molecular contributions to the taxic perspective outlined in this narrative came from T. Ryan Gregory (2004), who neatly linked his contemporary understanding of the structure of the genome with the prior work that his non-molecular predecessors had done with hierarchical systems at higher levels. At the top of his paper, Gregory cites yet another prescient statement from the pen of Stephen Jay Gould (1992): "Punctuated equilibrium [*sic*] is but one pathway to the elaboration of hierarchy, and probably not the best or most persuasive; that role will probably fall to our new understanding of the genome and the need for gene-level selection embodied in such ideas as 'selfish DNA.'" Using his own extensive data on relative genome sizes, Gregory explores the implications for understanding large-scale macroevolutionary events, concluding that "genome evolution cannot be understood without such a hierarchical approach" (179).

Since the early days of protein sequencing and similar approaches that antedated actual use of gene sequences, right on up through the use of increasingly detailed (if automated) gene sequences themselves, conflicts, rather than agreement, seem to be the norm when the results of studies of the fossil record are compared with the results emanating from the molecular biology lab.

And now more than ever, these conflicts nearly always involve not only whether the phylogenetic pattern inferred from fossils, recent morphological data, and molecular data agree, but also on the *timing* of evolutionary events. Timing conflicts stem, at base, from discrepancies on the age of fossils as determined by correlation with radiometrically dated rocks, versus the basic assumptions of the molecular clock, which has always been that mutations occur at a more or less statistically constant, monotonous rate. Thus genetic "distances" yield a rough estimate of time since divergence, assuming a certain rate of mutation, and results often differ sharply from the geological dates.

Oddly, the two dating techniques both depend on statistically constant, and in a sense empirically determinable, rates of change. Radioactive decay rates from parental to daughter isotopes can be measured in the lab. Likewise, mutation rates can be determined as well in the laboratory.

There are, of course, several lines of possible error—of measurement, of sampling, of assuming that a fossil, say, which may appear to be very old and primitive is really from the earliest stages of the evolution of a taxon, and the like. A recent spate of papers has revealed that many of these problems are being addressed, and new analytic techniques appear promising to resolve the persistent conflicts between molecular and fossil data on a more regular basis.

But conflicts persist. A recent multi-authored paper (O'Leary et al. 2013) analyzed a very large data base of living and fossil mammal morphology, comparing the results with prior molecular studies. The question was: When did the orders of placental mammals diversify? Ever since I've been around, the standard, and very sensible, answer was: non-placental mammals (of several lineages, many not surviving the Mesozoic) arose ca. 215 million years ago, right about at the same time as the dinosaurs. The dinosaurs radiated phylogenetically, while mammals were the "rats of the Mesozoic" and hardly evolved much at all. In the Middle to Upper Cretaceous Period, however, insectivores and possibly primates (both true placental mammals) first appeared. But the mammals did not radiate into major groups and ecological types until after the dinosaurs, already dwindling as the Cretaceous wore on, were finally totally annihilated in the end-Cretaceous extinction 66 to 65 million years ago.

Not so, says O'Leary and her colleagues. Their data suggest, instead, that all of the orders of placental mammals diversified relatively suddenly some 2 to 3 million years after the Cretaceous extinction event. Although some of their results propose different patterns of evolutionary relationships among placental mammals than have been proposed using molecular techniques, the real discrepancy is, once again, time.

The paleontologists point out that the fossil record of terrestrial vertebrates, including early, non-placental forms of mammals and other smaller forms of terrestrial life, is relatively good in the Upper Mesozoic.

The absence of placentals, let alone a diversified array of different orders of placentals, in the Mesozoic thus seems to be a real phenomenon. Rather than missing data, it seems pretty clear that the reason why diversified placental mammals have not been found in Mesozoic sediments is that they were not there: they simply had not evolved yet.

As a taxic-minded paleontologist, I was excited about these results. Radiations often appear (often with a bit of a lag after major extinctions events) to be "explosive." The Cambrian "explosion" (which may have taken some 10 million years to complete, starting around 530 million years ago) is nonetheless blindingly fast, separating as it does the absence of diversified metazoans (major, still-living marine invertebrate taxa [phyla] such as mollusks, arthropods, echinoderms, and the like) from the subsequent dense fossil record left by these groups from then on to the present day. There has been lots of extinction and further evolution within these phyla, but no new phyla have appeared since the Cambrian explosion. Here again, the fossils suggest an explosive pattern of evolutionary radiation, while molecular clock studies have suggested that the metazoans may have diversified over 1 billion years ago.

Accompanying O'Leary's paper, however, was a short report detailing how the paleontological results did not take into account the lengths of the branches of the molecular trees, which indeed suggest earlier dates for the radiation of the placental mammalian orders. The author Anne Yoder (2013) suggested that the data from all quarters had been slowly moving towards the "explosive radiation" model, but that it was too soon to throw out the contradictory molecular data and analyses that still point to an earlier, and in a sense more gradual, diversification of the placental mammalian orders.

I am neither a nuclear physicist nor a molecular geneticist. But clearly the problem lies in discrepant results in radiometric versus molecular clock dating. And so I close this narrative with a prediction, based on directions that I hope molecular evolutionary studies will eventually be able to take.

Several years ago, I read some of the papers of geneticist Mark Pagel and his colleagues. One paper in particular caught my eye: "Large Punctuational Contribution of Speciation to Evolutionary Divergence at the Molecular Level" (Pagel, Venditti, and Meade 2006; see also Venditti and Pagel 2009). Pagel and colleagues had found a way of testing the difference between the assumption that molecular substitutional change occurs in strictly a clock-like manner, or is, instead, sometimes concentrated at "nodes," which would be splitting events between species.

I will simply quote their abstract here, and then make a couple of comments. My goal is not so much advocacy, but to draw attention to what developments may lie in store vis à vis the juxtaposition of molecular biology with the taxic perspective:

> A long-standing debate in evolutionary biology concerns whether species diverge gradually through time or by punctuational episodes at the time of speciation. We found that approximately 22% of substitutional changes at the DNA level can be attributed to punctuational evolution, and the remainder accumulates from background gradual divergence. Punctuational effects occur at more than twice the rate in plants and fungi than in animals, but the proportion of total divergence attributable to punctuational change does not vary among these groups. Punctuational changes cause departures from a clock-like tempo of evolution, suggesting that they should be accounted for in deriving dates from phylogenies. Punctuational episodes of evolution may play a larger role in promoting evolutionary divergence than has previously been appreciated. (Pagel, Venditti, and Meade 2006:119)

In other words, given that a large percentage of substitutional changes indeed seem to be correlated with evolutionary "nodes" (episodes of splitting), and, assuming that virtually all lineages do indeed undergo speciation events, the simple assumptions of the molecular clock cannot possibly be consistently correct. Molecular clocks, instead, must routinely overestimate the length of time of divergence dates.

Pagel, Venditti, and Meade (2006) conclude their paper by saying:

> Whatever the mechanisms of the effects we have characterized, relatively rapid and punctuational bursts of evolution driven by speciation appear to make a substantial contribution to molecular divergence. By comparison, we found no molecular counterpart to the periods of stasis noted for morphological traits . . . , the other half of the conventional

> punctuated-equilibrium description of morphological evolution. There need not be any conflict between these two observations as it is well known that molecular change can occur independently of morphology. Punctuational effects are an area of great potential for future research on speciation combining functional-genomic, phylogenetic, physiological, behavioral, and paleontological investigations. (121)

Their comment that "molecular change can occur independently of morphology" helped assuage my concern that "only" (as I naively originally saw it!) 22 percent of molecular substitutional change occurs in association with nodes in their data set.

Which leads me to my closing prediction: I'm willing to bet it is the coding elements—the molecular elements that actually make products that end up being part of the phenotype—that differentially change in bursts at speciation events. In contrast, it is the non-coding elements—the "neutral" or "nearly neutral" elements—that, however important and whatever their functional role may be, are the predominant ones that go on ticking like a clock.

It would be nice to be able to tease these elements apart in these molecular data bases to see if, indeed, it is the coding elements that differentially change at speciation events, while the non-coding elements display true clock-like behavior. This would be my prediction for molecular evolutionary change given my taxic perspective.

In other words, it seems to me that the molecular clock concept is basically the gradualism perspective translated down to the molecular level. In terms of the production of morphological changes in evolution, it is probably as false at the molecular level as it has proven to be at the phenotypic/species level.

SUMMING UP: THE TAXIC PERSPECTIVE SINCE 1935

It is astonishing how many of the themes of the modern era of research and analysis of taxic themes were actually well known, at least in outline, to the pioneers in evolutionary biology between the years 1801 and 1859.

Giambattista Brocchi's (1814) analogy adumbrated hierarchy theory. It also anticipated the "species as individuals" discussions. Spatiotemporal patterns of species replacement as the species in biotas became progressively younger looking presaged much of the modern empirical work in speciation and punctuated equilibria theory. Adaptive change in isolated populations leading to the emergence of new species—explicitly present in Charles Darwin's notebooks of the late 1830s—was certainly there. Natural selection, one of the few laws relevant to both the taxic and transformational evolutionary perspectives, is of course still very much with us. Even species selection is hinted at: one need look no further than Darwin's subtitle of the *Origin: Or the Preservation of Favoured Races in the Struggle for Existence*. And finally, there are the patterns of cross-genealogical extinction and appearance of new species—patterns brought forth especially by Georges Cuvier (1812) and his student Alcide d'Orbigny (1842), both of whom were seemingly agnostic (rather than necessarily opposed) to the concept of evolution. Nonetheless Darwin saw d'Orbigny's work on *étages* and recognized them for what they were: patterns of multiple, concerted origins of species that were not rationally explicable through "my theory" of gradual change within different lineages.

Thus the basic elements of modern-day taxic thinking were all there in the initial cycle of evolutionary theory. But most were forgotten in the wake of Darwin's transformational thinking in the *Origin*, which was the book that did the incredibly important work, let us not forget, of finally bringing evolution out of the closet and conveying it the status of a proper, indeed monumental scientific theory.

There was no need to delineate a "taxic perspective" separate from other lines of thinking within evolutionary biology when I came on the scene as a greenhorn student in the 1960s. Ernst Mayr (1942, 1963) had written his important books on species and speciation, and I simply absorbed them and saw that the process of speciation was an integral part of evolutionary theory, which otherwise was pretty much a theory of adaptation through natural selection, integrated with the understanding of genetics (especially population genetics) current at that time. All this made perfect sense to me.

I had no idea that speciation theory was a relative newcomer on the block of conventional evolutionary theory—essentially the invention (despite his own intellectual forbearers) of the naturalist-turned-geneticist Theodosius Dobzhansky, as recently as the mid-1930s.

And the work of these pioneers in the 1930s, 1940s, and 1950s has long since mushroomed into the vibrant field of speciation studies today. I am particularly struck by the relatively recent incorporation of both ecological perspectives and the even more recent novel ability to integrate field studies with the techniques of molecular biology. It turns out that within-population variation is far more rampant at the genome level than it is at the phenotypic level—far more than even the early modern studies of genetic variation of several decades ago seemed to imply. After all, post-Darwinian biologists from time to time would worry about the morphological uniformity they would encounter in populations, even entire species, that many of their study organisms often seemed to present. Now, with concepts such as John Thompson's (2013) "relentless evolution," the picture is one of prolific variation at the genome level.

And yet, Thompson and his like-minded colleagues see that most of the genome-level variation and change seems to be important for the survival of local populations, but otherwise seldom "goes" anywhere. For lasting genetically based morphological evolutionary change, we mostly need to consider the higher level of speciation itself: the onset of genetic isolation, still seen as predominantly (if not exclusively) contingent upon geographic isolation. This is where concepts like "ecological speciation" are important: seeing the origin of new species, and the generation and conservation of adaptive change, as part and parcel a matter of reproductive isolation associated with adaptive change to "new" (to the species) ecological conditions. Indeed, I have always thought of the adaptive change in isolated populations, often on the periphery of an ancestral species' range, as being a chance for natural selection (given the requisite heritable variability) to, in effect, redefine the marginal (hence suboptimal) environment as the new "optimal."

Thus true speciation, involving actual genetic isolation, is an opportunity for the generation of adaptive change. It also provides the opportunity for the conservation of that genetic change, as the new adaptations won't be swamped and ultimately lost by subsequent interbreeding with the ancestral species—Dobzhansky's original point. Isolation is a necessary, though in itself not sufficient, cause for the origin of new species. And speciation literally does seem, just on the basis of modern studies of the living biota, to be the context of most of the lasting adaptive change in evolution.

Yet it was also true that other notables of the preceding generation, especially George Gaylord Simpson, remained aloof from the "biological species concept." Simpson and like-minded evolutionary biologists for the most part saw "speciation" as nothing more than the accumulation of sufficient genetically based morphological change through time to warrant recognizing the younger elements of a gradually evolving "species lineage" status as separate ("new") species.

But Simpson did see the relevance of patterns in the fossil record in truly long-term, real-time evolutionary/geological perspective. Specifically, his conviction that large-scale adaptive change tied up with the often sudden appearances of fully adaptively differentiated higher taxa ("taxa" being his later distinction from "categories"), was a clarion call to paleontologists of the next succeeding generation.

The next step seems, in retrospect, obvious. I, and later with my fellow student and colleague, Stephen Jay Gould, integrated Simpson's perspective on the fossil record with Dobzhansky and Mayr's ideas on the nature and origin of species.

But the really important innovation we brought to the table was the emperor's-new-clothes acknowledgment that little, if any, progressive anatomical change could be documented within lineages, even in exceptionally well-preserved examples of marine invertebrates. This was, and remains, an empirical claim. Stasis is not observed in 100 percent of cases. But species-wide patterns of morphological change through geological time are mostly oscillatory: rather like the chaotic genome-level within populations that is the observational basis of Thompson's "relentless evolution," the morphology of species through time varies, but seems to go "nowhere."

Stasis is not only the empirical underpinning of "punctuated equilibria," which is an amalgam of this critical empirical generalization plus speciation theory: what stasis also does is essentially falsify (as Darwin himself, as a younger man, realized would be the case) the idea that adaptive change is wholly, or at least mostly, a matter of natural selection modifying entire species through time in response to the inevitable environmental change that is always occurring. Punctuated equilibria gives the alternative, real-time temporal dimension to the picture that Dobzhansky had begun to develop less than forty years earlier.

So if ecological speciation looks at the importance of isolation for generating and conserving adaptive change from a within-species perspective, punctuated equilibria does the same thing from a within-to-among species perspective. Both approaches converge at seeing phylogenetically long-lasting adaptive change tied up largely, if not completely, in events of speciation through isolation. And, I must say, as someone who does have a pony in this race, the results from the molecular labs of Mark Pagel and colleagues (Pagel, Venditti, and Meade 2006; Venditti and Pagel 2009), finding essentially that a higher percentage of molecular elemental change is concentrated at splitting events ("nodes") than would be expected under the molecular clock model (the molecular version of gradualism), points the way to yet a third way of testing—and validating—the claim that speciation via isolation is the locus for much/most evolutionary change.

Starting with the paper on punctuated equilibria (Eldredge and Gould 1972a), paleontologists, evolutionary biologists, and philosophers began to pitch in and put some meat on the bare bones of Ernst Mayr's suggestion that species themselves play an active role in the generation of larger-scale patterns ("macroevolution") of the evolutionary process. That species can be differentially produced, and/or suffer differential deaths, in a manner at least superficially akin to natural selection itself, became a topic of great interest. Patterns such as trends and adaptive radiations came to be seen in this light. Whether or not such patterns are explicable through "species selection" (as Steve Gould [2002] maintained in his final book), or whether "species sorting" (Elisabeth Vrba's [1984] term) is the better characterization and sobriquet, remains a bit contentious. I prefer the latter.

Following biologist Michael Ghiselin's (1974) initial suggestion, species were begun—in some quarters at least—to be seen as "individuals" or as "particles." Such language emphasized the discreteness of species, with their own births, histories, and deaths. And it fit handily in with concepts of differential births and deaths of species within lineages.

It was thus perhaps inevitable that the hierarchical structure of biological systems in an evolutionary context would come to be examined in detail, following the initial lead of Theodosius Dobzhansky in the 1930s. And, again following Ghiselin, evolutionary biologists, paleontologists, and philosophers began parsing out the ontological status of various sorts of biological systems, from the molecular level up through the mega-levels of regional ecosystems and higher taxa.

One of the insights from this work, beginning with Niles Eldredge and Stanley Salthe (1984), was the recognition that biological systems formed by the matter-energy interactions among the next-lower-level parts (for example, local populations of different species within a local ecosystem) are fundamentally different from the informational systems of the germ line genome, species, and higher taxa. Thus two hierarchies are delineated: one the "ecological" or "economic" hierarchy (that is, somatic cellular processes-organisms-avatars [populations in an economic sense]-local ecosystems-regional ecosystems-the global system). The other is the "genealogical" hierarchy, consisting of the germ line genomes (and constituent parts)-organisms-demes (local populations in a reproductive sense)-species-monophyletic taxa. Interactions among components of the two hierarchies are seen, in this dual hierarchy model, as producing the phenomena of evolution.

Finally, with attention drawn to the reality and potential evolutionary implications of global mass extinctions in the early 1980s, and the work of Vrba (1985) delineating "turnovers," pulses of extinction that generate new species as well as eliminating others—the larger-scale connections between environmental perturbations began to become clear. Vrba explored how selection in isolation in disturbed habitats might lead to speciation as she conceptually linked processes in cross-genealogical ecosystems with the standard discourse in evolutionary theory, with its traditional focus on genealogical systems at the higher levels of organization (that is, species and higher taxa). Thanks to her work (and that of Brett and colleagues [Brett and Baird 1995]) in the 1980s and 1990s, we are, I believe, a lot closer to understanding the context of adaptive evolutionary change in the evolutionary process.

I tried to connect the processes of death and rejuvenation of such systems (at various levels, such as local ecosystem perturbations);

extinction and speciation at the regional level (that is, Vrba's "turnovers,"—as well as Brett and colleagues' "coordinated stasis" patterns); and global mass extinctions, to the dual hierarchy model, the better to show the relevance of hierarchy theory to evolutionary biology. This was the "sloshing bucket" of the late 1990s (Eldredge 2003).

Of course, a great counter-theme to the importance of speciation in isolation has been the persistence—in greatly modified form, reflecting the addition in knowledge, especially in the principles of heredity—of a theoretical perspective that considers only the process of adaptation: not only as the core of the evolutionary process, but nearly the only thing that evolution in its entirety consists of. This is, after all, the main theme that Darwin left us with when he founded the modern practice of evolutionary theory with the publication of *On the Origin of Species* (1859).

Many prominent evolutionary biologists in the latter decades of the twentieth century, in other words, considered only adaptation through natural selection. Richard Dawkins is especially important in this regard, as his notion of the "selfish gene" essentially redefined selection as a matter of competition of alternative forms of genes (that is, "alleles"). Much argumentation ensued over what really turned out to be a simple issue: Is adaptation through selection, in its myriad forms and guises, all there is to evolution? Or is context important? Is speciation through isolation important in specifying the locus and providing the causal context of the generation and accrual (preservation) of adaptive change in evolutionary history? Is speciation important beyond the fact that it is the way sexually reproducing lineages diversify—by being the locus of adaptive change beyond the simple generation of separate lineages?

I believe that knowing where, when, and why natural selection kicks in and effects adaptive change in the history of life, is indeed (as Dobzhansky more or less said in the 1930s) a very important component of evolutionary theory. And this is the reason that, by the late 1970s, I found it imperative to draw the distinction between this "taxic perspective" and the "transformationist perspective." I contrast this idea that species are real, have origins, histories, and deaths; and that their origins locate the time and place in which most lasting morphological, adaptive change actually occurs in evolution, with the traditional view that evolutionary adaptive change is purely a matter of natural selection modifying entire species gradually through time as environments inexorably change.

I have been asked, in effect: So what? Why should Niles Eldredge care about work from the first half of the nineteenth century so long forgotten that it basically had little to do with events starting with Dobzhansky in 1935?

My answer is simple: this entire double episode, in which the taxic perspective was invented to begin the very subject of serious scientific study of evolution, only to be forgotten until its resurrection, reminds me of convergent evolution, patterns of adaptive similarity in often exquisite detail, sometimes among collateral lineages, at others of far-distant lineages, whether phylogenetically, geographically, or temporally.

I have always thought of patterns of adaptive convergence to be the best single set of phenomena you can point to in order to explain to anyone why evolution must be true.

That the taxic perspective had a near-extinction event, but in due course had to be redeveloped in pretty much all its diversity of concepts, means to me that the taxic perspective is indeed a very good description of the structure of biological nature, and of the natural processes that have underlain the evolution of life on earth. And that means a lot to me.

Besides, it is irresistible to identify with the nearly forgotten past when it includes such gems as these words of Robert Grant (1829)—Giambattista Brocchi's analogy in near poetic form as he describes the world of "eternal ephemera":

> In this vast host of living beings, which all start into existence, vanish, and are renewed, in swift succession, like the shadows of the clouds in a summer's day, each species has its peculiar form, structure, properties, and habits, adapted to its situation, which serve to distinguish it from every other species; and each individual has its destined purpose in the economy of nature. Individuals appear and disappear in rapid succession upon the earth, and entire species of animals have their limited duration, which is but a moment, compared with the antiquity of the globe. Numberless species, and even entire *genera* and tribes of animals, the links which once connected the existing races, have long since begun and finished their career.

NOTES

In this section, I briefly cite some of the books, articles, manuscripts, and letters that either are discussed in the text or served as sources for my presentation of the specific contributions of the works of the scientists, philosophers, and historians who, in my opinion, were the most important shapers of evolutionary biology in general—and specifically of the taxic evolutionary perspective. Detailed citations of all these works are in the bibliography.

I also append some notes and reflections tied to the relevant portions of the text.

This book relies first and foremost on primary literature: it represents my considered understanding of what my predecessors and, in due course, my contemporaneous colleagues and my younger successors have to say about the evolutionary—especially taxic—issues discussed in the text. Perhaps the most difficult part of writing this narrative was trying to bring the same level of critical analysis to my own work as I sought to present for everyone else.

I am extremely grateful to Tom Baione, Mai Reitmeyer, and the entire staff of the library of the American Museum of Natural History—one of the great natural history libraries in the world—for all their help over the years. And I extend my thanks as well to the staff of the Rare Book Collection at Cambridge University Library for their hospitality and help when I read through Charles Darwin's Geological Diary and Geological Notes under the superb guidance of historian David Kohn early in 2006.

Nowadays, much (but not all!) of the critically important works by not only Darwin but also his predecessors, contemporaries, and successors is freely available online. Specific Web sites are cited in the bibliography, but I will mention as well Google Books and Google Scholar as treasure troves whose riches increase seemingly on a daily basis.

Except where noted, all the many quoted passages throughout the text are taken directly from the original (or reprinted) sources.

INTRODUCTION

The thoughts on taxic versus transformational approaches developed in my lecture at the University of Rochester were an outgrowth of thinking about the ramifications of, as well as the spirited objections to, the notion of "punctuated equilibria," as developed earlier in the 1970s (Eldredge 1971a; Eldredge and Gould 1972a, 1972b; Gould and Eldredge 1977). I remain grateful to colleagues H. B. Rollins and J. S. Schwartz for inviting me to participate in a symposium at the University of Pittsburgh, where I developed these thoughts further, leading to the publication of my paper "Alternative Approaches to Evolutionary Theory" (Eldredge 1979). All these papers (as well as many others of mine) can be found in downloadable format at www.NilesEldredge.com.

The voluminous notes, correspondence, and books by Charles Darwin relied on in this book are discussed in detail, and referenced, in chapters 2 and 3. Suffice it to say, at this point in the narrative, that the digital revolution has enabled Darwin historians such as David Kohn to publish scholarly transcriptions of many of Darwin's notes and unpublished (in his lifetime) manuscripts; it is now even possible to download a complete edition of *On the Origin of Species*. The most important Web sites for accessing this treasure trove of Darwiniana are the Darwin Manuscripts Project (darwin.amnh.org) and the Darwin Correspondence Project (www.darwinproject.ac.uk).

The most influential, and generally excellent, biographies of Darwin in recent times are Janet Browne's *Charles Darwin: Voyaging* (1995) and *Charles Darwin: The Power of Place* (2002), and Adrian Desmond and James Moore's *Darwin: The Life of a Tormented Evolutionist* (1991).

1. THE ADVENT OF THE MODERN FAUNA

Baron Georges Cuvier (1769–1832) was a French zoologist and paleontologist, sometimes called the "Father of Comparative Anatomy." Historian Martin Rudwick's account of Cuvier and his works, *Georges Cuvier, Fossil Bones, and Geological Catastrophes: New Translations and Interpretations of the Primary Texts* (1997), remains the key resource on this important figure. Cuvier was a colleague of Jean-Baptiste Lamarck (1744–1829) at the Jardin des Plantes, the Parisian center for research on and exhibitions of France's growing natural history collections. Although Cuvier was opposed to Lamarckian evolution, his influence on scientists such as Lamarck, Giambattista Brocchi, and Charles Darwin was enormous. Cuvier was a catastrophist, interpreting the fossil record of the history of life as showing numerous global extinction events, most of which he deemed to have been catastrophic, followed by the appearance, through unspecified means, of succeeding species. Cuvier established the reality of extinction to the satisfaction of most of his contemporaries. He published *Recherches sur les ossemens fossiles de quadrupèdes* in 1812; the "Preliminary Discourse" introducing that great work was widely read and was translated into English. Rudwick reports that Cuvier began pointing to the progressive "younging" of successions of species as early as 1812, in the "Preliminary Discourse."

Robert Jameson (1774–1854), as we shall see in some detail in this chapter, was arguably the most important proponent of transmutational ideas in Great Britain in the 1820s. He was a geologist but also a physician on the faculty of the medical school in Edinburgh. His translation of Cuvier's "Preliminary Discourse" into English was published as *Essay on the Theory of the Earth. With Geological Illustrations by Professor Jameson* (1813). Darwin took Jameson's course on natural history and is known to have read the fifth edition of *Essay on the Theory of the Earth* (1827).

Charles Lyell (1797–1875), often considered the founder of modern geology, wrote the three-volume *Principles of Geology* (1830–1833). The impact of these works (especially, though by no means exclusively, volume 2 on transmutation) on the young Charles Darwin emerges especially in chapter 2 of this book. For convenience, I have relied heavily on a single-volume edition of *Principles of Geology* (1997) edited by historian James Secord, whose introduction on Lyell and the *Principles* is invaluable.

John Herschel's letter to Lyell was published in Charles Babbage's *The Ninth Bridgewater Treatise* (1838), but the passage had appeared in print earlier. Darwin's exultant comment on the "mystery of mysteries" appears in Notebook E (1839) opposite page 58, on which is his succinct reduction of natural selection to three terse sentences.

Lyell specifically extols the importance and utility of focusing on fossil molluscan species to define the subdivisions, utilizing the collections of the French paleontologist Gérard Deshayes—but neither Lamarck's nor Brocchi's names appear in this section of Lyell's text. According to Rudwick in *Worlds Before Adam*, some of the names of the subdivisions had actually been coined by William Whewell. See also William B. N. Berry's informative *Growth of a Prehistoric Time Scale* (1968).

Brocchian Transmutation

I first encountered Brocchi's name and core ideas in volume 2 of Lyell's *Principles of Geology*. As discussed in chapter 2, Brocchi's analogy is clearly present in Darwin's essay "February 1835," in the latter half of his Red Notebook (1836–1837), in some of his Transmutation Notebooks, and in an especially revelatory letter to Leonard Jenyns (Darwin 1844b)—with vestiges remaining in his "Pencil Sketch" (1842), his "Essay" (1844a), and *On the Origin of Species* (1859). Yet Darwin never cites Brocchi by name. I linked Brocchi to Darwin via Lyell in "Experimenting with Transmutation: Darwin, the *Beagle*, and Evolution" (Eldredge 2009a). Stefano Dominici and I investigated the paper trail linking Brocchi to Darwin in the literature, largely published in Edinburgh between 1816 and the late 1820s. We reported the results in "Brocchi, Darwin, and Transmutation: Phylogenetics and Paleontology at the Dawn of Evolutionary Biology" (Dominici and Eldredge 2010), the basis of the somewhat expanded account presented in this book.

Brocchi appears prominently in the two magnificent volumes by Martin Rudwick: *Bursting the Limits of Time: The Reconstruction of Geohistory in the Age of Revolution* (2005) and *Worlds Before Adam: The*

Reconstruction of Geohistory in the Age of Reform (2008). In the former volume, Rudwick discusses Brocchi's analogy, saying that it "respected the reality of species as discrete entities or natural kinds, rather than dissolving them in an endless flux of transmutation. It also suggested, though less explicitly, that the *origin* of species, might have an equally natural, yet episodic, mechanism, analogous to the birth of individuals" (527).

The descriptions of plant and animal species by the Swedish naturalist Carolus Linnaeus (born Karl von Linné) set the standards for modern systematic biology. Although he appeared to waver somewhat toward the end of his life, Linnaeus was not an evolutionist (transmutationist). One hundred and one years after Linnaeus published *Systema naturae*. Darwin, who essentially founded the modern science of evolutionary biology, published *On the Origin of Species*. The question that drives this section of my narrative becomes, then, what happened between 1758 and 1859? Specifically, when, where, and by whom was the pursuit of a non-miraculous causal explanation for the origin of modern species first developed?

The details of Darwin's paleontological discoveries at Bahia Blanca in 1832 (and then again in 1833) are presented, with full citations, in chapter 2. The point here is that the early transmutationists such as Brocchi, Jameson (as we shall soon see), and Darwin used the persistence of genera—*as opposed to the persistence of their component species*—to broach the delicate subject of patterns of apparent births of descendant species from antecedent, congeneric species.

Robert Jameson, Robert Grant . . . and Charles Robert Darwin

Charles Darwin (1809–1882) wrote his *Autobiography* in 1876. According to his son Francis Darwin (1848–1925)—arguably the first of a still-growing horde of Darwin scholars, and the original editor of the *Autobiography*—Darwin intended his autobiographical sketch to be read only by members of his family. The version I have utilized was bundled with other Darwiniana by paleontologist George Gaylord Simpson and published as *Charles Darwin's Autobiography, with His Notes and Letters Depicting the Growth of the Origin of Species* (Darwin [1876] 1950), while

Darwin's granddaughter Nora Barlow edited another edition, *The Autobiography of Charles Darwin, 1809–1882* (1958).

The *Edinburgh New Philosophical Journal* has emerged as *the* hotbed of the publication of radical, transmutational thinking in Edinburgh—and thus in Great Britain. I was fortunate to discover that a complete run of these journals is housed in the library of the American Museum of Natural History, but they have yet to appear in their entirety on Internet sites such as Google Books.

Darwin's student book list is available at the Darwin Correspondence Project (www.darwinproject.ac.uk).

2. DARWIN AND THE *BEAGLE*

Chapter 2 is devoted entirely to the nearly five years (1831–1836) of Charles Darwin's experiences as naturalist and companion to Captain Robert FitzRoy on HMS *Beagle*, and to the relatively brief period on his arrival home when he was overtly (albeit to himself) developing a theory of transmutation from a primarily taxic perspective.

The account begins, substantively, with Darwin's initial geological and paleontological observations and deductions on the Atlantic island of St. Jago (in the Cape Verde Islands) in early 1832. It ends with his letter to Leonard Jenyns, in which Darwin (1844b) makes it clear that it was Giambattista Brocchi's analogy that first convinced him of transmutation.

The evidence is all in Darwin's own handwriting. There are several distinct categories of Darwin's manuscripts, books, letters, and notes that, together, yield the picture of his work on the *Beagle* expedition and shortly thereafter. I have (Eldredge 2009a; see also Dominici and Eldredge 2010) presented in abbreviated form my analysis based on the entire gamut of these documents:

- Correspondence. Darwin's letters, especially, though not exclusively, to his mentor, John Stevens Henslow, reveal valuable insights into Darwin's transmutational thinking while he was on the *Beagle*. Nora Barlow, Darwin's granddaughter, edited a useful compendium of Darwin's correspondence with Henslow: *Darwin and Henslow: The Growth of an Idea* (1967); a handy recent source is *The 'Beagle' Letters* (2008), edited by Frederick Burkhardt; and the correspondence is available at the Darwin Correspondence Project (www.darwinproject.ac.uk).

• Diary. Although the Diary that Darwin kept while on the *Beagle* is not particularly rich in purely scientific observations and conclusions, exceptions do occur—as in his mention of the snake at Bahia Blanca. Barlow edited and published *Charles Darwin's Diary of the Voyage of H.M.S. Beagle* (1934), and Richard Keynes, Darwin's great-grandson, edited another edition: *Charles Darwin's Beagle Diary* (1988), which is rendered more valuable by occasional notes from what Keynes calls "The Down House Notebooks": Darwin's pocket-size Field Notebooks, in the collection at Down House, Kent, which have been edited and published by Gordon Chancellor and John van Wyhe as *Charles Darwin's Notebooks from the Voyage of the 'Beagle'* (2009) and are available at Darwin Online (darwin-online.org.uk).

• Geological Diary/Geological Notes. These documents have yet to be fully published and, in any case, have never before been rigorously analyzed from a transmutational perspective for reasons discussed briefly in the text. The originals are in the Cambridge University Library; the portions most directly relevant to my text (those pertaining to South America) are catalogued as DAR 32–42. A transcription is available at Darwin Online (darwin-online.org.uk).

Historian Sandra Herbert is an exception to the generalization that scholars have almost completely ignored Darwin's Geological Diary. Her book *Charles Darwin, Geologist* (2005) covers many aspects of Darwin's geological work, including his paleontological experiences, especially in Patagonia. Herbert hints at the possible transmutational relevance of the Geological Diary, but stops short of concluding that it in fact shows the development of Darwin's transmutational thinking while he was on the *Beagle*.

Two essays in the Earthquake Portfolio (DAR 42) are catalogued after the Geological Diary. The first of these is "Reflection on Reading my Geological Notes" (DAR 42:93–96), which Herbert has transcribed and analyzed in "From Charles Darwin's Portfolio: An Early Essay on South American Geology and Species" (1995).

Herbert's transcription and annotation of the Red Notebook—in *The Red Notebook of Charles Darwin* (1980) and "Red Notebook" (1987)—is also critical, as it represents the most explicit rendering of Darwin's paleontological and zoological discoveries in South America in transmutational terms—albeit written some six months after his return to England.

The other, later essay in the Earthquake Portfolio is, along with Darwin's paleontological experiences at Bahia Blanca in 1832 (with a return trip and ruminations a year later), the most important component of the Geological Diary/Geological Notes. Called simply "February 1835" (DAR 42:97–99), it has been transcribed and analyzed by historian M. J. S. Hodge in "Darwin and the Laws of the Animate Part of the Terrestrial System (1835–1837): On the Lyellian Origins of His Explanatory Program" (1983), and transcribed by David Kohn (Darwin Manuscripts Project, darwin.amnh.org).

• Zoological Notes. Keynes, himself a zoologist, did much to explicate his great-grandfather's work—especially in the *Beagle* years. *Charles Darwin's Zoology Notes & Specimen Lists from H.M.S. Beagle* (2000), edited by Keynes, does for the Zoological Notes (DAR 29–31) what remains to be done for the Geological Diary/Geological Notes: a lucid transcription accompanied by many notes and penetrating observations. It is available at Darwin Online (darwin-online.org.uk).

• Rewritten and compiled Zoological Notes: Animal Notes and Ornithological Notes. Toward the end of the *Beagle* voyage, Darwin organized and compiled much of his zoological data. Barlow transcribed and annotated Ornithological Notes (DAR 29.2) and published them as "Darwin's Ornithological Notes" (1963), which is available at the Darwin Manuscript Project (darwin.amnh.org). Animal Notes (DAR 29.1:A1–19) can be found at Darwin Online (darwin-online.org.uk).

• Transmutation Notebooks (Notebooks B–E). Darwin's Transmutation Notebooks (DAR 121–124), dealt with at the end of this chapter and in chapter 3, record the development of his transmutational thinking and, by Notebooks D and E, his change of emphasis from a taxic to a transformational perspective. They were written between 1837 and 1839. Embryologist and Darwin enthusiast Gavin de Beer published a version of these notebooks as "Darwin's Notebooks on Transmutation of Species: Parts I–IV" (1960–1961) and, with M. J. Rowlands, "Darwin's

Notebooks on Transmutation of Species: Addenda and Corrigenda" (1960–1961), while the definitive transcription and scholarly annotation is by David Kohn: "Notebook B–Notebook E" (1987).

• *Journal of Researches*. In 1839, Darwin combined his Diary with his scientific observations into this monumental, informative, and downright delightful book, which has come to be known as *The Voyage of the Beagle*. The book summarizes his experiences in the field, with but a few hints of the transmutational thoughts that Darwin harbored. To disguise his ideas on the births and deaths of species through natural causes, as he compared his fossils with elements of the living South American fauna, Darwin developed his Law of Succession, which boils down simply to the taxic pattern of clearly related, endemic species replacing one another up through the stratigraphic record to modern times.

Yet Darwin, in the unanimous opinion of all Darwin scholars, was a convinced transmutationist at least by early 1837, based on evidence in the Red Notebook (as the text shows, I am convinced that he was at least taking a hard look at transmutation as early as 1832), and had discovered natural selection at least a year before *The Voyage of the Beagle* appeared. This disingenuous approach in public, in my opinion, is a mere extension of his secretive style developed in his notes while on the *Beagle*.

The second edition of *Journal of Researches*, with still more sly hints on evolution, appeared in 1845. Unlike the Diary, neither edition of the *Voyage* is to be taken as a literal sequential narrative of that journey: for example, Darwin does not reveal that his initial visit to Bahia Blanca—when he recovered his critically important fossils—actually occurred in 1832, not in 1833, when he returned for a second look. Readers of the *Voyage* will come away thinking that Darwin did not get to Bahia Blanca until 1833.

Bahia Blanca

The Fossils—and Shades of Giambattista Brocchi

George Gaylord Simpson's first book *Attending Marvels: A Patagonian Journal* (1934) is a good adventure story—interspersing accounts of his paleontological experiences with vignettes of gunfire in the rebellious streets of Buenos Aires. Although he says little about evolution—and nothing about Darwin—in the pages of this book, Simpson does reveal his predilection for gradualist, transformational evolutionary thinking in several striking passages, discussed in "A Question of Individuality: Charles Darwin, George Gaylord Simpson, and Transitional Fossils" (Eldredge 2009b).

For an interesting recent look at Darwin's (vertebrate) paleontological work in South America, and its relevance to the formation of Darwin's evolutionary views, see Paul Brinkman's "Charles Darwin's *Beagle* Voyage, Fossil Vertebrate Succession, and 'The Gradual Birth & Death of Species'" (2009).

In *Journal of Researches* (1839:209), Darwin, following Richard Owen's identifications, changed the identification of the small rodent fossils found at Monte Hermoso from *Cavia* (the fossil and recent cavy species) to *Ctenomys* (the fossil and recent tucutucu species), using them as an example of his Law of Succession—and the probable source of Simpson's grudging admission that these fossils may have played a role in the emergence of Darwin's evolutionary thinking.

The portion of the Geological Notes dealing with the rocks and fossils at Bahia Blanca (Punta Alta and Monte Hermoso), as well as the notes from Darwin's return trip in 1833 (labeled "Appendix") are catalogued as DAR 32.1.61–74. All quotes in this section are from those pages.

Darwin and the Geographic Replacement of Closely Allied Species in Patagonia

In "Darwin's Ornithological Notes" (1963:273–278), Nora Barlow follows up, closing with an appendix that is an even more comprehensive summary of Darwin's writings on the rheas. She includes Darwin's account of eating the specimen of the smaller, southern species of rhea shot for Christmas dinner—before he realized what it must be.

Crossing the Andes, Then on to the Galápagos

The bicentennial of Darwin's birth in 2009 was marked, of course, by many lectures, symposia, articles, and books (even museum exhibitions!). One of the most useful is "Darwin en Argentina," a special issue, available in English as well as Spanish, of the *Revista de la Asociación Geológica Argentina* (Aguirre-Urreta, Griffin, and Ramos 2009). Among many gems, there are accounts of Darwin's trek across the Andes and of his work there and its relation to modern understanding of Andean uplift. The outcrops (including the newly discovered human and megafaunal footprint sites) at Bahia Blanca are also included in this wonderful compendium.

3. ENTER ADAPTATION AND THE CONFLICT BETWEEN ISOLATION AND GRADUAL ADAPTIVE CHANGE

It is tempting to think that Darwin was being a bit sarcastic, linking William Paley with Euclid as a source of "delight." Yet Paley's book was taken sufficiently seriously that, as I propose, his "argument from design" pretty much inhibited explanation of adaptations through natural causes for the first two or three decades of the nineteenth century.

Adaptation and Natural Selection in Speciation and Gradual Phyletic Change

David Kohn's "Notebook B–Notebook E" (1987) is the source of all citations in this chapter. Given the large number of quoted passages in my text, I have inserted the numbers *of the relevant Notebook page* (*not* of Kohn's transcription), directly in association with each quotation. Kohn's introductory essay to each of the four Transmutation Notebooks serves as an invaluable consideration of the rich content of these notebooks, which more than any of Darwin's other writings together reveal the order and logic underlying Darwin's creation of the structure and content of his later two essays in the 1840s—"Pencil Sketch" (1842) and "Essay" (1844a)—and so, too, of the unfinished "Big Species Book" of the mid-1850s and, ultimately, of *On the Origin of Species* (1859) itself. I learned how to read Darwin's handwriting by working with Kohn in the rare book room at the Cambridge University Library in 2006, reading Darwin's Geological Notes and other manuscripts. Kohn is a master at deciphering Darwin's handwriting, but in the transcription of the passage from page 155 of Notebook B, I beg to differ on one word: I read the word "spots" where Kohn has rendered it "sports." Though either word is plausible, "spots" make more sense in this geographic context. The reader can judge by examining the image of this passage at the Darwin Manuscripts Project (darwin.amnh.org).

Darwin's Evolutionary Texts

"Pencil Sketch"

Francis Darwin—Charles and Emma Darwin's seventh child, and a noted zoologist in his own right—was arguably the first Darwin scholar of worth. In 1909, he published *The Foundations of the Origin of Species: Two Essays Written in 1842 and 1844* in honor, simultaneously, of the hundredth anniversary of his father's birth and the fiftieth anniversary of the publication of *On the Origin of Species*. The book consists of Francis Darwin's transcriptions of the "Pencil Sketch" and the "Essay."

These transcriptions, with a few editorial changes in the "Essay," were republished under the editorship of embryologist Gavin de Beer, in *Evolution by Natural Selection [by] Charles Darwin and Alfred Russel Wallace* (Darwin 1958), and are available at Darwin Online (darwin-online.org.uk). I have relied on Francis Darwin's transcription, as edited by de Beer (Darwin 1958), for the purposes of this narrative, and the page numbers for the quoted text refer to this edition. A new transcription, by Kohn, of these two manuscripts is available at the Darwin Manuscripts Project (darwin.amnh.org).

"Essay"

The "Fair Copy" of the manuscript of the "Essay" (Darwin 1844a) was written out by an amanuensis and intended, possibly, for future publication. Darwin made some annotations, including the important one singled out here, published as a footnote by Francis Darwin in 1909. To my knowledge not yet transcribed and published, the original is available at the Darwin Manuscripts Project (darwin.amnh.org).

Volume 1 of Alcide d'Orbigny's (1802–1857) monumental *Paléontologie française* (1842) is dedicated to Jurassic cephalopods (mainly ammonoids) and is famous as his first articulation of the concept, and delimitation, of sequential *étages* in the Jurassic rocks of France and surrounding regions. The version consulted here, ironically, is Darwin's very own copy, which is available at the Biodiversity Heritage Library (biodiversitylibrary.org/item/106856#page/7/mode/1up). For more on d'Orbigny's life and work, see Marie-Thérèse Vénec-Peyré, "Beyond Frontiers and Time: The Scientific and Cultural Heritage of Alcide d'Orbigny (1802–1857)" (2004).

Darwin's Principle of Divergence

In 1975, historian R. C. Stauffer published *Charles Darwin's Natural Selection: Being the Second Part of His Big Species Book Written from 1856 to 1858*. The only published source of Darwin's manuscript, which was not published in his lifetime, the book is available at Darwin Online (darwin-online.org.uk).

Darwin's *On the Origin of Species* was, of course, first published in 1859. The sixth edition (1872) is important because it is the version that most subsequent readers had available to them, roughly until 1959—the centennial of the original publication of the *Origin of Species*. Evolutionary biologist Ernst Mayr wrote the introduction to a facsimile edition of the first edition, published in 1959. Both editions (plus others) are available at Darwin Online (darwin-online.org.uk) and the Darwin Manuscripts Project (darwin.amnh.org).

I have discussed all the texts, in a somewhat broader context than in this chapter narrative focusing on the taxic perspective in evolutionary biology, in *Darwin: Discovering the Tree of Life* (Eldredge 2005).

PART II. REBELLION AND REINVENTION

I find it deeply ironic that, until very recently, it was next to impossible for someone not affiliated with a university or another form of research institution to have access to most of the important early writings on evolutionary biology. Indeed, in some instances, such as Darwin's early notes, one had to travel to the rare book room of Cambridge University Library and to Down House. All that has changed, as my text and the accompanying notes for the introduction and chapters 1 to 3 of this narrative have made clear: the Internet has made virtually everything of importance available to anyone with a computer and Internet service.

Not so, ironically, of the literature of the twenty and twenty-first centuries! To be sure, copies of Theodosius Dobzhansky's, Ernst Mayr's, and George Gaylord Simpson's most important books are usually available, often at no great expense, through used book dealers.

That leaves scientific, historical, and philosophical papers published in technical journals. Some of these are available through "open access" platforms, such as all the back issues of *Evolution: Education and Outreach*, where Stefano Dominici's "Brocchi's Subapennine Fossil Conchology" (2010) and Dominici and Niles Eldredge's "Brocchi, Darwin, and Transmutation: Phylogenetics and Paleontology at the Dawn of Evolutionary Biology" (2010), as well as other papers by me and other authors cited in the bibliography are freely available. But otherwise, and for the most part, one needs to have access to a major library to find articles published in such journals as *Evolution*, *American Naturalist*, *Science*, and *Nature*. Even though all of these journals are now archived and continuously available online, potential readers must be affiliated with institutions that provide access to archival organizations such as JSTOR. Otherwise, there is a hefty cost to download each pertinent article from the evolutionary literature from 1935 (where this narrative resumes) right down to the present moment.

I can but cite these papers and hope that readers can find them without too much difficulty.

4. SPECIES AND SPECIATION RECONSIDERED

In my opinion, Theodosius Dobzhansky's "A Critique of the Species Concept in Biology" (1935), discussed extensively in this chapter, triggered the renaissance in taxic evolutionary thinking.

In *Genetics and the Origin of Species* (1937), Dobzhansky remarks that if this "maxim is taken too literally, it overshoots the mark," as G. J. Romanes was apparently suggesting that adaptive change is indeed causally connected with speciation-in-isolation—the strong version of the importance of isolation in the generation of adaptive change that Darwin himself saw in the late 1830s, but later abandoned. Stephen Jay Gould and I teamed up in 1982 to bring out reprints of Dobzhansky's *Genetics and the Origin of Species* (1937) and Ernst Mayr's *Systematics and the Origin of Species* (1942). Gould wrote the editorial introduction to Dobzhansky's book, while I did the same for Mayr's.

The literature of modern studies of speciation is voluminous. For a recent review, and guidance to this literature, James Sobel and colleagues' "The Biology of Speciation" (2009) provides a good start, as does Francesco Santini, Maria Pia Miglietta, and Anuschka Faucci's "Speciation: Where Are We Now?" (2012). And I highly commend John N. Thompson's *Relentless Evolution* (2013), which examines evolutionary processes, including speciation (especially "ecological speciation") from the perspective of within- and among-genetic variation in environmental and geographic contexts.

5. PUNCTUATED EQUILIBRIA

When George Gaylord Simpson declined to include my editorial introduction to a reprint of *Tempo and Mode in Evolution* (1944), I can't say that I was surprised, nor did I particularly hold it against him. I published it, instead, in *Unfinished Synthesis: Biological Hierarchies and Modern Evolutionary Thought* (Eldredge 1985b), along with my detailed examination of the three editions of Theodosius Dobzhansky's *Genetics and the Origin of Species* (1937, 1941, 1951) and of Ernst Mayr's *Systematics and the Origin of Species* (1942).

Simpson's clarification of the ontological status of taxa and categories appeared in "The Meaning of Taxonomic Statements" (1963). In my opinion, this distinction is a perfect example of the sort of ontological analysis that Michael Ghiselin called for in his paper "A Radical Solution to the Species Problem" (1974), which is discussed in chapter 6.

The Genesis of Punctuated Equilibria

Both Steve Gould and I have given our own separate accounts of the history, nature, and meaning of "punctuated equilibria"; component terms (for example, "stasis"); and extended concepts (such as "species selection," "hierarchy theory," and the like). Steve's last technical book, *The Structure of Evolutionary Theory* (2002), has much to say about these matters—and contains an extensive pertinent bibliography. For my own part, I wrote *Time Frames: The Rethinking of Darwinian Evolution and the Theory of Punctuated Equilibria* (Eldredge 1985a) as a brief account of the history and content of punctuated equilibria, and have since written a general audience account: "The Early 'Evolution' of 'Punctuated Equilibria'" (2008a). In addition, I have contributed the chapter "Stephen Jay Gould in the 1960s and 1970s and the Origin of 'Punctuated Equilibria'" (2013) to a volume on Steve's scientific legacy. We also severally, and at times together, wrote on punctuated equilibria—some of which papers are alluded to in appropriate places as this chapter unfolds.

The best, independent account of the history of punctuated equilibria is David Sepkoski's *Rereading the Fossil Record: The Growth of Paleobiology as an Evolutionary Discipline* (2013b). Sepkoski had full access to my own files, and those of others, in preparing what I think is a highly accurate and dispassionate account of the events leading up to the publication of "Punctuated Equilibria: An Alternative to Phyletic Gradualism" (Eldredge and Gould 1972a). In addition, see Sepkoski's "Punctuated Equilibria" (2013a).

"February 1965" and "April 1968"

The essays comprising my personal paper trail on the development of punctuated equilibria, as well as many others pertaining to punctuated equilibria, are housed in the library of the American Museum of Natural History, whose staff has scanned and posted them online. They are either handwritten or typescripts, and include what I have called "February 1965" (much more modest, with far less import, than Darwin's "February 1835," but nonetheless my oldest surviving evolutionary essay) and "April 1968" ("Some Aspects of Species-level Evolution in Paleontology"). I omit (to everyone's relief, I am sure) my actual doctoral dissertation (Eldredge 1969, 1972).

"The Allopatric Model and Phylogeny in Paleozoic Invertebrates"

In 1974, I published a follow-up paper more fully developing the model of abrupt replacement of stable species in the fossil record: "Stability, Diversity and Speciation in Paleozoic Epeiric Seas."

Stasis: The Big Gorilla in the Room

Our biggest critic—who showed up with original, and published, data and analysis—was Philip Gingerich, in his articles "Stratigraphic Record of Early Eocene *Hyopsodus* and the Geometry of Mammalian Phylogeny" (1974) and "Paleontology and Phylogeny: Patterns of Evolution at the Species Level in Early Tertiary Mammals" (1976). Phil had good data, but, as we pointed out, he was not looking at the entire geographic distribution of the species under study. (Phil also privately admitted to me many years later that "you guys were right.") There were many other critics, some of whom also showed up with data purporting to support a gradualist perspective. We conceded, in "Punctuated Equilibria: The Tempo and Mode of Evolution Reconsidered" (Gould and Eldredge 1977), for example, that haploid, or alternating sexual/asexually reproducing organisms—such as marine planktonic single-celled organisms—often do show patterns of gradual evolutionary change over wide geographic areas. On the evolution of the radiolarian genus *Pterocanium*, see David Lazarus, Reed P. Scherer, and Donald Prothero's "Evolution of the Radiolarian Species-complex *Pterocanium*: A Preliminary Survey" (1985).

Internet search engines, of course, will turn up many more discussions, pro and con, of punctuated equilibria.

6. SPECIATION AND ADAPTATION

Elisabeth S. Vrba

Steve Gould, in a striking passage in *The Structure of Evolutionary Theory* (2002), wrote of species selection that "for no other problem have I made so many published mistakes, and undergone so many changes of viewpoint" (670), to which I briefly allude in the text. I have already given the reasons for my preference for Elisabeth Vrba's position on species selection/species sorting.

The late philosopher David Hull made many important direct contributions to evolutionary theory, especially in the 1970s and 1980s. As we shall see in the following section, Hull was especially important in furthering and promoting biologist Michael Ghiselin's notion that species are in fact "individuals" rather than "classes," presented in his paper "A Radical Solution to the Species Problem" (1974). Indeed, it was in "Individuality and Selection" (1980) that Hull developed his notion that, for selection to obtain, the "individuals" involved (at whatever level) must be both "replicators" and "interactors." I always found this distinction compelling, and as we shall see a bit later in this narrative, Hull's distinction between interactors and replicators dovetails perfectly with the distinction, first drawn by me working with biologist Stanley N. Salthe, between the dual ecological hierarchy (composed of nested sets of interacting individuals) and the genealogical hierarchy (composed of nested sets of reproducers, or "more-makers," the extension of Hull's category of "replicators").

Bruce S. Lieberman's work on macroevolutionary patterns, in conjunction with cladistic analyses integrated with paleobiogeographic analyses, and his work on stasis have been prolific. In addition to E. O. Wiley and Lieberman's *Phylogenetics: The Theory and Practice of Phylogenetic Systematics* (2011), which presents an evolutionary concept of species ("species form lineages and are the largest tokogenetic arrays in which reproduction predominates" [65]), he has written *Paleobiogeography* (2000). His work on speciation rates, such as those obtained among the earliest trilobites of the basal Cambrian evolutionary "explosion," has taken the study of speciation, as seen in the fossil record, to new levels.

Hierarchy Theory

In *Interactions: The Biological Context of Social Systems* (1992), Marjorie Grene and I discuss the ecological (economic) and genealogical hierarchies, and extend the analysis of interactions between components that, in our view, are critical to understanding the origin, structure, and internal dynamics of social systems.

Mass Extinctions and Macroevolutionary Rebounds in Higher Taxa

The literature on the modern-day, human-engendered mass extinction event currently engulfing our planet is voluminous. As an entrée into this subject, I'll cite one of my own works: *Life in the Balance: Humanity and the Biodiversity Crisis* (1998). A more recent, equally accessible book is Elizabeth Kolbert's *The Sixth Extinction: An Unnatural History* (2014).

BIBLIOGRAPHY

Aguirre-Urreta, Beatriz, Miguel Griffin, and Victor A. Ramos, eds. 2009. Darwin en Argentina. Special issue, *Revista de la Asociación Geológica Argentina* 64:1–180.

Alvarez, Luis W., Walter Alvarez, Frank Asaro, and Helen V. Michel. 1980. Extraterrestrial cause for the Cretaceous-Tertiary extinction. *Science* 208:1095–1108.

Babbage, Charles. 1838. *The Ninth Bridgewater Treatise*. 2nd ed. London: John Murray.

Barlow, Nora, ed. 1934. *Charles Darwin's Diary of the Voyage of H.M.S. Beagle*. Cambridge: Cambridge University Press.

Barlow, Nora, ed. 1958. *The Autobiography of Charles Darwin, 1809–1882*. New York: Harcourt, Brace & World. [Available at http://darwin-online.org.uk]

Barlow, Nora, ed. 1963. Darwin's Ornithological Notes. *Bulletin of the British Museum (Natural History)*, Historical Series 2, no. 7. [Available at http://darwin-online.org.uk]

Barlow, Nora, ed. 1967. *Darwin and Henslow: The Growth of an Idea. Letters, 1831–1860*. London: John Murray.

Bayón, Cristina, Teresa Manera, Gustavo Politis, and Silvia Aramayo. 2011. Following the tracks of the first South Americans. *Evolution: Education and Outreach* 4:205–217. [Available at http://link.springer.com/journal/12052]

Berry, William B. N. 1968. *Growth of a Prehistoric Time Scale*. San Francisco: Freeman.

Brett, Carlton E., and Gordon Baird. 1995. Coordinated stasis and evolutionary ecology of Silurian to Middle Devonian faunas in the Appalachian Basin. In *New Approaches to Speciation in the Fossil Record*, edited by Douglas H. Erwin and Robert L. Anstey, 285–315. New York: Columbia University Press.

Brinkman, Paul D. 2009. Charles Darwin's *Beagle* voyage, fossil vertebrate succession, and "the gradual birth & death of species." *Journal of the History of Biology*. [Available at DOI 10.1007/s10739-009-9189-9]

Brocchi, Giambattista. 1814. *Conchiologia fossile subapennina*. Milan: Stamperia Reale.

Brooks, Daniel R., and E. O. Wiley. 1986. *Evolution as Entropy: Toward a Unified Theory of Biology*. Chicago: University of Chicago Press.

Browne, Janet. 1995. *Charles Darwin: Voyaging*. Vol. 1 of *A Biography*. New York: Knopf.

Browne, Janet. 2002. *Charles Darwin: The Power of Place*. Vol. 2 of *A Biography*. New York: Knopf.

Bunge, Mario. 1977. *Treatise of Basic Philosophy*. Vol. 3, *The Furniture of the World*. Dordrecht: Reidel.

Burkhardt, Frederick, ed. 2008. *The 'Beagle' Letters*. Cambridge: Cambridge University Press. [Available at http://www.darwinproject.ac.uk]

Burkhardt, Frederick, et al., eds. 1985–2014. *The Correspondence of Charles Darwin*. 21 vols. to date. Cambridge: Cambridge University Press.

Carruthers, R. G. 1910. On the evolution of *Zaphrentis delanouei* in Lower Carboniferous times. *Quarterly Journal of the Geological Society of London* 66:523–538.

Chancellor, Gordon, and John Van Wyhe, eds. 2009. *Charles Darwin's Notebooks from the Voyage of the 'Beagle.'* Cambridge: Cambridge University Press. [Available at http://darwin-online.org.uk]

Clarkson, Euan N. K. 1966. Schizochroal eyes and vision in some Silurian acastid trilobites. *Palaeontology* 9:1–29.

Crampton, H. E. 1916. Studies on the variation, distribution, and evolution of the genus *Partula*. The species inhabiting Tahiti. *Carnegie Institute of Washington Publication* 228:1–311.

Crampton, H. E. 1932. Studies on the variation, distribution, and evolution of the genus *Partula*. The species inhabiting Moorea. *Carnegie Institute of Washington Publication* 410:1–335.

Cuvier, Georges. 1812. *Recherches sur les ossemens fossiles de quadrupèdes* [including the "Preliminary Discourse"]. Paris: Deterville.

Darwin, Charles. 1820s–. Miscellaneous writings. Darwin Manuscripts Project, edited by David Kohn, American Museum of Natural History, http://darwin.amnh.org; Darwin Correspondence Project, directed by James Secord, Cambridge University, http://www.darwinproject.ac.uk; Darwin Online, edited by John van Wyhe, http://darwin-online.org.uk.

Darwin, Charles. 1831–1836a. Diary. Down House, Kent. [See Barlow 1934; Keynes 1988; http://darwin-online.org.uk]

Darwin, Charles. 1831–1836b. Field Notebooks. Down House, Kent. [See Chancellor and Van Wyhe 2009; http://darwin-online.org.uk]

Darwin, Charles. 1832–. Correspondence. Cambridge University Library. [See Barlow 1967; Burkhardt 2008; Burkhardt et al. 1985–2014; http://www.darwinproject.ac.uk]

Darwin, Charles. (1832) 2008a. Letter to Caroline Darwin, October 24–November 24. In *The 'Beagle' Letters*, edited by Frederick Burkhardt, 166–170. Cambridge: Cambridge University Press. [Available at http://www.darwinproject.ac.uk]

Darwin, Charles. (1832) 2008b. Letter to John Henslow, May 18–June 16. In *The 'Beagle' Letters*, edited by Frederick Burkhardt, 126–130. Cambridge: Cambridge University Press. [Available at http://www.darwinproject.ac.uk]

Darwin, Charles. (1832) 2008c. Letter to John Henslow, July 23–August 15. In *Charles Darwin. The 'Beagle' Letters*, edited by Frederick Burkhardt, 141–143. Cambridge: Cambridge University Press. [Available at http://www.darwinproject.ac.uk]

Darwin, Charles. (1832) 2008d. Letter to John Henslow, October 26–November 24. In *The 'Beagle' Letters*, edited by Frederick Burkhardt, 170–173. Cambridge: Cambridge University Press. [Available at http://www.darwinproject.ac.uk]

Darwin, Charles. 1832–1836a. Geological Diary/Geological Notes. DAR 32–42, Cambridge University Library [includes the Earthquake Portfolio (DAR 42), containing two essays: "Reflection on Reading my Geological Notes" (DAR 42:93–96) and "February 1835" (DAR 42:97–99)]. [See http://darwin-online.org.uk]

Darwin, Charles. 1832–1836b. Zoological Notes. DAR 29–31, Cambridge University Library. [See Keynes 2000; http://darwin-online.org.uk]

Darwin, Charles. 1834. Reflection on reading my Geological Notes. DAR 42:93–96, Cambridge University Library. [See Herbert 1995; http://darwin-online.org.uk]

Darwin, Charles. (1834) 2008. Letter to John Henslow, March. In *The 'Beagle' Letters*, edited by Frederick Burkhardt, 259–263. Cambridge: Cambridge University Press. [Available at http://www.darwinproject.ac.uk]

Darwin, Charles. 1835. February 1835. DAR 42:93–96, Cambridge University Library. [See Hodge 1983; http://darwin.amnh.org (transcription)]

Darwin, Charles. (1835) 2008a. Letter to Caroline Darwin, July 19–August 12. In *The 'Beagle' Letters*, edited by Frederick Burkhardt, 349–351. Cambridge: Cambridge University Press. [Available at http://www.darwinproject.ac.uk]

Darwin, Charles. (1835) 2008b. Letter to John Henslow, July 12–August 12. In *The 'Beagle' Letters*, edited by Frederick Burkhardt, 353–354. Cambridge: Cambridge University Press. [Available at http://www.darwinproject.ac.uk]

Darwin, Charles. (1835) 2008c. Letter to William Fox, August 9–August 12. In *The 'Beagle' Letters*, edited by Frederick Burkhardt, 351–352. Cambridge: Cambridge University Press. [Available at http://www.darwinproject.ac.uk]

Darwin, Charles. 1836a. Animal Notes. DAR 29.1: A1–19, Cambridge University Library.

Darwin, Charles. 1836b. Ornithological Notes. DAR 29.2, Cambridge University Library. [See Barlow 1963; http://darwin-online.org.uk]

Darwin, Charles. 1836–1837. The Red Notebook. Down House, Kent. [See Herbert 1980, 1987; http://darwin.amnh.org (original and transcription)]

Darwin, Charles. 1837–1839. Transmutation Notebooks B–E. DAR 121–124, Cambridge University Library. [See Kohn 1987; http://darwin.amnh.org (transcription); http://darwin-online.org.uk]

Darwin, Charles. 1839. *Journal of Researches into the Geology and Natural History of the Various Countries Visited by H.M.S. Beagle, Under the Command of Captain FitzRoy, R.N., from 1832 to 1836* [*The Voyage of the Beagle*]. London: Henry Colburn. [Available at http://darwin-online.org.uk]

Darwin, Charles. 1842. Pencil Sketch. DAR 6, Cambridge University Library. [See Darwin 1909:1–53; http://darwin.amnh.org (transcription); available http://darwin-online.org.uk]

Darwin, Charles. 1844a. Essay. DAR 7, Cambridge University Library. [See Darwin 1909:57–255; http://darwin.amnh.org (transcriptions of original and "Fair Copy"); http://darwin-online.org.uk]

Darwin, Charles. 1844b. Letter to Leonard Jenyns, November 25. [See http://www.darwinproject.ac.uk]

Darwin, Charles. 1845. *Journal of Researches into the Natural History and Geology of the Various Countries Visited by H.M.S. Beagle, Under the Command of Captain FitzRoy, R.N., from 1832 to 1836*. 2nd ed. London: John Murray. [Available at http://darwin-online.org.uk]

Darwin, Charles. 1856–1858. *Natural Selection*. DAR 7–15, Cambridge University Library. [See Darwin (1856–1858) 1975; http://darwin-online.org.uk]

Darwin, Charles. 1859a. Letter to Charles Lyell, December 27. [See http://www.darwinproject.ac.uk]

Darwin, Charles. 1859b. Letter to Richard Owen, December 13. [See http://www.darwinproject.ac.uk]

Darwin, Charles. 1859c. *On the Origin of Species by Means of Natural Selection, or the Preservation of Favoured Races in the Struggle for Life*. London: John Murray. [Available at http://darwin.amnh.org; http://darwin-online.org.uk]

Darwin, Charles. 1872. *On the Origin of Species*. 6th ed. London: John Murray. [Available at http://darwin.amnh.org; http://darwin-online.org.uk]

Darwin, Charles. (1876) 1950. *Charles Darwin's Autobiography, with His Notes and Letters Depicting the Growth of the Origin of Species*. Edited by Sir Francis Darwin. Introduction by George Gaylord Simpson. New York: Schuman. [See also Barlow 1958; http://darwin-online.org.uk]

Darwin, Charles. 1958. *Evolution by Natural Selection [by] Charles Darwin and Alfred Russel Wallace*. Edited, with a foreword, by Sir Gavin de Beer. Cambridge: Cambridge University Press.

Darwin, Erasmus. 1794, 1796. *Zoonomia; or, the Laws of Organic Life*. 2 vols. London: Printed for J. Johnson.

Darwin, Francis, ed. 1909. *The Foundations of the Origin of Species: Two Essays Written in 1842 and 1844*. Cambridge: Cambridge University Press.

Dawkins, Richard. 1976. *The Selfish Gene.* Oxford: Oxford University Press.

de Beer, Gavin, ed. 1960–1961. Darwin's notebooks on transmutation of species: Parts I–IV. *Bulletin of the British Museum (Natural History)*, Historical Series 2, nos. 2–5:23–183. [Available at http://darwin-online.org.uk]

de Beer, Gavin, and M. J. Rowlands. 1960–1961. Darwin's notebooks on transmutation of species: Addenda and corrigenda. *Bulletin of the British Museum (Natural History)*, Historical Series 2, no. 6:185–200. [Available at http://darwin-online.org.uk]

Desmond, Adrian, and James Moore. 1991. *Darwin: The Life of a Tormented Evolutionist.* New York: Norton.

Dobzhansky, Theodosius. 1935. A critique of the species concept in biology. *Philosophy of Science* 2:344–355.

Dobzhansky, Theodosius. 1937. *Genetics and the Origin of Species.* New York: Columbia University Press.

Dobzhansky, Theodosius. 1940. Speciation as a stage in evolutionary divergence. *American Naturalist* 74:312–321.

Dobzhansky, Theodosius. 1941. *Genetics and the Origin of Species.* 2nd ed. New York: Columbia University Press.

Dobzhansky, Theodosius. 1951. *Genetics and the Origin of Species.* 3rd ed. New York: Columbia University Press.

Dominici, Stefano. 2010. Brocchi's subapennine fossil conchology. *Evolution: Education and Outreach* 3:585–594. [Available at http://link.springer.com/journal/12052]

Dominici, Stefano, and Niles Eldredge. 2010. Brocchi, Darwin, and transmutation: Phylogenetics and paleontology at the dawn of evolutionary biology. *Evolution: Education and Outreach* 3:576–584 [Available at http://link.springer.com/journal/12052]

d'Orbigny, Alcide. 1842. *Paléontologie française.* Vol. 1, *Terrains oolitiques ou Jurassiques.* Paris: Arthus Bertrand, Masson, Paris. [For Darwin's copy, see http://biodiversitylibrary.org/item/106856#page/7/mode/1up]

d'Orbigny, Alcide. 1845–1846. Letters to Charles Darwin. Cambridge University Library. [See http://www.darwinproject.ac.uk]

Eldredge, Niles. 1965. February 1965. American Museum of Natural History Library. [See http://dx.doi.org/10.5531/sd.paleo.1]

Eldredge, Niles. 1968a. Some aspects of species-level evolution in paleontology ["April 1968"]. American Museum of Natural History Library. [See http://dx.doi.org/10.5531/sd.paleo.2]

Eldredge, Niles. 1968b. Convergence between two Pennsylvanian gastropod species: A multivariate mathematical approach. *Jour. Paleontology* 42:186–196.

Eldredge, (Robert) Niles. 1969. Geographic variation and evolution in *Phacops rana* (Green, 1832) and *Phacops iowensis* Delo 1935, in the Middle Devonian of North America. Ph.D. diss., Columbia University. [See http://digitallibrary .amnh.org/dspace/handle/2246/1095]

Eldredge, Niles. 1970a. The allopatric model and phylogeny in Paleozoic invertebrates—final draft. American Museum of Natural History Library. [See http:// dx.doi.org/10.5531/sd.paleo.8]

Eldredge, Niles. 1970b. The process of speciation and interpretation of the fossil record [initial draft of Eldredge and Gould 1972a]. American Museum of Natural History Library. [See http://dx.doi.org/10.5531/sd.paleo.3]

Eldredge, Niles. 1971a. The allopatric model and phylogeny in Paleozoic invertebrates. *Evolution* 25:156–167.

Eldredge, Niles. 1971b. Letter to Thomas J. M. Schopf, June 21. American Museum of Natural History Library. [See http://dx.doi.org/10.5531/sd.paleo.6]

Eldredge, Niles. 1972. Systematics and evolution of *Phacops rana* (Green, 1832) and *Phacops iowensis* Delo, 1935 (Trilobita) from the Middle Devonian of North America. *Bulletin of the American Museum of Natural History* 147:45–114. [Available at http://digitallibrary.amnh.org/dspace/handle/2246/1095]

Eldredge, Niles. 1974. Stability, diversity, and speciation in Paleozoic epeiric seas. *Journal of Paleontology* 48:540–548.

Eldredge, Niles. 1979. Alternative approaches to evolutionary theory. In "Models and Methodologies in Evolutionary Theory," edited by J. H. Schwartz and H. B. Rollins. Special issue, *Bulletin of the Carnegie Museum of Natural History* 13:7–19.

Eldredge, Niles. 1982a. Introduction to *Systematics and the Origin of Species*, by Ernst Mayr. 1942. Reprint. New York: Columbia University Press.

Eldredge, Niles. 1982b. Phenomenological levels and evolutionary rates. *Systematic Zoology* 31:338–347.

Eldredge, Niles. 1985a. *Time Frames: The Rethinking of Darwinian Evolution and the Theory of Punctuated Equilibria*. New York: Simon and Schuster.

Eldredge, Niles. 1985b. *Unfinished Synthesis: Biological Hierarchies and Modern Evolutionary Thought*. New York: Oxford University Press.

Eldredge, Niles. 1998. *Life in the Balance: Humanity and the Biodiversity Crisis*. Princeton, N.J.: Princeton University Press.

Eldredge, Niles. 2003. The Sloshing Bucket: How the physical realm controls evolution. In *Evolutionary Dynamics: Exploring the Interplay of Selection, Accident, Neutrality, and Function*, edited by James P. Crutchfield and Peter Schuster, 3–32. New York: Oxford University Press.

Eldredge, Niles. 2005. *Darwin: Discovering the Tree of Life*. New York: Norton.

Eldredge, Niles. 2008a. The early "evolution" of "punctuated equilibria." *Evolution: Education and Outreach* 1, no. 2:107–113. [Available at http://link.springer.com/ journal/12052]

Eldredge, Niles. 2008b. Some thoughts on "adaptive peaks," "Dobzhansky's dilemma"—and how to think about evolution. *Evolution: Education and Outreach* 1, no. 3:243–246. [Available at http://link.springer.com/journal/12052]

Eldredge, Niles. 2009a. Experimenting with transmutation: Darwin, the *Beagle*, and evolution. *Evolution: Education and Outreach* 2:35–54. [Available at http://link.springer.com/journal/12052]

Eldredge, Niles. 2009b. A question of individuality: Charles Darwin, George Gaylord Simpson, and transitional fossils. *Evolution: Education and Outreach* 2:150–155. [Available at http://link.springer.com/journal/12052]

Eldredge, Niles. 2013. Stephen Jay Gould in the 1960s and 1970s and the origin of "punctuated equilibria." In *Stephen J. Gould: The Scientific Legacy*, edited by Gian Antonio Danieli, Alessandro Minelli, and Telmo Pievani, 3–20. Milan: Springer. [Available at http://link.springer.com/chapter/10.1007/978-88-470-5424-0_1

Eldredge, Niles, and Stephen Jay Gould. 1970. Speciation and punctuated equilibria: An alternative to phyletic gradualism [draft with "inverted authorship" ("Gould, Stephen Jay and Niles Eldredge"), with pencilled notation of rearrangement of authors]. American Museum of Natural History Library. [See http://dx.doi.org/10.5531/sd.paleo.4]

Eldredge, Niles, and Stephen Jay Gould. 1972a. Punctuated equilibria: An alternative to phyletic gradualism. In *Models in Paleobiology*, edited by Thomas J. M. Schopf, 82–115. San Francisco: Freeman, Cooper. [Available at www.NilesEldredge.com; http://www.blackwellpublishing.com/ridley/classictexts/eldredge.pdf]

Eldredge, Niles, and Stephen Jay Gould. 1972b. Speciation and punctuated equilibria: An alternative to phyletic gradualism [final (third) draft]. [See http://dx.doi.org/10.5531/sd.paleo.7]

Eldredge, Niles, and Marjorie J. Grene. 1992. *Interactions: The Biological Context of Social Systems*. New York: Columbia University Press.

Eldredge, Niles, and Stanley N. Salthe. 1984. Hierarchy and evolution. *Oxford Reviews in Evolutionary Biology* 1:182–206.

Eldredge, Niles, John N. Thompson, Paul M. Brakefield, Sergey Gavrilets, David Jablonski, Jeremy B. C. Jackson, Richard E. Lenski, Bruce S. Lieberman, Mark A. McPeek, and William Miller III. 2005. The dynamics of evolutionary stasis. In *Macroevolution: Diversity, Disparity, Contingency: Essays in Honor of Stephen Jay Gould*, edited by Elisabeth S. Vrba and Niles Eldredge, 133–145. Supplement to *Paleobiology* 31, no. 2. Lawrence, Kans.: Paleontological Society.

Fisher, Ronald A. 1930. *The Genetical Theory of Natural Selection*. Oxford: Clarendon Press.

Ghiselin, Michael. 1974. A radical solution to the species problem. *Systematic Zoology* 23:536–544.

Ghiselin, Michael. 1987. Species concepts, individuality, and objectivity. *Biological Philosophy* 2:127–143.

Gingerich, Philip. 1974. Stratigraphic record of Early Eocene *Hyopsodus* and the geometry of mammalian phylogeny. *Nature* 248:107–109.

Gingerich, Philip. 1976. Paleontology and phylogeny: Patterns of evolution at the species level in Early Tertiary mammals. *American Journal of Science* 276:1–28.

Gould, Stephen Jay. 1966. Allometry and size in ontogeny and phylogeny. *Biological Reviews* 41:587–640.

Gould, Stephen Jay. 1969. An evolutionary microcosm: Pleistocene and Recent history of the land snail *P. (Poecilozonites)* in Bermuda. *Bulletin of the Museum of Comparative Zoology* 138: 407–532.

Gould, Stephen Jay. 1977. *Ontogeny and Phylogeny*. Cambridge, Mass.: Harvard University Press.

Gould, Stephen Jay. 1982. Introduction to *Genetics and the Origin of Species*, by Theodosius Dobzhansky, xvii–xli. 1937. Reprint, New York: Columbia University Press.

Gould, Stephen Jay. 1992. Punctuated equilibrium in fact and theory. In *The Dynamics of Evolution: The Punctuated Equilibrium Debate in the Natural and Social Sciences*, edited by Albert Somit and Steven A. Peterson, 54–84. Ithaca, N.Y.: Cornell University Press.

Gould, Stephen Jay. 2002. *The Structure of Evolutionary Theory*. Cambridge, Mass.: Harvard University Press.

Gould, Stephen Jay, and Niles Eldredge. 1977. Punctuated equilibria: The tempo and mode of evolution reconsidered. *Paleobiology* 3:115–151.

Gould, Stephen Jay, and Elizabeth A. Lloyd. 1999. Individuality and adaptation across levels of selection: How shall we name and generalize the unit of Darwinism? *Proceedings of the National Academy of Sciences USA* 96:11904–11909.

Gould, Stephen Jay, and Elisabeth S. Vrba. 1982. Exaptation—A missing term in the science of form. *Paleobiology* 8:4–15.

Grant, Peter, and Rosemary Grant. 2011. *How and Why Species Multiply: The Radiation of Darwin's Finches*. Princeton, N.J.: Princeton University Press.

Grant, Robert E. 1826. Notice of a new zoophyte (Cliona celata, Gr.) from the Firth of Forth. *Edinburgh New Philosophical Journal* 1:78–81.

Grant, Robert E. 1829. *Essay on the Study of the Animal Kingdom* [inaugural address of 1828]. London: John Taylor.

Gregory, T. Ryan. 2004. Macroevolution, hierarchy and the C-value enigma. *Paleobiology* 30:179–202.

Grene, Marjorie J. 1958. Two evolutionary theories. *British Journal for the Philosophy of Science* 9:110–127; 185–193.

Grene, Marjorie J. 1987. Hierarchies in biology. *American Scientist* 75:504–510.

Griffith, Edward, Charles Hamilton Smith, and Edward Pidgeon, eds. 1827. *The Animal Kingdom, Arranged in Conformity with Its Organization, by the Baron Cuvier*. Vol. 2. London: George B. Whittaker.

Gulick, J. T. 1905. Evolution, racial and habitudinal. *Carnegie Institute of Washington Publication* 25:1–269.

Haffer, Jürgen. 1974. *Avian Speciation in Tropical South America: With a Systematic Survey of the Toucans (Ramphastidae) and Jacamars (Galbulidae)*. Publications of the Nuttall Ornithological Club 14. Cambridge, Mass.: Nuttall Ornithological Club.

Haldane, J. B. S. 1932. *The Causes of Evolution*. New York: Harper.

Harris, Marvin. 1968. *The Rise of Anthropological Theory: A History of Theories of Culture*. New York: Crowell.

Herbert, Sandra, ed. 1980. *The Red Notebook of Charles Darwin*. Ithaca, N.Y.: Cornell University Press.

Herbert, Sandra, ed. 1987. Red Notebook. In *Charles Darwin's Notebooks, 1836–1844: Geology, Transmutation of Species, Metaphysical Enquiries*, edited by Paul H. Barrett, Peter J. Gautrey, Sandra Herbert, David Kohn, and Sydney Smith, 17–81. Ithaca, N.Y.: Cornell University Press.

Herbert, Sandra. 1995. From Charles Darwin's portfolio: An early essay on South American geology and species. *Earth Sciences History* 14:23–36.

Herbert, Sandra. 2005. *Charles Darwin, Geologist*. Ithaca, N.Y.: Cornell University Press.

Herschel, John. 1830. A *Preliminary Discourse on the Study of Natural Philosophy*. London: Longman, Rees, Orme, Brown, Green, and Taylor.

Herschel, John. 1836. Letter to Charles Lyell. [See Kohn 1987:413 (note 59-2 to transcription of Notebook E, 59); http://darwin.amnh.org]

Hodge, M. J. S. 1983. Darwin and the laws of the animate part of the terrestrial system (1835–1837): On the Lyellian origins of his explanatory program. *Studies in the History of Biology* 6:1–106.

Horner, Leonard. 1816. Review of *Conchiologia Fossile Subapennina*, by Giambattista Brocchi. *Edinburgh Review* 26:156–180.

Hull, David. 1976. Are species really individuals? *Systematic Zoology* 25:174–191.

Hull, David. 1978. A matter of individuality. *Philosophy of Science* 45:335–360.

Hull, David. 1980. Individuality and selection. *Annual Reviews of Ecology and Systematics* 11:311–332.

Jackson, Jeremy B. C., and Alan H. Cheetham. 1999. Tempo and mode of speciation in the sea. *Trends in Ecology and Evolution* 14:72–77.

Jameson, Robert. 1813. *Essay on the Theory of the Earth. With Geological Illustrations by Professor Jameson*. Edinburgh: Blackwood.

Jameson, Robert. 1826. Observations on the nature and importance of geology. *Edinburgh New Philosophical Journal* 1:293–302.

Jameson, Robert. 1827a. Death of Professor Brocchi. *Edinburgh New Philosophical Journal* 3:383.

Jameson, Robert. 1827b. *Essay on the Theory of the Earth. With Geological Illustrations by Professor Jameson*. 5th ed. Edinburgh: Blackwood.

Jameson, Robert. 1827c. Of the changes which life has experienced on the globe. *Edinburgh New Philosophical Journal* 3:298–301.

Jameson, Robert. 1829. Of the continuity of the animal kingdom by means of generation, from the first ages of the world to the present times: On the relations of organic structure and parentage that may exist between the animals of the historic ages and those at present living, and the antediluvian and extinct species. *Edinburgh New Philosophical Journal* 7:152–155.

[Jameson, Robert, trans.]. 1836. Biographical memoir of M. de Lamarck. By the Baron Cuvier. *Edinburgh New Philosophical Journal* 20:1–22.

Johnson, Ralph Gordon. 1972. Conceptual models of benthic marine communities. In *Models in Paleobiology*, edited by Thomas J. M. Schopf, 148–159. San Francisco: Freeman, Cooper.

Jordan, David Starr. 1905. The origin of species through isolation. *Science* 22:545–562.

Keynes, Richard, ed. 1988. *Charles Darwin's Beagle Diary*. Cambridge: Cambridge University Press. [Available at http://darwin-online.org.uk]

Keynes, Richard, ed. 2000. *Charles Darwin's Zoology Notes & Specimen Lists from H.M.S. Beagle*. Cambridge: Cambridge University Press. [Available at http://darwin-online.org.uk]

Kohn, David, ed. 1987. Notebook B–Notebook E. In *Charles Darwin's Notebooks, 1836–1844: Geology, Transmutation of Species, Metaphysical Enquiries*, edited by Paul H. Barrett, Peter J. Gautrey, Sandra Herbert, David Kohn, and Sydney Smith, 167–455. Ithaca, N.Y.: Cornell University Press. [Available at http://darwin.amnh.org]

Kohn, David, Gina Murrell, John Parker, and Mark Whitehorn. 2005. What Henslow taught Darwin. *Nature* 465:643–45.

Kolbert, Elizabeth. 2014. *The Sixth Extinction: An Unnatural History*. New York: Holt.

Kurtén, Bjorn. 1965. Evolution in geological time. In *Ideas in Modern Biology*, edited by John A. Moore, 329–354. Garden City, N.Y.: Natural History Press.

Lamarck, J.-B. 1801 *Systême des animaux sans vertèbres*. Paris: Deterville.

Lamarck, J.-B. 1809. *Philosophie zoologique*. Paris: Deterville.

Lamarck, J.-B. (1809) 1984. *Zoological Philosophy by J. B. Lamarck*. Translated by Hugh Elliot. Chicago: University of Chicago Press.

Lazarus, David, Reed P. Scherer, and Donald Prothero. 1985. Evolution of the radiolarian species-complex *Pterocanium*: A preliminary survey. *Journal of Paleontology* 59:183–220.

Lieberman, Bruce S. 2000. *Paleobiogeography*. New York: Kluwer Academic.

Lieberman, Bruce S. 2001. A test of whether rates of speciation were unusually high during the Cambrian radiation. *Proceedings of the Royal Society, Biological Sciences, Series B*, 268:1707–1714.

Lieberman, Bruce S. 2003. Taking the pulse of the Cambrian radiation. *Journal of Integrative and Comparative Biology* 43:229–237.

Lieberman, Bruce S., Carlton E. Brett, and Niles Eldredge. 1995. A study of stasis and change in two species lineages from the Middle Devonian of New York State. *Paleobiology* 21:15–27.

Linnaeus, Carolus [Carl von Linné]. 1758. *Systema naturae*. 10th ed. Stockholm: Salvii.

Lyell, Charles. (1830–1833) 1997. *Principles of Geology*. 3 vols in 1. Edited by James Secord. London: Penguin.

MacGillavry, H. J. 1968. Modes of evolution mainly among marine invertebrates. *Bijdragen tot de dierkunde* 38:69–74.

Malthus, Thomas. 1826. *An Essay on the Principle of Population*. 6th ed. 1798. London: John Murray.

Margulis, Lynn. 1974. Five-kingdom classification and the origin and evolution of cells. *Evolutionary Biology* 7:45–78.

Matthew, Patrick. 1831. *On Naval Timber and Arboriculture*. Edinburgh: Adam Black.

Mayr, Ernst. 1940. Speciation phenomena in birds. *American Naturalist* 74:249–278.

Mayr, Ernst. 1942. *Systematics and the Origin of Species*. New York: Columbia University Press.

Mayr, Ernst. 1963. *Animal Species and Evolution*. Cambridge, Mass.: Harvard University Press.

Mayr, Ernst, and Provine William B., eds. 1988. *The Evolutionary Synthesis: Perspectives on the Unification of Biology*. Cambridge, Mass.: Harvard University Press.

Miller, William, III. 2006. What every paleontologist should know about species: New concepts and questions. *Neues Jahrbuch für Geologie und Palaontologie Monatshefte* 2006:557–576.

Nehm, Ross, and Ann F. Budd, eds. 2008. *Evolutionary Stasis and Change in the Dominican Republic Neogene*. New York: Springer.

Newell, Norman D. 1967. Revolutions in the history of life. *Geological Society of America Special Paper* 89:63–91.

O'Leary, Maureen A., et al. 2013. The placental mammal ancestor and the post–K-Pg radiation of mammals. *Science* 339:662–667.

Pagel, Mark, Chris Venditti, and Andrew Meade. 2006. Large punctuational contribution of speciation to evolutionary divergence at the molecular level. *Science* 314:119–121.

Paley, William. 1813. *Natural Theology; Or Evidences of the Existence and Attributes of the Deity, Collected from the Appearances of Nature*. 14th ed. 1802. London: Faulder.

Pancaldi, Giuliano. 1991. *Darwin in Italy: Science Across Cultural Frontiers*. Translated by Ruey Brodine Morelli. Bloomington: Indiana University Press.

Paterson, Hugh E. H. 1985. The recognition concept of species. In *Species and Speciation*, edited by Elisabeth S. Vrba, 21–29. Transvaal Museum Monograph 4. Pretoria: Transvaal Museum.

Rampino, Michael R. 2011. Darwin's error? Patrick Matthew and the catastrophic nature of the geologic record. *Historical Biology: An International Journal of Paleobiology* 23:227–230.

Rowe, A. W. 1899. An analysis of the genus *Micraster*, as determined by rigid zonal collecting from the zone of *Rhynchonella Cuvieri* to that of *Micraster cor-anguinum*. *Quarterly Journal of the Geological Society of London* 55:494–547.

Rudwick, Martin J. S. 1997. *Georges Cuvier, Fossil Bones, and Geological Catastrophes: New Translations and Interpretations of the Primary Texts*. Chicago: University of Chicago Press.

Rudwick, Martin J. S. 2005. *Bursting the Limits of Time: The Reconstruction of Geohistory in the Age of Revolution*. Chicago: University of Chicago Press.

Rudwick, Martin J. S. 2008. *Worlds Before Adam: The Reconstruction of Geohistory in the Age of Reform*. Chicago: University of Chicago Press.

Salthe, Stanley N. 1975. Problems of macroevolution (molecular evolution, phenotype definition, and canalization) as seen from a hierarchical viewpoint. *American Zoologist* 15:295–314.

Salthe, Stanley N. 1985. *Evolving Hierarchical Systems: Their Structure and Representation*. New York: Columbia University Press.

Santini, Francesco, Maria Pia Miglietta, and Anuschka Faucci. 2012. Speciation: Where are we now? An introduction to a special issue on speciation. *Evolutionary Biology* 39:141–147.

Schaeffer, Bobb. 1965. The role of experimentation in the origin of higher levels of organization. *Systematic Zoology* 14:318–336.

Secord, James. 1991. Edinburgh Lamarckians: Robert Jameson and Robert E. Grant. *Journal of the History of Biology* 24:1–18.

Sepkoski, David. 2013a. Punctuated equilibria. In *Oxford Bibliographies: Evolutionary Biology*, edited by Jonathan Losos. [Available at http://www.oxfordbibliographies.com/obo/page/evolutionary-biology]

Sepkoski, David. 2013b. *Rereading the Fossil Record: The Growth of Paleobiology as an Evolutionary Discipline*. Chicago: University of Chicago Press.

Sepkoski, J. John. 1978. A kinetic model of Phanerzoic taxonomic diversity I. Analysis of marine orders. *Paleobiology* 5:223–251.

Simpson, George Gaylord. 1934. *Attending Marvels: A Patagonian Journal*. London: Macmillan.

Simpson, George Gaylord. 1944. *Tempo and Mode in Evolution*. New York: Columbia University Press.

Simpson, George Gaylord. 1953. *The Major Features of Evolution*. New York: Columbia University Press.

Simpson, George Gaylord. 1961. *Principles of Animal Taxonomy*. New York: Columbia University Press.

Simpson, George Gaylord. 1963. The meaning of taxonomic statements. In *Classification and Human Evolution*, edited by Sherwood L. Washburn, 1–31. Chicago: Aldine.

Simpson, George Gaylord. 1970. Uniformitarianism. An inquiry ino principle, theory and method in geohistory and biohistory. In *Essays in Evolution and Genetics in Honor of Theodosius Dobzhansky*, edited by M. K. Hecht and W. C. Steere, 43–96. New York: Appleton-Century-Crofts.

Simpson, George Gaylord. 1982. *Attending Marvels: A Patagonian Journal*. Rev. ed. [with afterword by Simpson]. Chicago: University of Chicago Press.

Simpson, George Gaylord. 1984. *Discoverers of the Lost World: An Account of Some of Those Who Brought Back to Life South American Mammals Long Buried in the Abyss of Time*. New Haven, Conn.: Yale University Press.

Sobel, James M., Grace F. Chen, Lorna R. Watt, and Douglas W. Schemske. 2009. The biology of speciation. *Evolution* 64:295–315.

Stanley, Steven M. 1975. A theory of evolution above the species level. *Proceedings of the National Academy of Sciences* 72:646–650.

Stanley, Steven M., and X. Yang. 1987. Approximate evolutionary stasis for bivalve morphology over millions of years: A mutivariate, multi-lineage study. *Paleobiology* 13:113–139.

Stauffer, R. C., ed. 1975. *Charles Darwin's Natural Selection: Being the Second Part of His Big Species Book Written from 1856 to 1858*. Cambridge: Cambridge University Press. [Available at http://darwin-online.org.uk]

Sulloway, Frank J. 1979. Geographic isolation in Darwin's thinking: The vicissitudes of a crucial idea. *Studies in the History of Biology* 4:23–65.

Sulloway, Frank J. 2009. Tantalizing tortoises and the Darwin-Galápagos legend. *Journal of the History of Biology* 42:3–31.

Thompson, John N. 1994. *The Coevolutionary Process*. Chicago: University of Chicago Press.

Thompson, John N. 2005. *The Geographic Mosaic of Coevolution*. Chicago: University of Chicago Press.

Thompson, John N. 2013. *Relentless Evolution*. Chicago: University of Chicago Press.

Trueman, A. E. 1922. The use of *Gryphaea* in the correlation of the Lower Lias. *Geological Magazine* 59:256–268.

Van Dam, Jan A., Hayfaa Abdul Aziz, M. Ángeles Álvarez Sierra, Frederik J. Hilgen, Lars W. van den Hoek Ostende, Lucas J. Lourens, Pierre Mein, Albert J. van der Meule, and Pablo Pelaez-Campomanes. 2006. Long-period astronomical forcing of mammal turnover. *Nature* 443:687–691

Venditti, Chris, and Mark Pagel. 2008. Speciation and bursts of evolution. *Evolution: Education and Outreach* 1, no. 3:274–280. [Available at http://link.springer.com/journal/12052]

Venditti, Chris, and Mark Pagel. 2009. Speciation as an active force in promoting genetic evolution. *Trends in Ecology and Evolution* 25:14–20.

Vénec-Peyré, Marie-Thérèse. 2004. Beyond frontiers and time: The scientific and cultural heritage of Alcide d'Orbigny (1802–1857). *Marine Micropaleontology* 50:149–159.

Vrba, Elisabeth S. 1980. Evolution, species and fossils: How does life evolve? *South African Journal of Science* 76:61–84.

Vrba, Elisabeth S. 1984. What is species selection? *Systematic Zoology* 33:318–328.

Vrba, Elisabeth S. 1985. Environment and evolution: Alternative causes of the temporal distribution of evolutionary events. *South African Journal of Science* 81:229–236.

Vrba, Elisabeth S., and Niles Eldredge. 1984. Individuals, hierarchies and processes: Towards a more complete evolutionary theory. *Paleobiology* 10:146–171.

Vrba, Elisabeth S., and Stephen Jay Gould. 1986. The hierarchical expansion of sorting and selection: Sorting and selection cannot be equated. *Paleobiology* 12:217–228.

Waterhouse, George. 1837. List of Edentates. DAR 39:133, Cambridge University Library. [See http://darwin.amnh.org]

Whewell, William. 1837. *History of the Inductive Sciences, from the Earliest to the Present Times*. 3 vols. London: Parker.

Wiley, E. O. 1978. The evolutionary species concept reconsidered. *Systematic Zoology* 27:17–26.

Wiley, E. O., and Bruce S. Lieberman. 2011. *Phylogenetics: The Theory and Practice of Phylogenetic Systematics*. Hoboken, N.J.: Wiley.

Wright, Sewall. 1931. Evolution in Mendelian populations. *Genetics* 16:97–159.

Wright, Sewall. 1932. The roles of mutation, inbreeding, crossbreeding, and selection in evolution. *Proceedings of the Sixth International Congress of Genetics* 1:356–366.

Wright, Sewall. 1941. The material basis of evolution. *Scientific Monthly* 53:165–170.

Wright, Sewall. 1945. *Tempo and Mode in Evolution*: A critical review [review of *Tempo and Mode in Evolution*, by George Gaylord Simpson]. *Ecology* 26:415–419.

Wright, Sewall. 1967. Comments on the preliminary working papers of Eden and Waddington. In *Mathematical Challenges to the Neo-Darwinian Interpretation of Evolution*, edited by Paul S. Moorehead and Marvin M. Kaplan, 117–120. Philadelphia: Wistar Institute Press.

Yoder, Anne D. 2013. Fossils versus clocks. *Science* 339:656–658.

INDEX